THE MYSTERIOUS SCIENCE OF THE SEA,
1775–1943

HISTORY AND PHILOSOPHY OF TECHNOSCIENCE

Series Editor: Alfred Nordmann

TITLES IN THIS SERIES

1 Error and Uncertainty in Scientific Practice
Marcel Boumans, Giora Hon and Arthur C. Petersen (eds)

2 Experiments in Practice
Astrid Schwarz

3 Philosophy, Computing and Information Science
Ruth Hagengruber and Uwe Riss (eds)

4 Spaceship Earth in the Environmental Age, 1960–1990
Sabine Höhler

5 The Future of Scientific Practice 'Bio-Techno-Logos'
Marta Bertolaso (ed.)

6 Scientists' Expertise as Performance: Between State and Society, 1860–1960
Joris Vandendriessche, Evert Peeters and Kaat Wils (eds)

7 Standardization in Measurement: Philosophical, Historical and
Sociological Issues
Oliver Schlaudt and Lara Huber (eds)

FORTHCOMING TITLES

Reasoning in Measurement
Nicole Mößner and Alfred Nordmann (eds)

Research Objects in their Technological Setting
*Bernadette Bensaude-Vincent, Sacha Loeve, Alfred Nordmann and
Astrid Schwarz (eds)*

THE MYSTERIOUS SCIENCE OF THE SEA, 1775–1943

BY

Natascha Adamowsky

Routledge
Taylor & Francis Group

LONDON AND NEW YORK

First published 2015 by Pickering & Chatto (Publishers) Limited

Published 2016 by Routledge
2 Park Square, Milton Park, Abingdon, Oxon OX14 4RN
605 Third Avenue, New York, NY 10017

First issued in paperback 2021

Routledge is an imprint of the Taylor & Francis Group, an informa business

BRITISH LIBRARY CATALOGUING IN PUBLICATION DATA

Adamowsky, Natascha, author.
The mysterious science of the sea, 1775–1943. – (History and philosophy of
technoscience)
1. Ocean. 2. Oceanography – History.
I. Title II. Series
551.4'6'09-dc23

ISBN 13: 978-1-03-209846-3 (pbk)
ISBN 13: 978-1-84893-532-7 (hbk)

Typeset by Pickering & Chatto (Publishers) Limited

CONTENTS

Acknowledgements   vii

List of Figures   ix

Introduction: Perspectives on the Epistemology and Aesthetics of
     Oceanic Wonders   1

1 Wondrous and Terrible Sea – Stations of an Unknown Modernity   13

2 Wonder and Mystery   37

3 *Twenty Thousand Leagues under the Sea*: The Overture to a Passion   73

4 *Mise-en-Scène*: Invented Realities, or the Mediality of Wonders –
     The Sea in the Aquarium   101

5 *Mise-en-Action*: Asleep in the Deep   145

Epilogue   175

Works Cited   177

Notes   209

Index   245

ACKNOWLEDGEMENTS

This book has been written in German. Special thanks to Henry Erik Butler and Michelle Miles for their wonderful translation.

LIST OF FIGURES

Figure 2.1: *Pelikanaal* 64
Figure 2.2: *Chaunangium Crater* 68
Figure 2.3: *Aphrocallistes beatrix J. E. Gray* 69
Figure 3.1: *Frontispiece*, J. Verne, *Vingt Mille Lieues Sous Les Mers* 74
Figure 3.2: *Aquarium*, J. Verne, *Vingt Mille Lieues Sous Les Mers* 77
Figure 5.1: *Observation Pond Empty* 155
Figure 5.2: *Pike Following up a Roach* 156

INTRODUCTION: PERSPECTIVES ON THE EPISTEMOLOGY AND AESTHETICS OF OCEANIC WONDERS

Wonders of the Sea and their Media

The sea conceals arcana in its depths which no glance can penetrate, which no genius can depict except with the help of imagination. In the aërial and terrestrial worlds, and even in the celestial space, Nature liberally unrolls before our eyes her marvellous pictures. From one pole to the other we may explore all the parts of our domain; ... but of this ocean, this thin stratum of water a few thousand yards in thickness, stretched over our little planet, we know by sight only the surface and the borders.[1]

How many has the enormous Sphinx devoured of those who have attempted to divine its enigmas, to pierce its mysteries! What matters it? The work goes and goes forward. The human eye has penetrated that formidable night.[2]

Secret, enigma and night – the ocean, its mysteries and wonders: the French natural scientist Arthur Mangin (1824–87) describes a mood that pervades the nineteenth century as a whole in his extraordinarily successful work, *Les Mystères de l'Océan* (1864). The sea's riches had been discovered; now, extravagant efforts were undertaken to achieve dominion over its inhabitants and terrains. In the early phase of the modern age, a panoply of passions – whether scientific or artistic, reserved for initiates or the seemingly mad – emerged and enriched the popular dramaturgy of riddling and cryptic phenomena with a demimonde of the hitherto unknown. To be sure, talk of wondrous and strange secrets occurred everywhere in works of amateur and popular science, to say nothing of scientific literature that prized the aesthetic dimensions of nature. But the sea, with its unfathomable depths and extraordinary recesses, represented a particularly significant focal point for both epistemology and media aesthetics. From the standpoint of popular science, accounts of wondrous marine phenomena had the peculiarity of being based 'upon another prodigy not less astonishing than the former'.[3] 'We live in an age of miracles', Jules Michelet (1798–1874) wrote,

> this is not to astonish us. Any one would formerly have been laughed at who had
> ventured to say that some animals, disobedient to the general laws of nature, take the
> liberty to breathe through their paws.[4]

Epistemologically speaking, the realm of the sea lay beyond human reach and was accessible only through considerable technological effort – this universe must be experienced by way of artifacts in artificial worlds. In their influential book on oceanographic research, *The Machine in Neptune's Garden*, Helen Rozwadowski and David van Keuren write:

> [T]he oceans are a forbidding and alien environment inaccessible to direct human
> observation. They force scientist-observers to carry their natural environment with
> them ... [O]ceanography's necessary dependence upon technology ... create[s] a per-
> suasive argument that the machine is the garden. That is, what oceanographers have
> learned about the ocean has been based almost exclusively on what various technolo-
> gies, or machines, have taught them.[5]

Marine worlds have to be experienced in mediated form, then. The relation between mankind and the sea is fundamentally based on technology that transforms it so that it may be grasped by human senses and understanding. For some time now, historians of science have focused their attention on the emergence and constitution of new objects of investigation. For all that, it is only recently that insights from the cultural history of media have also been incorporated – for example, the fact that generating scientific objects follows from processes of medialization, that is, from various transformations that occur by artificial means. Hereby, a three-dimensional object yields a two-dimensional view, a colourful continuum turns into photographic image marked by contrast and so on. And since, as is well known, media 'das zu Vermittelnde immer unter Bedingungen stellen, die sie selbst schaffen und sind'[6] – or, in other words, since the medium is always inscribed in what it lends mediatized form – it is necessary to pay attention to the means by which the sea is made available to us. This holds especially for the conditions and structures of non-discursive media and the deictic orders they involve.[7] Put still differently: for a history of the 'wonders of the sea' – or, alternately, their 'exploration', 'unveiling' or 'decipherment' – it is essential to consider the constellations of media that enable a given sphere to be known: what aesthetic practices, technologies and instruments were employed, by whom and in what cultural-historical context.

It may seem anachronistic to call the sea a treasure house of modern wonders. Don't monsters, sea serpents and leviathans number among the fears of unenlightened times? Wasn't it precisely the nineteenth century that, with its great passion for amateur science, diffused knowledge about marine life forms, fostered the systematic study of the ocean and witnessed the founding of the field of oceanography? Don't we now know that sea serpents do not exist? The possibilities are not mutually exclusive; all the same, the connections prove complex and intertwined.

On Wonders: Approaches via the History of Media and Knowledge

Reality crushes fiction, Félix Belly observed in 1855 apropos of the first world's fair in Paris, and science and technology bestow more splendor and marvel upon everyday reality than the poet's imagination has ever dreamed of.[8] A firm belief in the wonders of science and technology characterized incipient modernity, which never tired of celebrating its successes as sure signs that a paradisiacal future was imminent. At the same time, however – and as is well known – science and technology rely on a rationalistic dispositive that excludes the existence of wonders categorically. Accordingly, the question arises about the strangely anachronistic combination of progress and myth; after all, wonders usually count as *premodern*. What we normally mean when we speak of 'wonders' stands diametrically opposed to the modernist paradigm – points of disagreement about periodization and dominant themes notwithstanding – which includes the Cartesian worldview, secularization, rational methods of thought, the empiricism and experimentation of the natural sciences, and the thoroughgoing mechanization in all spheres of life. Wonders have little in common with any of these things. Indeed, quite the opposite holds. Wonders are abnormal, irrational, criminal, metaphysical, esoteric, naïve, beautiful, sublime and terrible – in brief, they constitute a sphere of the extraordinary and the marginal, which is multifarious and hardly consistent.

Max Weber's (1864–1920) diagnosis of modernity and its conditions of emergence – the 'Entzauberung der Welt' (disenchantment of the world)[9] – has become proverbial. The phrase refers to a process that pushed away 'belief in magic' and the 'magischen Sinngehalt' (magic sense) of the cosmos – especially after the Renaissance when modern science and the rationalism, experimentation and empiricism it entailed transformed the world into a 'kausalen Mechanismus' (causal mechanism).[10] This perspective on the 'impoverishing' effects of social evolution stands opposed to the Enlightenment view, which interprets modernization as mankind's progressive emancipation from religious and metaphysical constructs (which were supposed to have progressively yielded to the explanations of natural science). Eighteenth-century philosophes inaugurated this tradition; for them, illumination symbolized a new age of Reason; in contrast, medieval theology and metaphysics represented the gloomy miscarriages of a dark age.[11] Either way, consensus holds that an epochal shift occurred. Its defining feature is the end of wonders.

Oddly enough, the state of wonder and wondrous phenomena often provided a subject of debate in modernity. Artifacts and events were granted this title, and expeditions counted as opportunities to experience them. If the pages that follow take up a few of these 'marvels', the point is not to show countercurrents to 'rationalistic' modernity that are, say, occultist, esoteric, mystical or superstitious in origin; nor is the purpose to uncover a 'reenchantment of the world' that supposedly took place.[12] Instead, the aim is to show that both the history of modern success as well as the history of modern loss – more specifically, the triumphs and failures of modern natural science and technology – prove inadequate on closer inspection. Hereby, nineteenth-century 'wonders of technology' and 'wonders of life' occupy the central position. This focus makes it possible to study, in particularly concrete fashion, the modes of apparition of wonders and the wondrous – what histories of rupture and decline cannot account for inasmuch as they follow from the interplay of continuity and change.

To be sure, the *Great Debate on Miracles*[13] took place; all the same, research has demonstrated that enormous influence was exercised by traditions that had nothing to do with modernity in genealogical terms. Likewise, it is true that Enlightenment science identified the causal relations underlying natural phenomena and advertised its success at 'unmasking' putative wonders. Finally, 'wonder' indeed ceased to provide a rubric under which objects of scientific investigation were examined. Nevertheless, it was precisely 'long-known' objects such as fossils, monstrosities, machines, automata, microscopic close-ups and astronomic perspectives that commanded scientific attention in the nineteenth century and inspired the staging of wondrous states and facts.

There can be no doubt that a transformation occurred around 1800. All the same, the victory march of modern natural science hardly exhausted the significance of wonders. Instead, it was precisely here, in the realm of science and technology, that they flourished – in terms of content, structure, aesthetics, as well as rhetoric. Wonders did not represent foreign bodies or carryovers from premodern times; instead, they constituted a key feature of modernity. The Enlightenment may have managed to exclude 'wonder' as an explanatory category in the accelerating sciences, but the magical-mythical and aesthetic-poetic dimensions defining it as a dispositive of experience remained unaffected.

To account for modern wonders' modes of apparition, some preliminary clarification is required. What is it that defines them – especially in contrast to what is normally taken to characterize modernity? A distinction must be observed between 'wonder' and 'the wondrous'. The point is to avoid the common mistake of essentializing or ontologizing wonders. Needless to say, the study at hand does not seek to answer the question, posed time and again, whether miracles really do or do not occur; instead, it stands as historical fact that wonders are *cultural* phenomena. The matter involves representations and practices that achieve expression – or, alternately, are given form – in texts, images, artifacts, performances and stagings. Accordingly, our point of departure is that human beings experience wonder and voice their amazement. Thus, in this study, 'wonder' refers to a category that classifies perceptions and experiences culturally and determines discursive positions. 'The wondrous', on the other hand, is understood in terms of media-based productivity that is supposed to lend expression to wonderment, convey it or evoke it. The wondrous is distinguished by self-reflection and the creative ambition of furthering intensive experiences. Taken together, 'wonder' and 'the wondrous' – which condition each other mutually – are meant to enable a form of historical observation that does not view them as the opposite of 'rationality', but instead demonstrates how they are constructively arranged in the aesthetic field. The focus, then, is the interplay of concepts, discourses and cultural and medial practices where historically variable modes of encounter with, and valuations of, the experience of wonder emerge.

On the Difficulties Posed by Grand Narratives for History

In recent research, consensus largely holds that the history of wonders cannot be conceived as a linear narrative. The objection also concerns one of the most influential works in the field, Hans Blumenberg's *Der Prozeß der theoretischen Neugierde* (The Process of Theoretical Curiosity) (1966).[14] In Blumenberg's account, the legitimacy of theoretical engagement with the world on the model of ancient philosophy – chiefly through the works of Plato, Aristotle and the Stoics – was followed by the Church Fathers' condemnation of curiosity as a vice. Then, after moral reservations had weakened in the course of the sixteenth century, the 'scientific revolution' finally swept them away. The Enlightenment inaugurated the systemic investigation of nature and introduced the world of possibilities offered by modern science.

This perspective – which is presented in radically abbreviated form here, and strictly for heuristic purposes – proves very seductive. All critique notwithstanding, it remains widespread today inasmuch as it provides the mental framework for modern understandings of history. The heroes of science are supposed to have freed natural, human curiosity from the strictures imposed by dogmatic clerical authority, thereby inaugurating the progressive unfolding of superior intellectual understanding. The point is not to bring up the crude and schematic nature of this model yet again. Instead, it is a matter of investigating the possibility of continuity. As William Bouwsma aptly notes in an essay on Blumenberg: '[A]lthough most of us have had to five up on the assurance of the Enlightenment, it remains a temptation to which we would still like to believe we can yield'.[15] Accordingly, any self-reflective study must soon recognize that, even if the 'grand narrative' no longer proves appropriate in a methodological sense, it hardly seems possible to do without it when writing the history of wonder or curiosity.[16] For this reason, Neil Kenny, the English historian of language, calls for two kinds of historiography: one that concentrates on short periods of time, or *histoire événementielle*, and another that focuses on the *longue durée* and pursues relatively stable relations of meaning over the centuries.[17] Daston and Park offer similar considerations and express doubt whether an account of curiosity as a story of success – how mortal sin transformed into a relatively insignificant form of human weakness – can be told. For them, wonder is a flexible concept: a set of 'sensibilities that overlapped and recurred', 'shifting its contents and its meaning in innumerable ways'.[18] At the same time, however, one cannot overlook the fact that their account, which focuses on elite culture, displays pronounced narrative elements. Daston and Park do not just describe the ascent of wonders as an integral component of the study of nature; they also present them as 'cherished elements of élite culture' at the end of the eighteenth century. 'When marvels themselves became vulgar, an epoch had closed'.[19] Ever since, an 'odor of the popular' has clung to them.[20]

Needless to say, such a summary and elitist assessment is problematic. Above all, it cannot hold for the people who were responsible for nineteenth- and twentieth-century works of wonder and enthusiastically celebrated them. To be sure, the culture of European elites offers an interesting site for exploring the relative value attached to marvels. That said, if one focuses exclusively on the members of the *Royal Society* and *Académie Royale des Sciences,* one leaves out an array of important actors who played key roles in the movement and modes of apparition of wondrous objects: craftsmen, technicians, artists, collectors, merchants, traders, travellers, conquerors and amateurs. The many excellent studies exploring the avenues of exchange between elite and popular cultures include the collection of essays edited by Pamela Smith and Paula Findlen, *Merchants and Marvels: Commerce, Science and Art in Early Modern Europe* (2002); Stephen Greenblatt's *Marvelous Possessions: The Wonder of the New World* (1991) still represents a milestone in literature on the topic. The latter work provides the point of departure for the argument to follow, which qualifies the claim that wonder – that is, the culture of wonders – wound up tainted, after 1800, by the 'odor of the popular'.

The Discovery and Conquest of the New World

The discovery of America, according to Greenblatt, witnessed the birth of the wondrous in the modern world. The event decisively changed European notions about what is possible and impossible, what counts as marvellous and what is natural. Accordingly, the following does not address what 'really' happened, but the 'representational practices that the Europeans carried with them to America ... when they tried to describe to their fellow countrymen what they saw and did'.[21] Front and centre stands the figure of *amazement,* which served to convey the experience of 'discovery' in both emotional and intellectual terms and, for this reason, gave rise to a particular practice of representation that enabled wonders to be approached with a certain kind of instrumentally rational interest.[22] With the 'wondrous riches' that subsequently flowed to Europe, reports from the New World also spread; they invoked the experience of amazement so frequently it seems they radiated an almost irresistible force. The manifold aspects of this process receive abbreviated treatment in the book at hand, but paradigmantic instances serve to illustrate symbolic thresholds in a wide-ranging history. Throughout, the aim is to show how, when the New and Old Worlds met, a 'neue Nomenclatur des Wunderbaren' (new nomenclature of the wondrous)[23] took the place of images of ancient, Christian, medieval and courtly wonder – as known from Pliny's natural history, the travel reports of Marco Polo and Mandeville, chivalric romances and pious hagiographies.

For the Christian west, the voyage of Cristobal Colón (1451–1506) marks a caesura in many respects. When Columbus sighted one of the islands of the Bahamas on 12 October 1492 and thought he was in the Far East, the European *'Weltbild'* (world picture) was convulsed, in the experience of the traveller and explorer, by amazement on a scale hitherto unknown. The wondrous encounter initially took form as lasting enchantment and Columbus repeatedly remarks in his journal, 'yo no sé ya cómo lo escriba' and 'digo que es verdad que es maravilla' (I don't know anymore how I should describe it. I say it is a true miracle).[24]

The entries of Columbus, the rapt discoverer of worlds, offer far more than a description of his inner spiritual landscape. His accounts of wonder opened the space for a new discourse of wonder – indeed, at its core, almost everything he recorded seems to be concerned with the miraculous nature of the New World.[25]

The ways that Europeans found to confront their amazement proved radical and irreversible. The very first wonder that Columbus invoked did not concern the land itself so much as possessing it. This perspective was new: in the Middle Ages, the experience of wonder had still been tied to a feeling of resourcelessness – acknowledgement of the world's manifold and impenetrable nature.[26] According to Jacques LeGoff, medieval wonders can be understood to represent a certain kind of cultural resistance to appropriation and exploitation.[27] For Columbus, on the other hand, amazement does not simply mean acknowledging the unusual; it also connotes 'a certain excess, a hyperbolic intensity, a sense of awed delight'.[28]

The conqueror's amazement seemed to extend beyond all measure. It entailed boundless mania for possession that could only be relieved by destroying the newly discovered realms of wonder and subjugating their inhabitants. In 1492, Europe initiated a campaign of discovery that colonized the world of wonders both materially and discursively. Taking possession of the marvellous did not occur only through violence; it also took place by naming and classification – confiscating, cataloguing and inventorying objects within the parameters of European-modern science. According to standard accounts, through the conquest and colonization of the New World and its treasures, Europe entered a new age of wonder. From the inception, it implied a revolution:

> New objects, new people and new knowledge flooded into people's consciousness ... The great period of wonder of the sixteenth and seventeenth centuries AD came into being as a result of an excess of novelty, and was brought to an end, as nature prescribes and Bacon and Descartes had foreseen, by a wave of explanation and classification ... Not only were the tools for preserving and developing knowledge vastly improved, but as explorers travelled to all continents the limits of knowledge of this planet were being reached. In a sense the age of wonder of the sixteenth and seventeenth centuries brought wonder to an end.[29]

This view must be challenged. When *conquistadores* shared their amazement, it did not provoke the loss of wonder, which then vanished more and more as Europeans claimed the world for themselves. On the contrary, in the centuries that followed, the astonishment that Europeans experienced shook standing categories and coordinates time and again. Wonders did not cease; rather, the wondrous proliferated. As Greenblatt observes, amazement did not yield 'knowledge of the other' in European representational practice, but its medial objectification; 'the principal faculty involved in generating these representations is not reason but imagination'.[30] Here began a varied aesthetic praxis of representing wonders, lending them a degree of visibility unknown before this time. The results were not Baroque models that then ceased production; instead, they represented a constant feature of European world conquest, renewed over and over by the influx of shiploads of exotic and wonderfully strange objects and organisms until the early twentieth century.

Wonders of the Modern World

The nineteenth century also discovered new worlds and relations no one would have dared dream of before: the 'lost worlds' of prehistory and the beginnings of time, the endless abysses of the ocean deep, invisible waves and rays and the microscopic realm of bacteria, viruses and other enigmas. The new orders of knowledge that convulsed the age fill whole libraries – most notably, perhaps, works on evolution, medicine and thermodynamics. In Vienna, the institute for biological research became known as the haunt of magicians; in Berlin, Rudolf Virchow's pathological museum was commonly referred to as the Kaiser Virchow Memorial Church almost as soon as it opened.[31] Modernity counts as the age of technological innovation, pure and simple; for the first time in the history of culture, the world was conceived – and planned – 'technomorphically'. The steam engine, railways, telegraphy, electricity, photography, cinematography, world's fairs, transatlantic cables, scientific progress and vast engineering projects attest to the ambition of obliterating all scale and dimension. Whatever heights, depths, sizes or speeds had held until now were surpassed, almost beyond measure.

Needless to say, this state of affairs differed from circumstances in the fifteenth century. At the same time, the question arises whether the modern age took a different approach or, instead, still sought out 'marvelous possessions' at the outer reaches of the known world. Either way, one cannot speak of an 'end of wonders':

> Because there are many kinds of marvelous, or at least many different ways in which it manifests itself, it cannot be said that the marvelous as a cultural force expired in the 18th century thanks to the ascendency of neoclassicism or the Enlightenment. It is true that the former established new rules of taste and that the latter preferred not to see the world as full of marvels and mysteries, wondrous to contemplate, but as a site of social, political, and economic problems to be solved, forces to be tamed, including

what it regarded as religious superstition and absolutist tyranny.

> The marvelous is ... not a transient moment, an aberration of the baroque, but a continuing quest to explore and express the essential joy and pain of existence in a still mysterious world.[32]

Continuity and Change

The work at hand shares the view, also presented in other studies, that modernity abounds in talk of wonders and marvels.

> Yet, when we survey our own culture and consider its fascination with things both large and small (skyscrapers and microchips), with spectacular human accomplishment (the landing on the moon or the prodigious talent of a child violonist), with the discoveries made by the space probe *Voyager*, with the special effects of movies, and with fireworks, parades, and extraordinary technological inventions, than it seems that a love for the marvelous has not been extinguished, but is rather, still very much with us.[33]

Compared to those of earlier times, modern wonders involve the transformation and recombination of elements. For the seventeenth century, the glowing calf's head described by Robert Boyle (1627–91) represents a typical object of wonder. The item in question was altogether familiar; nor is it likely that this was the first specimen displaying a strange radiance. 'Suddenly', however, Boyle and his employees had become aware of the phenomenon. In the mid-nineteenth century, something similar occurred when sea anemones and algae were declared wondrous. These forms of life had led a 'quiet' existence for centuries on the coasts of England – now, the accounts offered by natural scientists made them into marvels to be sought out and collected.

To understand the qualification of 'wonder', it is important to consider the different media employed, which aestheticize – that is, make available to the human senses – a natural world which would otherwise prove inaccessible. Both in the seventeenth and nineteenth centuries, for instance, looking through microscopes disclosed marvellous new realms. In equal measure, Baroque observers and later nature-lovers would speak of the wonders they beheld in the tiniest dimensions.

Of course, transformation and new contextualizations occurred alongside continuities. In the nineteenth century, objects that had long counted as wondrous – e.g. fossils and monsters – still featured in the canon of scientific marvels. Even if opinion no longer held that they were the skeletons of giants or or cyclops-skulls, but instead the remains of dinosaurs and mammoths, they continued to be presented at world's fairs and as 'Wonders of the Natural History Museum'.[34] In the following, it is less a matter of taking stock of modern wonders than of describing exemplary sites and techniques of display, which bear on the forms wonders assumed after they supposedly vanished around 1800.

Vertical and Temporal Dimensions of Wonder

To account for the particular inflection of wonder in modern society calls for a history of medially supported stimuli and displacements of attention. Such an approach brings into view a new type of wonder that no longer occupies the cartographic 'margins'. Modern marvels are not necessarily situated at the ends of the earth. They do not need to be rare – only to reveal what remains ungraspable in surroundings that appear familiar and well-known.

An initial overview of modern wonders yields two complementary forms. Items and beings of wonder are no longer defined in terms of faraway or exotic provenance. To be sure, shiploads of strange people, animals and plants continued to arrive from distance places. But even if they played an important role in the nineteenth and early twentieth centuries, they did not define the novelty of the age. Indeed, a feature of the times was the disappearance of 'blank spaces' on maps. By the end of the eighteenth century, the outlines of the world had largely been charted, and most of its parts were already 'discovered' and claimed. Less and less room seemed available for expansion. In consequence, the nineteenth century witnessed struggles among empires competing to carve up the few regions that remained unoccupied and practically unexplored. Above all, such lands were found at the poles, in mountain ranges and parts of Africa and Asia. Here, the foreign and the unusual beckoned – but not the radically other or the completely impossible. As a corollary of the penetration of the world by science and technology, the idea of mysterious regions and marvels of nature hitherto unseen exercised more and more fascination. The chief spaces promising a 'fresh' perspective on the unknown were connected to elements inhabited by other species – the fish in the water and the birds in the sky. As focus shifted from the horizontal axis to the vertical, a new dimension of wonder was disclosed. No longer reaching 'far afield' or to the 'end of the world', wonder was now a matter of expanding potential experience as much as possible. 'Wie ein Fisch sein, wie ein Vogel sein, wie Gott sein (eine extraterrestre Position einnehmen): das Ziel der Anstrengung ist es, sich das ureigene Element einer anderen Art anzueignen' (Like a fish, like a bird, like God (to swing up to an extraterrestrial viewpoint): the struggle aims at adopting the innate element of another mode of life)[35] – and, in the process, to change the conditions of one's own existence. The marvels of modernity – especially, scientific inventions, 'epistemological breakthroughs' and the mise-en-scène of the 'wonders of technology and nature' (whereby 'decoding' amounted to dominion, of course) – asymptotically approached what remained unattainable as a matter of principle. 'Movement along the depth axis is consistently associated with a transformation, be it a change of worlds or condition'. Accordingly, voyage into the sky or the ocean depths does not simply offer a ticket to a foreign world – it leads to a different ontological status, another 'state of mind'.[36]

The centrepiece of the study at hand consists of technological artifacts and media-aesthetic strategies that provide access to the vertical dimension. Attention falls on an element in which invention and inventiveness are inscribed from the inception – one that involves temporalising the marvellous, as it were. The unknown realms of the ocean depths represent the horizon where these wonders appear. Here, one encounters what Rosalind Williams has remarked of the 'boom' in subterranean worlds that occurred in the nineteenth century:

> And as imagination reached further into the past, it also leapt further into the future. The sight of the ruins of past civilizations inevitably suggested what the present would be in the future, when what was now on the surface would become part of the buried past.[37]

A widespread perception precipitated from this solution:

> [T]hose who venture into subterranean or submarine regions are both time travellers and spiritual pilgrims. They plunge below surface material reality to the truth that lies hidden below. The lower world is, paradoxically, identified with higher truth.[38]

Where the wonders of the sea stand at issue, discourse both stretches into the past, back to when life first emerged and assumed its manifold forms, and toward the future, which promises improvement – if not a return to paradise.[39] The marvels of the deep provide aesthetic evidence for the 'law of progress' in evolutionary biology, which applies to the history of peoples and cultures as much as to that of animals and plants.

Ocean-Dwellers: Miraculous Life

Verticalization and temporalization, as the *novum* of modern wonder, concern selection. The point is not to define all marvels after 1800 in this way; as indicated, many 'old-fashioned', 'horizontal' wonders occurred in changed form and new medial arrangements. However, the intellectual panorama at issue may be seen in the arc extending from wonders in the sky to those in the sea. Since time immemorial, the two spheres have been related: an oceanic dimension has been inscribed in the discourse of flight from the inception; by the same token, voyages by sea are often described as flying in the heavens above:

> It has often been said that studying the depths of the sea is like hovering in a balloon high above an unknown land which is hidden by clouds, for it is a peculiarity of oceanic research that direct observations of the abyss are impracticable. Instead of the complete picture which vision gives, we have to rely upon a patiently put together mosaic representation of the discoveries made from time to time by sinking instruments and appliances into the deep, and bringing to the surface material for examination and study.[40]

This study follows a monograph on the conquest of aerial space and its oceanic dimensions;[41] now, attention goes in the opposite direction: by way of exemplary cases, it describes the emergence of a wondrous science of the sea that, metaphorically, constitutes an entire cosmos. Comprehensiveness is not the aim. When discussion concerns, e.g. shells, fossils or oceanographic research, the point is not to narrate *the* history of shell-collecting, fossil-interpretation or marine research – especially since excellent historical treatments aleady exist on the emergence and development of paleontology, geology and oceanography. Instead, the items named are conceived as catalysts which, in different forms of medialization, came to provide objects of curiosity and amazement. Standing in for the sea, they represent the varying forms of mankind's paradoxical and ambivalent relation to the ocean – which, until now, scholars have not discussed in terms of themes of wonder. Accordingly, the focus is how the perception and experience of the sea, in the west, are interwoven with a particular history of knowledge, imagination and the medial and aesthetic practices they involve. As in the earlier study on the 'wonders of technology' that permitted the 'conquest of the the skies', the concepts of wonder and ignorance are not equated. Just as the 'wondrous works of technology' do not imply that the technology in question has not been understood, the 'marvels of nature' hardly mean that nature defies comprehension. Instead, the experience of amazement and admiration means an intensified encounter with nature in its vast and extravagant dimensions, which only grows as more knowledge is gained (e.g. in terms of evolutionary biology). In the words of Matthias Jakob Schleiden (1804–81), the nineteenth century found in the sea 'eine so unerschöpfliche Zeugungskraft der Nautur, wie sie uns in gleicher Fülle nirgends sonst auf Erden entgegentritt' (an inexhaustible procreativeness of nature which cannot be found anywhere else on earth).[42] Especially in this context, this means rejecting historiographic clichés of mankind's primal fear of the ocean and its forbidding foreignness – which are supposed to have dominated European culture for millennia before finally being dispelled in the nineteenth century. After all, Egyptian and Greco-Roman cosmogonies already featured the idea of a creative, primal ocean and ascribed to it opposing metaphysical qualities, viewing it as both masculine and feminine (e.g. in the Greek gods Okeanos and Thetys), gently lapping and furiously raging, a mirror-like surface and a black abyss, beautiful and terrifying and fertile and fatal.[43] The aim, then, is to observe transformation as well as continuity in the way submarine wonders were encountered and represented in the nineteenth century; despite all the discoveries the age witnessed, the mobility of marine objects of wonder hardly vanished. That is: even though, following the Enlightenment and in a time of progress, the sea began to change, in the words of Jean Delumeau, from a kind of taboo and site of fear *par excellence*[44] into a landscape of scientific research and investigation, this did not mean that it ceased to be perceived as a realm of wonders and secrets; if anything, this dimension thrived in ways unimaginable until then – as it still does now.

1 WONDROUS AND TERRIBLE SEA – STATIONS OF AN UNKNOWN MODERNITY

For ages, the sea was seen as a fluid surface stretching over an abyss as dark as the night. Skill and understanding might enable one to travel the realm of waves, but it seemed that only death lay in the depths. Although cultural entertainments included grottoes, fountains and artificial lakes, the general view of the world had not yet incorporated narrative or visual forms of the deep sea. Historians largely agree that the underwater realm did not make a large impression until around 1800. Conquerors, discoverers and researchers held to a terrestrial orientation.[1] 'Maritime realities were not easily understood in the early modern era because philosophical and scientific concepts that made the sea appear foreign and hostile enjoyed great prestige and authority'.[2] Then, in the early nineteenth century, enthusiasm for the ocean penetrated all social orders. The public discovered the seaside, plunged into the water and gathered anything and everything they could for examination.

Standard accounts stress the relationship between the discovery of the coast and the emergence of railways in the first half of the nineteenth century. For all that, however, it was already fashionable in mid-eighteenth-century England to spend holidays at the shore. As early as 1660, a certain 'Dr. Wittie of Scarborough'[3] had extolled the salubrious effects of a seaside stay with moderate enjoyment of the waters; by 1735, Scarborough had come to be known as a popular resort town. The writings of the English physician Richard Russell (1687–1759) – *De tabe glandulari* (1759) and *A Dissertation on the Use of Sea-Water* (1752) – had helped 'beachcombing', 'sea-bathing' and knowledgeable discourse about algae and seaweed to become preferred pastimes of the upper classes.[4] Finally, in 1787, the Prince of Wales, Prince Regent and future King George IV (1762–1830) commissioned the architect John Nash (1752–1835) to construct a 'royal pavilion' in Brighton. The same year, William Cowper (1781–1800) composed an ironic encomium of a seaside holiday:

> In coaches, chaises, caravans and hoys
> Fly to the coast for daily, nighlty joys,
> And all, impatient of Nylond, agree
> With one consent, to rush into the sea.[5]

While French emigrants viewed the young British *macaronies* (dandies) in horror – they threw themselves into the waves even at ebb tide and when weather conditions were terrible – artists and writers discovered the comical side of resort-life. Theatre pieces such as *A Trip to Scarborough*,[6] drawings and caricatures by James Gillray (1757–1815), George Cruikshank (1792–1878) and Robert Seymour (1798–1836) and novels like *The Expedition of Humphry Clinker*[7] portray an age in which swimming in the sea counted as a comical and extravagant matter, yet by no means an infrequent occurrence.

Still, by making travel to the sea easier and facilitating the transport of marine life back to the metropolis and salons, the rails exercised a strong influence on the evolution of marine biology; public interest was either roused or heightened to examine creatures of the coasts and the seas.[8] The same holds for the research instruments that were invented. Scholars agree that the introduction of improved and, more importantly, affordable microscopes in the 1830s stimulated amateur scientific expeditions to the shore and even out onto the water. Within a short time, examining the tiniest aquatic organisms came to provide a universally prized amusement, and images of the microscopic world became a central component of visual culture. The new microscopes undoubtedly played a part in the emergence of greater interest for natural science – especially as they became increasingly available to the middle classes. But for all that, the microscopic viewpoint, taking account of even the smallest phenomena, had already proven fashionable at eighteenth-century courts, where microscopes provided an instrument of amusement for the loftier estates.[9] It is surely an exaggeration to contend that a sudden and fundamental change in ways of seeing prompted the sea to transform from a hideous abyss into a site for romantic reverie because new modes of transportation and achromatic lenses had now been introduced.[10] Could an 'aversion to the sea reaching back millennia' dissolve into thin air simply because it was possible to travel faster, and in greater comfort, into this deadly realm and study its horrors magnified to the -*nth* degree? The claim seems just as unlikely as the (related) notion that a paradigmatic change of habits of perception might have occurred 'suddenly', owing simply to technological developments. At any rate, Alain Corbin's extensive study, *The Lure of the Sea: Discovery of the Seaside in the Western World 1750–1840,* affirms that in the course of roughly one hundred years, mortal fear of forbidding, black waters transformed into rich and multifaceted fascination. Between 1750 and 1840, Corbin maintains, there emerged a yearning for the sea and shore that replaced a knotted mass of representations which had reinforced mankind's age-old horror of the oceanic depths for centuries.[11] But for all that, the history of such 'horror' is hardly continuous. Even though negative associations predominated from antiquity until the Enlightenment, there had always been admiring and curious views of the ocean and marine life; they still have not received due attention.

The thesis of the work at hand is that the relationship between humankind and the sea did not simply change from bad to good in the course of enlightenment; from the inception, it was – and remained – ambivalent and paradoxical. This is not to deny that the nineteenth century became enamoured of the ocean in a way without precedent. Rather, the point is to bring out, against the background of changing attitudes, the many shades of meaning, metamorphoses and aesthetic effects arising from contradictory conceptions of the sea as the utterly Other.

The Sea Before 1800

Creation, Disaster and Safe Passage

'In the beginning God created the heaven and the earth. Now the earth was formless and empty, darkness was over the surface of the deep, and the Spirit of God was hovering over the waters'. The opening verses of Genesis – the first Book of Moses – form the core of canonical representations of Creation for the Western world. However, they do not answer one question: Did God also create the waters upon which His Spirit then moved?[12] Verse six reads: 'Let there be a vault between the waters to separate water from water'. The next four verses describe how God created such a firmament and called it 'Heaven', made the waters underneath gather into bodies called 'Seas', and named the dry regions that resulted 'Earth'. Nothing is said about the origin of water. As Wolfgang Detel has observed, in the first two verses of the Genesis there are two different versions of the world's creation: the younger one imagines God as a pure spiritual being who creates the world out of nothing; in the older one the water is already existing prior to the act of creation and God is not its creator, but the one who shapes and enlivens it.[13]

The older version contains an echo of the essential meaning of water: the most primordial element of the Creation-myth, pre-existing Biblical cosmogony. The more recent version offers greater appeal in soteriological terms: not only does it give God dominion over the course of the waters, it puts power over the *being* of water in His hands. Accordingly, when describing the heavenly state of redemption in the Book of Revelation, John declares that he witnessed 'there was no more sea'.[14]

Thereby, we have also reached the dark side of water imagery in the West. Despite the positive qualities that water possesses in Christian worship – as a means of purification, atonement and rebirth – the sea long counted as the abode and medium of uncanny, ravenous demons. In the Christian imaginary, water is the means of the most terrible of punishments, culminating in God's reversal of Creation when he opened the floodgates of the sky and subterranean depths. All life that did not find refuge in the ark was consigned to doom. The history of Biblical motifs has never been able to separate itself from this catastrophe. God may well have created the rainbow to signify reconciliation and love for mankind, and the Bible may abound in miracles such as events at the Sea of Reeds and wells springing from barren rocks, yet the Christian understanding of water remains fundamentally shaped by the myth of the Deluge. In consequence, the sea has represented 'the remnant of the disaster', 'an instrument of punishment' and 'an abyss full of debris'.[15] Its constant restlessness seems to warn of another Flood. The 'apocalyptic age' of the Reformation was gripped by the particularly intense fear of destruction in watery form. Even though the final cataclysm of the Apocalypse was supposed to occur through divine fire, it seemed most likely that the catastrophe would begin with the rage of the sea. *Artes moriendi*, which circulated widely from the fifteenth century on, granted a significant role to water among the fifteen premonitions heralding the 'Return of Our Lord'. Waves will flood the mountains; fish and other sea monsters will rise from the deep with ghastly howls. The ocean and all waters will blaze and bellow at fire falling from the sky.

In myths of the flood, a universal experience transforms into an image: archaic human fear is concentrated in the indomitable sea. The sea conveys what is inherently uncanny and prodigious about the element of water.[16] One way to free oneself from the fear inspired by the sea was to conquer it. Accordingly, the struggle between 'man' and the sea numbers among the archetypal narratives of seafaring peoples. The wanderings of Odysseus first lent it the form of heroic epic, giving catastrophic experiences at sea – storms, shipwrecks and death – the shape of collective song. At the same time, general fear of destructive forces was countered by the seafarers' pride and the value they attached to nautical travel as the basis for all the achievements of civilization. 'Many wonders there be', the chorus declares in Sophocles's *Antigone*, 'but naught more wondrous than man: Over the surging sea, with a whitening south wind wan, Through the foam of the firth'[.][17]

In Poseidon, the sea takes the stage as a mighty adversary. Its floods harbour horrors: the terrible Keto, mother of the viper-haired Gorgons, the *halioi gerontes*, 'Old Men of the Sea' and other *monstra marina*. Even so, the deep is also home to the divine Nereids, whom Hesiod describes in his *Theogony* as rosy and delightful, graceful and kindly, beautifully proportioned and alluring. They stand at the beginning of the tradition of a marine world outside of time, where sons and daughters of the deep kiss and embrace, and splendid chariots fashioned of shells and drawn by majestic hippocamps race over a sea bathed in bright flames. This dance of mythological figures remains just as present in Western visual culture today as the Roman fondness for views of the open sea. Excavations at Pompeii and Herculaneum, for example, have revealed countless terraces, pavillions, gardens and colonnades attesting to the pleasure residents took in contemplating the sea; two thousand years later, their sentiments echo in the sighs of visitors, who still flock to these sites.

The ancient view of the sea as part of an Arcadian landscape resembles the paradisiacal images that attended Christian voyages to sea. It is no accident, Horst Bredekamp observes, that Columbus was obsessed by the idea that the mastering of the sea would be identical with the revelation of the Eschaton.[18] Consequently Columbus considered himself the instrument of divine revelation and Christ's second coming. When he reached the West Indies, he thought he was beholding what was left of earthly paradise: the Orinoco, for example, seemed to be one of the four mythical rivers running through the Garden of Eden. Fittingly, his contemporaries celebrated the discovery of the New World as an event equalling the world's creation at the beginning of time. With utmost clarity, such eschatologically tinged pathos reflected belief that the ancient terror of the sea had been overcome – a sense of triumph due to technological mastery. Even though fear of the unfathomable and seemingly infinite dimensions of the sea would persist until the late-twentieth century, Columbus's crossing of the Atlantic inaugurated a new epoch in the relationship between man and the sea.

For all that, the first journey to the New World did not allay fears of the ocean. Discoveries of the alien expanses overseas were attended by polyphonous echoes: sailors' accounts of savage cannibals and nightmarish marine life fed into the collective imaginary. Over the course of the sixteenth century, the sea remained a no-man's-land of untamed and untameable elements, beasts and demonic beings. Countries might fight for sovereignty over marine realms, but the sea itself escaped all governmental control. Contemporary maps present the ocean as inhabited by murderous whales, flying fish, enormous crabs and huge serpents; here and there, they show ships falling prey to the abominations' insatiable hunger. Then, during the seventeenth century, the scenario began to change slowly. Instead of monsters representing the sea's indomitable nature, more decorative and fanciful figures emerged, which seemed to follow human efforts with curiosity and interest.

> In this shift from the ocean's representation as a terrifying wild wherein societies and nature interact to its representation as an empty space to be crossed by atomistic ships, one can see the beginnings of the ocean-space construction (and the scientific outlook) that was to predominate during the following centuries of industrial capitalism.[19]

Territorialization, Gernot and Hartmut Böhme observe, put an end to the age-old history of fear of the sea. Achieving mastery, by technical and technological means, is a matter of neutralizing the fear that always remains active in the affective ambivalence of elements understood in mythological terms.[20] Needless to say, such designs do not unfold as readily as one is led to believe by the claim that the seas were gradually 'colonised' from 1500 onwards. Pathos and triumphant eruptions of sentiment at new dominions acquired in vast oceanic space are recorded well into modern times. Still in 1859, Edward Forbes (1815–54) – one of the most well-known representatives of the first generation of deep-sea explorers – adopted the perspective of a Spanish conquistador when recounting the discovery of a new world:

> When a whole dredgeful of living creatures from the unexplored depth appeared, it was as if we had alighted upon a city of the unknown people ... And when, at close of day

our active labours over, we counted the bodies of the slain, or curiously watched the proceedings of those whom we had selected as prisoners, and confined in crystal vases, filled
with limited allowance of their native element, our feelings, of exultation were as vivid,
and surely as pardonable, as the triumphant satisfaction of some old Spanisch 'Conquistador', musing over his siege of a wondrous Astlan city, and reckoning the number
of pointed Indians he had brought to the ground by the prowess of his stalwart arm.[21]

Treasures of the Sea

The consensus among historians is that the sea counted as an accursed empire of
darkness in early modernity. Agitated by devilish powers, it offered doomed creatures a place where each of the inhabitants consumed the other in turn. As Corbin
puts it, the ocean represented the 'world turned upside-down' and the epitome of
unreason.[22] To be sure, this view of the sea was widespread, yet it did not prevail
everywhere or at all times. In the fifteenth century, at the latest, all European courts
knew that the ocean concealed inconceivable treasures: pearls, coral, shells and snails.

Already in antiquity, coral had held exceptional significance: its many
forms represented the world as a whole, symbolizing the overabundance of artful Nature, which here revealed to mortals its full beauty in petrified form.[23]
Baroque cabinets of wonder practically overflowed with the strangest marine
organisms: dried seahorses, turtle shells, shark teeth, narwhal horns, crabs and
the spears of swordfish. Such specimens aroused intense curiosity. Even if the sea
was not deemed a 'cabinet of wonder' itself, it represented the treasure house of
a miraculous and exceptional world. The sea contained 'masterworks' of both
beauty and terror – but above all, an inconceivable multiplicity of forms of life.

Fittingly, Conrad Gesner's (1516–65) *Fisch-Buch* (Book of the Fishes),
which appeared in 1558, lists almost 800 different marine organisms.[24] In the
foreword, the author refers to the multiplicity of life-forms he has gathered as
proof for the miracles God has wrought in the depths of the sea; the same sense
of wonder runs through the book as a whole. Gesner's *Fisch-Buch* was the most
comprehensive work on marine life compiled until this point. It built on the
works of Adam Lonicer (1528–68), Guillaume Rondelet (1507–66) and Pierre
Belon (1517–64),[25] and, more still, on Aristotle's studies of marine biology in
De generatione animalium (which was widely read in the sixteenth century).
Gesner had drawings and specimens sent to him from all over Europe, and he
spent several months at the fishmarket in Venice studying many a Mediterranean 'catch of the day'. At the same time, his book compiled all the beasts and
fantastic creatures of the deep that were known at the time – whatever imagination, sailors' yarns and clever combinations of fish skin and seal bone presented.
The *Fisch-Buch* mirrors an age when geographical discoveries were increasing the
domain of natural history exponentially; each year, shiploads of remarkable and
altogether improbable animals and plants arrived in the harbours of Europe. (In

turn, the great expeditions of the seventeenth, eighteenth and nineteenth centuries brought forth an entirely new picture of the globe; it ultimately prevailed, not least of all, because of the curiosity and amazement generated by its convincing aesthetics, exceptional sense of beauty and marvelous splendour). The illustrations and reports reflect contemporary enthusiasm for encounters with a world that was new – still unexplained and not yet fully sounded; evidently, many people expected that discoveries would liberate them from established ways of thinking. In this context, the natural sciences considered it a matter of course not to sift out reports which seemed incredible; rather, they should be published for the purpose of clarification. Accordingly, it does not speak for Gesner's 'ignorance' or 'naivete' so much as his circumspection and attentiveness that his works include everything about sea monsters and demons that stirred the early-modern soul. One reads of 'serpent whales' (which still ghosted about the sea as late as the twentieth century),[26] turtles and other 'most miraculous' inhabitants of the deep – not just Gesner's favourites, but also those of the Baroque Age, which was enamored at all things wondrous – including 'flowers of the sea' and 'blossoms from the earth's core': shellfish and fossils.

The World in an Oyster: The Mysteries of Shells and Snails

Molluscs, it seems, exercise a singular aesthetic fascination for the European eye. The immeasurable diversity of shape and colour in their housing, to say nothing of the inhabitants dwelling within, casts a spell that elicits amazement to this day. In *Poetics of Space,* Gaston Bachelard lends it expression in an apt question: 'Is it possible for a creature to remain alive inside stone, inside this piece of stone?' Conchylian wonder, according to Bachelard, represents an enduring topos in the history of the Western imagination.[27] Time and again, it has witnessed astonishing beings emerge from shells.

As Jurgis Baltrusaitis has observed, ancient gemstones already featured the motif of the spiral abalone bringing forth the most unexpected creatures all at once. A dog jumps out of a snail shell, a carnivore is hiding in it.[28] Indeed, Aphrodite herself came into the world in this manner, rising up out of a violet snail or mussel 'womb'. Venus emerged from the shell that bears her name – as did Eros (Amor), although he sometimes came out of an abalone. The genesis of gods in the watery element concealed a deeper meaning. The sea counted as the point of origin for all forms of life, whether they then dwelt in the water or on land; the horse in Poseidon's train came from the sea, as well.

The Middle Ages took up the motifs and symbols of conchylian nativity. Hares, birds, stags and dogs spring from shells as if from a magician's hat, transforming the ancient image into a scene belonging to the visual culture of medieval Europe. By the same token, rosettes, fashioned from shells and jewelry, shaped like the conch of Venus featured prominently in all designs and appointments of luxury and refinement. Such enthusiasm reveals a mode of perception that saw something miraculous at work in these fascinatingly regular, smooth and shiny formations from the ocean depths. After all, didn't conches magically preserve the sound of the rushing waves of the faraway sea from which they had come?[29]

Shellfish were strange creatures that held the Baroque Age spellbound. Starting in the seventeenth century, at the latest, the whole of Europe showed signs of 'conchyliomania' (although fascination had already gripped natural historians and artists at the middle of the preceding century).[30] Large and important collections of shells were made, some of which still constitute a significant portion of the holdings of museums of natural history (to say nothing of countless, private collections). *The Universal Conchologist* (1784), by Thomas Martyn, includes a brief list of the shells accumulated by the Countess of Bute. The author remarks: 'These private collections often lacked the classified rigour of the didactic ones formed by serious scientists, but they showed how fashionable an interest in natural history had become'.[31] The boom that ensued occurred above all because shells united, in a single form, all the oppositions between art and nature that had come to be important for the Baroque Age.[32] Moreover, their marine origin recalled the origin of all life in the primordial ocean; thereby, it connected with one of the greatest mysteries that the new sciences eagerly sought to 'unveil'.

As we have observed, the sea had already represented the source of life in antiquity. Aristotle, for example, in his work on reproduction, *De generatione animalium*, had used mussels to illustrate the process of *creatio ex nihilo*.[33] He viewed these creatures as the products of primordial generation pure and simple, for they emerged spontaneously when the sun warmed the dense mud of the ocean floor.

Over the following centuries, the idea that life had started in water resurfaced, but it achieved real weight beginning with early modern studies of the natural world. In the early seventeenth century, a veritable uproar attended the publication of *Telliamed,* a work by the French consul Benoît de Maillet (1659–1738). Starting with the basic idea that water must once have covered the entire globe, de Maillet articulated the main considerations and lines of thinking for living organisms' capacity for metamorphosis. Over the course of enormous stretches of time, the sea level had sunk. De Maillet concluded that all organic beings – plants and animals, including human beings – must trace their origins back to the water; it was necessary, then, to assume that a general transmutation of species had occurred (on the basis of factors such as environment, climate, and the like), leading water-dwellers to become land creatures. Needless to say, the claim commanded attention and elicited almost as many howls of protest.[34]

However, the early modern way of seeing brought new nuances to an image of the ocean already viewed in positive terms. Voices that spoke of the seas' unique fertility multiplied. Aristotle had already declared that aquatic life presents a richer array of species and shapes inasmuch as the element they inhabit generally proves 'more quickening' and is better suited to producing physical form.[35] In a certain sense, this ancient line of thinking had never broken off entirely. In *De Beata Vita*, for example, Augustine refers to the sea as bearing the seed of life, even though he also sees it as a mirror of death.[36]

Then, toward the end of the seventeenth century, the adherents of Natural Theology discovered the riches of the sea. Their hymns of praise presented a sea glowing with what remained of earthly paradise, a blessed treasure house of immeasurable abundance. The celebrated physico-theologian Johann Albert Fabricius (1668–1736), for instance, considered the sea a central element in both aquatic and geological cycles; it was not devoid of meaning or life, but teemed with the most interesting organisms.[37] The infinite ocean depths, he averred, contain mountains and valleys, open landscapes and plains, flowers and orchards and forests and meadows. The works of John Ray (1627–1705) also declare that the forms sea creatures can assume are boundless. The gleam of shells, the colourful pomp of corals, and the incomparable lustre of pearls reveal the majesty of God's Creation. Gradually, the picture emerged of submarine nature as the secret refuge of the wonders of the created universe, where unfathomable, enigmatic Providence hides its treasures.[38]

Increasingly, the multiplicity of marine life-forms – or, alternately, the realm of molluscs – provided the object of aesthetic appreciation and discourse. Gesner's *Fisch-Buch* was hardly the sole work to feature illustrations of marine organisms. Others included those of Lonicer, Belon, Rondelet, Adriaen Coenen (1514–87), and Ulisse Aldrovandi – as well as the *Libri Picturati (c.* 1560).[39] In the seventeenth and eighteenth centuries, majestic works exercised widespread fascination: books by Nicolaes de Bruyn (1571–1656), Adrien de Collert (1560–1618), Albert Flamen (1620–93), Crispin van de Pas (1598–1670), John Johnston (1603–75), Francis Willughby (1635–72), Louis Ferdinand Marsili (1658–1730), Jakob Theodor Klein (1685–1759), Edmé Louis Billardon de Sauvigny (1736–1812), Karl Freiherr von Meidinger (1750–1820), Marcus Elieser Bloch (1723–99) and Bernhard Germaine de Lacépède (1756–1825).[40] Significant studies of conchyliology included illustrated volumes by Fabio Colonna (1567–1650) and Basil and Rupert Besler (1561–1629 and 1607–1661);[41] more notable still were *Ricreatione dell' occhio e della mente* (1681) by Filippo Buonannis (1638–1725), *Historia Conchyliorum* (1685–92) by Martin Lister (1638–1712), and Georg Eberhard Rumphius's (1627–1702) *Rariteitkamer* (1705). Eighteenth-century standouts include Antoine Joseph Dézallier d'Argentville's (1680–1765) *Conchyliologie* (1780) and, to be sure, *Conchylien-Cabinet* (1786–95), which the authors, Friedrich Martini and Johann Chemnitz, designed as a virtual museum on paper. The list is far from exhaustive. These and other illustrated works of natural history demonstrate that marine organisms occupied a secure position in the visual culture of the day and enjoyed great aesthetic popularity.

Well into the nineteenth century, the descriptive and illustrative style of compendia of this sort expressed the fond gaze of collectors more than it followed the systematizing logic of natural scientists. Each individual specimen should receive its due; every single shell was to be presented in its particular physiognomy. Accordingly, the books present row after row of miniature portraits showing incomparably unique creatures; the new breed of 'conchylia-writers' approached the individual texture of each specimen with a soft touch.

In the texts accompanying the images, one encounters flaps and wings, fabrics the colour of gingerbread, crooked muzzles, ladies' gowns, princely raiment and Berber tents. No limit was imposed on the analogies and comparisons that authors' descriptive fancy might make; indeed, it seems that writers sought to outdo nature itself, competing by means of poetry.[42] The shell posed a riddle – and not just in its array of thousands of colours. A particular mystery radiated from its biological design, the architecture of its turns and twists, the morphological contrasts between delicate and solid aspects, and the aesthetic play alternating between clarity and melting colours, capricious inventions and elegant simplicity. Writers explored how one form developed out of another; in the process, urgent questions arose about the orientation of time and space. As Karin Leonhard has demonstrated in an outstanding essay on the shell as a symbolic form, the spiral of the nautilus shaped the Baroque world-picture. It 'became a paradigm for the passage of time and historical growth – space-time made visible'.[43]

The shell became both the aesthetic and the epistemic marvel of Baroque curiosity. Its prominent position in the cosmic order of seventeenth-century natural science derives from its spiral form, which expresses its connection to the expanses of the sea, space and time – which, ultimately, are related. In historical terms, the point of departure was the examination of marine life-forms, and especially shellfish, which also stood at the centre of encounters with fossils – a petrified age's representatives in the future.

A Sea of Stone: Fossils as Messages in a Bottle

Fossils were objects in which the most varied forms of insight intersected. Into the seventeenth century, early modern natural history still viewed them as lithic formations, above all: interesting mineral forms that one collected as pieces of primordial generation that had not advanced and managed to achieve life.[44] As such, they were not seen as the remains of organic beings from an earlier time, but rather as 'jokes of nature', *ludi naturae*.[45] That is, they represented transitional forms between the mineral, vegetable and animal realms. However, the perspective started to change in the mid-sixteenth century, as efforts were made to depict the caprices of nature within a system of classification. Marine fossils formed the focus; they counted as the simplest objects because they displayed similarity to living species. In the course of these scientific encounters, petrified organisms came to unite the newly emergent figures of biological and geological knowledge in an aesthetic riddle; thereby, the space where they originated, the sea, was brought into the consciousness of the age as an epistemic dimension to be unveiled.

The earliest known illustrations of fossils were two shells in *De Re Metallica* (1557), by Christopher Encelius (Christoph Entzelt, 1517–83).[46] For all that, it was Gesner, yet again, who authored the first important work on the subject. *De Rerum fossilium* (1565) incorporated illustrations in systematic fashion and inaugurated an iconographic tradition reflecting one of the weightiest problems facing researchers: how to understand fossils' enigmatic form.[47] Undoubtedly, many fossils resembled organic forms of life – teeth, bones or shells; but even so, plenty of reasons for 'misunderstanding' existed. What is more, a great portion of the petrified organisms were not yet known in the Renaissance. Against this backdrop, it is perhaps more than a 'happy accident' that most of the fossils Gesner presented came from marine animals, many of which he had described seven years earlier in his book on fish.

Gesner was likely the first to picture fossils in biological relation to living organisms. In *Historia animalium* (1558), he had already remarked the similarity between familiar *glossopetrae* (literally, 'tongue-teeth') and shark teeth, which he had presented side by side. Gesner, like all his contemporaries, was far from understanding fossils in terms other than those of Aristotelianism or Neo-Platonism; he did not make the leap to viewing them as organic remainders of animals and plants that had died (or, indeed, gone extinct). But for all that, his books offered a condensed, aesthetic expression of the relationship between marine creatures, petrification and living organisms; in the centuries to follow, it would represent the epicentre of epistemological earthquakes that shook the West.

In particularly concrete and immediate fashion, and as if by visual analogy, petrified shells seem to embody the modern interpretation of fossils. In antiquity, isolated references had already intimated the organic origin of petrified molluscs. The first extensive reflection was made by Leonardo da Vinci; in his notebooks, he observed that the similarity between fossilized and living shells is too exact for a causal relation not to exist.[48] But in the sixteenth century, few natural historians thought the same – e.g., Girolama Fracastoro (1478–1553) or Bernard Palissy (1510–90). Until late in the seventeenth century, the idea of fossils' organic origin did not manage to find broad support. A decisive turn occurred when the renowned shell-expert Fabio Colonna (1567–1640) published *Aquatilium, et Terrestrium aliquot Animalium* (1616). Like Gesner, Colonna presented a precise and elaborate nomenclature and, above all, offered an aesthetic mode of argument through visual illustrations. He was the first to view fossils in biological terms instead of a mineralogical context. His depiction of fossils and living species alongside each other afforded a model that, in years to come, would focus attention in a new way.

It may seem surprising, then, that shells caused the biggest headaches for experts – after all, they are supposed to number among the 'simplest' fossils. There are two main reasons this occurred. For one, shells were found in places that beggared belief: on mountain peaks or hidden in deep bedrock. How did they get there? Until the nineteenth century, answers involved variations on the theme of the Deluge. They ranged in colouration from the Biblical and miraculous to the natural-philosophical; finally, the ground opened and an abyss of time was revealed – vast beyond all that had been imagined until then. The second reason was given exemplary formulation by one of the foremost seventeenth-century experts on shells, Martin Lister (1638–1712). Lister doubted that fossils had organic origins. He observed that fossilized shells resemble those of living creatures, but are not identical with them; inasmuch as there are no living examples of petrified specimens, they cannot derive from an organic source. Lister's friend John Ray continued this line of reasoning and declared that any such creatures would have had to become extinct.[49] This, however, was inconceivable in terms of the Christian understanding of divine creation: why should animals and plants that could exist, and obviously also had existed, have vanished from the earth? Where were these organisms reproduced in stone now? Those who preferrd an organic origin for fossils saw only one way out of the problem: somewhere in God's wide world, the descendants of these species must still be hiding. Given that mainly exotic marine animals and plants stood at issue and little was known about their living conditions, it seemed the most plausible explanation. Indeed, in the mid-eighteenth century the first 'lost' species emerged from the depths of the West Indian Ocean – among others, the 'sea lily' (*crinoidea*) and ammonites, deap-sea molluscs.[50] With that, one of the most alluring spells fell over the Western perception of submarine space; in the nineteenth century, in particular, it unfolded a rich cosmos of images: the sunken world and voyages to the beginning of time. In the space of the imagination, the depths of the ocean gradually transformed into one of the greatest spheres of natural-historical mystery, a shrine of revelation that seemed to house all the answers to the question of life itself. Still at the end of the eighteenth century, Jean-Guillaume Bruguière (1750–98), one of the leading naturalists and a prominent expert on fossilized shells, argued that ammonites, belemnites and many other organisms had not died out, but had migrated into the dark night of the deep: 'The vast floor of the sea', he wrote, 'could still be paved with them.'[51]

The Sea as a Place of Curiosity

The sea has represented a site of curiosity since well before the founding of oceanography in the second half of the nineteenth century:

> [I]t is certain that the modern science of oceanography has grown out of concepts which began to take shape in Europe during the Renaissance era. The development was a slow process, chequered by interuptions and reverses, but it is often possible to see an underlying continuity of ideas and methods. Yet man's interest in the sea is much older than this ... Attempts to give a rational account of the origin and nature of the sea ... go back at least to the Greek scientists of Asia Minor who, in the sixth century BC, abandoned the mythological interpretations of the universe ... in favour of explanations based on the operation of natural causes.[52]

A philosophy of water, writes Hartmut Böhme, inaugurated European philosophy in the first place, 2,500 years ago.[53] It began with Thales of Miletus, who declared water the primordial element and principle of all things. Subsequently, ancient philosophers devoted great attention to water: hydrological cycles, the relationship between evaporation and rain, the question of subterranean connections between bodies of water, the salinity of the ocean, the workings of ebb and tide and the directions of ocean currents.[54] Although new 'discoveries' hardly occurred in the centuries between Pliny's natural history and the Renaissance, the questions posed by water and the sea were frequently discussed. The Middle Ages were interested primarily by the tides, but the Renaissance added the matter of the ocean's depth. Although the idea can be traced back to the ninth century, the reflections on possible methods of measurement offered by Nicholas of Cusa (1401 – 64), Leon Battista Alberti (1404–72), Marin Mersenne (1588–1648) and Bernhard Varen (1622–50/51) made the topic exercise particular fascination.

Early Oceanography, or The Art of Living Underwater

The Renaissance made the ocean a fixed component of the *new sciences*. '[T]he sea and its behavior had become an object of study in its own right'.[55] The question of tides quickly opened onto the core debates of cosmology; here, one beheld a drama of nature in which the invisible forces that hold the universe together unfold in spectacular fashion.[56] It is worth noting that the 'science of the sea' – which, at the time, did not exist as an independent discipline – already possessed a clearly structured and extensively ramified research programme. The catalogue of tasks for one the first sea expeditions (1661–2), led by Edward Montagu, Earl of Sandwich (1625–72), specified: 'The six topics of inquiry were to be the depths of the sea, horizontal and vertical variations in its salinity, the pressure of sea water, tides and currents in the Strait of Gibraltar and luminescence'.[57] Investigating events underwater was also a common theme of discussion. In *The Art of Living Underwater* (1716), Edmond Halley (1656–1741) notes at the outset: 'There have been many Methods proposed, and Engines contrived, for enabling Men to abide a competent while under Water'.[58]

According to Margaret Deacon, the first heyday of marine research occurred in 1660 and revolutionized the scientific study of the sea. That said, one should not mistake this fact for the predominance of marine subjects in scientific enterprise as a whole – much less for mounting enthusiasm for the sea among researchers in general. It may be assumed that three central figures of the period – Sir Robert Moray (1608–73), Henry Oldenburg (1618–77) and Lawrence Rooke (1622–62) – had a special passion for maritime matters. However, other luminaries such as Robert Boyle certainly had no soft spot for the ocean. Indeed, the example of Boyle proves especially instructive in light of the prevalent notion that natural scientists of the nineteenth century received the gift of an ocean that had remained practically unexplored. Boyle's four essays on maritime research summarize all the methods and ideas available toward the end of the seventeenth century. *Other Inquiries Concerning the Sea*[59] sketched the most concise research programme for investigating the submarine world to date. It had been preceded, a year earlier, by *New Experiments and Observations touching Cold* (1665), and it was followed by *Tracts about the Cosmical Qualities of Things. The Temperature of the Submarine Regions. The Temperature of the Subterraneall Regions. The Bottom of the Sea* (1671) and *Observations and Experiments about the Saltness of the Sea* (1673).

Boyle was no sea-lover. He lived in London and suffered from poor health; consequently, his comprehensive studies on the ocean were based on information provided by a broad network of informants. Although Boyle did not deign to name most of them, there is no doubt they existed. Deacon concludes:

> [W]hile the main achievements in marine science during the seventeenth century were the work of comparatively few people, the extent of interest and involvement was very much wider than this. Certainly without these largely unknown workers and their enthusiasm for scientific research at sea, Boyle's work on marine science would have been largely impossible.[60]

Boyle's example is telling about the multifaceted views that people held of the sea as the seventeenth century drew to a close. It makes it clear that a general aversion to the sea did not prevail in the Baroque Age; rather, general interest held for matters pertaining to ocean exploration and marine phenomena. Whenever amateurs or experts made a new discovery, basic curiosity assured them of the public's full attention.

The Whale as a Wonder of Nature and Renewed Interest in the Sea

In 1675, a book was published in Hamburg that described, with enthusiasm matched only by its detail and precision, the whales of the Arctic Sea.[61] Until now, cetological knowledge had been defined by the writings of the Scandinavian natural historian Olaus Magnus (1490–1557), who had recorded encounters on the 'Midnight Sea': enormous monsters bearing horns and flashing teeth upon a fiery countenance casting fierce and cunning glares.[62] Aldrovandi's *Monstrorum historia* (1613) and Gesner's *Historia animalium* (1551–8) had included whales with the faces of lions, panthers, rams, apes and women; their eyes were said to be so large that fifteen human beings might have fitted inside. The ship-barber Friderich Martens's *Spitzbergische oder Groenlandischen Reise Beschreibungen* (Travel Accounts of Spitsbergen and Greenland) drew a wholly different picture.[63] He, too, contemplated 'God's exceptional Providence', yet he portrayed beings wholly different in kind. With equal measure of fascination and tenderness (as it were), he described the form, colouration and dimensions of Arctic whales – their migrations, dietary habits, predators and, above all, behaviour when encountering human beings. The accompanying engravings of a bowhead and a finback whale were especially breathtaking for the times. All of a sudden, the uncanny and monstrous beasts turned into true-to-life, fascinating beings of wondrous beauty. On this score, Martens has no equal among his contemporaries. His descriptions combine attentive observation of nature, whose results are recognized in cetological research to this very day,[64] with the contemporary aesthetics of natural wonder. Under his pen, even the skin of whales appears as a *lusus naturae* and he praises the beautiful creatures as peaceful and retiring.

Martens's account of his travels generated great notice not only in Germany, but in other countries, too. It was translated into Italian, French, Dutch and English and published in Venice, Bologna, Amsterdam and London. The book's success reflects a time which had great interest in marine life, which is, in sum, precisely what one should bear in mind: the Baroque perception of the sea and its creatures was thoroughly ambivalent. Worries, fears and reservations were voiced, but so were curiosity and enthusiasm. Members of the Royal Society researched the composition of seawater, its weight and the reasons for the movement of tides – exceptional scientists include, in addition to Moray, Oldenburg, Rooke and Boyle, figures such as John Wallis (1616–1703), Robert Hooke, Isaac Newton (1642–1727) and Halley. At the same time, the prominent Anglican cleric Thomas Burnett (1635–1715) reflected on how it was possible for the earth to have drowned in the Deluge, given that its water supply was limited. Halley spent hours in a diving bell at the floor of the Thames, investigating the change of the water's colours as light decreased with progressing depth. He remarked that the upper surface of his hand, where the sun shone, appeared red, while the underside glowed green; this led Newton to conclude 'that the Sea-Water reflects back the violet and blue-making Rays more easily, and lets the red-making Ray pass most freely and copiously to great Depths'.[65] Around the same time, Louis Ferdinand Comte de Marsili (1658–1730) set out to explore the floor of the Mediterranean.[66] His *Histoire physique de la Mer,* which appeared in 1725, was the first book devoted entirely to marine research. Eight years later, a wonderfully illustrated volume, *Ueber die Asterien* (On Asterias), was published in Leipzig, the first book on sea stars exclusively.[67] The same period witnessed people descend upon the English coast to go swimming, collect shells or paint the shifting border where land and water touched. Amidst the busy clamour, the sea emerged as a site of popular diversion in the course of the eighteenth century; by its end, the shore was full of human activity. Scholars and amateur researchers went about roaming and combing for the wondrous phenomena on the exposed seabed, thrilled at the sight of algae, shells and zoophytes. Drawn by the vast emptiness and the hazy light, some wandered onto the tidal flats and kept the surprised fishermen company. Finally, the shore offered Romantic spirits a place of sentimental reverie, where the rushing ebb and tide echoed the depths of their own souls.[68]

Ultimately, whales migrated into show business-and world literature, too. In 1825, a dead, 28.5 metre long blue whale was found off the shore of Ostend; it was promptly defleshed and sent on a European tour. The famous 'Ostend Whale' travelled for seven years through the Netherlands, England, and France; it was one of the most popular attractions of its day, eternalized by numerous descriptions and illustrations:

> In British Columbia in 1835, the Pacific Whaling Company toured a 55-foot, 66-ton fin whale on a railroad flatcar, calling it 'A Mystery of This Age' and 'Playmate of Dinosaurs and Mastodons, Last of a Race of Towering Giants ... A fin whale had washed ashore near the Swedish city of Malmö in 1866, and when it was exhibited in Stockholm, the mouth was propped open so wide that spectators could wander in and out.[69]

The nineteenth century witnessed a mounting wave of enthusiasm for whales. Ever since, it has been impossible to imagine modern visual or literary culture without them. Admiration for the animals' beauty mixes with a deep shudder at their incomparable dimensions and the mightiness of their expressions of life. Thereby, the wondrousness they incarnate is magnified by their absolute foreignness, both in life and in death; the reality of whales' submarine existence proves just as impossible for human beings to grasp as the tortured silence of their mass death.

Time and the Sea

In the course of the eighteenth century, the coast was increasingly perceived as a privileged site for understanding the riddles of the world. At the shore, questions surfaced about the earth's past and the origins of life; it seemed to point to the inconceivable temporal dimensions discussed by geologists during the same period.[70]

> As the history of humankind became disconnected from that of the planet, the idea of a very ancient earth, indifferent to human presence, achieved sublimity. Observers began to perceive the coast, rocks, and cliffs as products of age-old wear. The shore simultaneously held images of past, present, and future. Cliff views offered three-dimensional spectacles that satisfied observers and readers who were becoming accustomed to new three-dimensional representations. People came to the coasts to browse in the archives of the earth. Reflecting on endless waves and the indistinct shoreline, they could envision the eternity of the world.[71]

A revolution had occurred in the way temporal duration was conceived, and the water of the ocean now played a decisive role in the earth's history. In this context, the shore represented a logical destination and object of study. Nowhere else was it so easy to take stock of the rhythms of nature, the duration of geological periods, the indeterminacy of biological borders and the spheres of life with all the astonishing metamorphoses they presented. *Libido sciendi* gradually gave rise to a mounting desire for the shore; it entailed the cultivation of practices that combined aesthetic enjoyment, the pleasure of scientific observation, and the satisfaction of physical exertion. As Corbin writes, 'various ways of appreciating ... forms of contemplation, and habits sprung up ... and formed a system': a 'laboratory' for a 'cluster' of practices and experiences to be had with, underneath, and on the sea.[72]

The Mysterious Arrow of Time: Temporalized Wonders in the Abyss

It seemed that the coasts – even more than mountain peaks or cliff – would afford insight into the geological archives of the world.[73] The decisive factor, thereby, was the role the sea had played in the earth's formation. Two traditions, which were different but not necessarily contradictory, converged here. On the one hand, geological and oceanological interests had long formed an alliance to understand the face and form of the earth. Marsigli had already set to sea because he was interested in geological formations; he was convinced that investigating the reliefs on the ocean floor represented the sole means of understanding how landscapes change. On the other hand, many people of the day held that the jagged shapes in the earth's profile could only be explained in terms of the Mosaic Deluge.[74]

   In the eighteenth century, only a minority of scholars doubted the historicity of the cataclysm described in the Bible. Increasingly critical perspectives and the difficulty of reconciling empirical observation with Holy Writ gave rise to new explanations of, and theories about, the earth's history. Although they took distance from traditional conceptions of the Deluge in part, the sea – with its currents, tides and varying water levels – still occupied the center of reflection. 'Biblical imagery', Corbin writes, 'persisted in the figure of the primordial or 'primitive' ocean, which cast its shadow over the emerging science of geology'.[75]

   In this context, one of the most significant finds was undoubtedly the *homo diluvii testis*, 'witness of the Deluge', which Johann Jakob Scheuchzer (1672–1733) described in 1726.[76] Some hundred years later, Georges Cuvier (1769–1832) would identify this 'human being' as a giant salamander.[77] A particular irony of this effort at clarification is that it did not make the public less inclined to believe in the Biblical Flood, but more willing to do so. How else, if not in the course of a giant catastrophe, had these strange animals vanished from the face of the earth? What is more, a giant salamander proved no less wondrous than a human being who had been the 'witness of the Deluge' – it roused no less amazement than the mammoth or the marine 'Lizard of Maastricht', whose enormous jaw was discovered in 1780 in a cave near the Dutch city.

   As this example illustrates, a great difficulty encountered by Enlightenment-era scholars involved understanding geological processes and conditions, their absolutely incomprehensible dimensions. Animals, or, as the case may be, bones, are objects; they share their physiological characteristics – incomparable form and gigantic size – in a material way. Temporal duration, however, requires the mediation of culture: both a logical model and a mode of visualization adequate to the contemporary world. This fact became particularly evident when, in the mid-seventeenth century, a new conception of time began to mature, which separated the history of the earth from that of humankind.[78] It became increasingly difficult for philosophers to lend credence to astronomical calculations – for example, the English Archbishop James Usher's (1581–1656) claim that the world had been created on 23 October 4004 BC at nine o'clock in the morning.[79] Now, incomprehensible geological time joined the brief duration of days, years and even the historical time of succeeding centuries; but for all that, no points of reference existed for integrating this vast dimension into mankind's picture of itself and the world. Tentative divisions between epochs, periods and phases were proposed, yet the depth of time stretched far beyond human experience. To this day, its ungraspable immeasurability represents something fundamentally beyond all understanding – as one comes to appreciate when considering the multiplicity of explanatory illustrations and comparisons offered for concepts such as geological miles, the Cosmic Calendar and the Cosmic Clock.

As Paolo Rossi observes, people in Hooke's day understood that the past reached back six thousand years. When Kant lived, they were aware of a few million years.[80] Then, James Hutton (1726–97) introduced the notion of deep time with *Theory of the Earth* (1795). In a well-known text, John Playfair (1748–1819) described the enormous geological unconformity that his friend presented to him in 1788:[81]

> On us who saw these phenomena for the first time, the impression made will not easily be forgotten ... What clearer evidence could we have had of the different formation of these rocks, and of the long interval which separated their formation, had we actually seen them emerging from the bosom of the deep? We felt ourselves necessarily carried back to the time when the schistus on which we stood was yet at the bottom of the sea, and when the sandstone before us was only beginning to be deposited ... Revolutions still more remote appeared in the distance of this extraordinary perspective. The mind seemed to grow giddy by looking so far into the abyss of time.[82]

Over a thousand pages, Hutton's *Theory of the Earth* develops no theme as fully as, or at greater length than, the author's *astonishment* at the immense expanse of time. With due logical rigor, Charles Lyell (1797–1875) connected the new depth of time with the other immeasurable vastness that had 'just' shaken the scientific community's ways of thinking: the extension of space in the Newtonian cosmos:

> Such views of the immensity of past time, like those unfolded by the Newtonian philosophy in regard to space, were too vast to awaken ideas of sublimity unmixed with a painful sense of our incapacity to conceive a plan of such infinite extent. Worlds are seen beyond worlds immeasurably distant from each other, and beyond them all innumerable other systems are faintly traced on the confines of the visible universe.[83]

In this way, geology 'struck the imagination with the endless renewal of its combi-nations'.[84] In consequence, a time at which the human race had not even existed yet came to house an array of images of continents and the ocean for natural scientists. Such images make plain that the immeasurability of time was projected onto the sea and into its unknown depths. As the greatest unknown of the day, oceanic space came to represent the incommensurable – the vertiginous 'abyss of time'. It was inhabited by the fabulous creatures that, at the beginning of the nineteenth century, emerged from the depths of the ages on the south coast of England at Lyme Regis.

Fantastic Sea of Stone

A series of spectacular fossil discoveries marked the early nineteenth century. They hardly occurred by chance, for scientists and amateurs had grown increasingly vigilant as they exchanged information and insights in journals such as *Organic Remains* (1804–11) and the superbly illustrated *Mineral Conchology* (from 1829 onwards). The market for fossils flourished. Especially on the coasts of England, talented *fossilists* – fossil-seekers from the ranks of 'common people' – roamed the fields and shores.[85] Popular accounts of the newest geological theories and dis-coveries were *en vogue* – for example, *Wonders from Geology* (1838) by Gideon Algernon Mantell (1790–1852) (who also wrote other extremely successful titles). These works fanned enthusiasm for collecting petrified specimens even more.[86]

Wonders from Across the Sea: The Discovery of Marine Reptiles

As mentioned, Lyme Bay in the southwest of England, off Dorset, was one of the most important sites of discovery. Sealed in layers of slate and chalk (also known as 'black jura' [limestone]) lay the secrets of an enormous sea from time immemorial which had turned to stone – the gate to an unknown world.[87] In 1811, young Joseph Anning, the son of a carpenter who was a fossil-collector himself, made a remark-able discovery on the beach of Lyme Regis: the gigantic head (1.2 metres long) of a creature with elongated jaws full of pointy teeth and boney eyesockets as large as saucers. A year later, his sister Mary found what seemed to be the animal's trunk; together they yielded a complete, coherent skeleton over five metres in length.[88]

The strange being was displayed publically for the first time at Bullock's Museum in Piccadilly, London, and caused a sensation throughout the land. The anatomy of the weird animal – its incredibly long jaws drew back into an uncanny smile and it stared through huge eyesockets – defied explanation. The long snout running to a point resembled a dolphin's, but the teeth seemed to belong to a crocodile; finally, the vertebrae were slim, like a fish's spine. The leading geologists of the time, William Buckland (1784–1856), William Conybeare (1787–1857), and Henry De la Beche (1796–1855) examined the animal thoroughly and conducted vigor-ous correspondence with Cuvier and his assistant Joseph Pentland (1767–1873);

finally, they concluded that it was a kind of 'sea lizard'. In the meanwhile, many counties in southwest England had provided fossil remains of the same kind of animal. When De la Beche and Conybeare finally published their findings in 1821, they presented the picture of a primordial sea in which reptiles over seven metres in length had plowed the waves.[89] They called them *ichthyosaurs*, 'fish-lizards'.

Two years later, on the evening of 10 December 1823, Mary Anning discovered the skull of a strange 'turtle' with a long, snakelike neck at the foot of the Black Ven cliff. After hours of excavation, a fantastic creature 2.7 metres in length emerged from an ancient grave; instead of legs or fins, it had paddles made of thin bones countless in number. News of the animal, which Conybeare christened *plesiosaur* ('reptile-like') spread like wildfire. Here again was an absolutely improbable form of life; its long neck seemed to contradict all the laws of anatomy. Accordingly, Cuvier suspected a hoax, and at first the researchers of the *Geological Society* held that the strange creature would have been utterly incapable of living. Within a brief span of time, however, further finds proved the authenticity of the Anning fossil; despite the fact that their brains had been tiny and their necks obscenely long, plesiosaurs had flourished across a broad range. On 20 February 1824, Conybeare presented the 'wonder from across the sea' to the *Geological Society* in London as the 'most "monstrous" [thing] that had yet been found amid the ruins of the former world'. 'To the head of the lizard, it unites the teeth of the crocodile; a neck of enormous length, resembling the body of a serpent; a trunk and tail having the proportions of an ordinary quadruped, the ribs of a chameleon, and the paddles of a whale'.[90] As Mantell wrote soon afterward, this enormous reptile would only have needed wings to be a dragon.[91] And so, two bizarre beasts had roamed the primordial seas. What kind of world had this been?

Strange fossils on the beach of Lyme Regis also included numerous dark-grey formations, shaped like longish pebbles and displaying a flaxcomb pattern . The objects were often found inside ichthyosaur specimens. Chemical analysis revealed them to be fossilized clumps of faeces – for the most part, the bones of other organisms. Gradually, the unsettling picture emerged of a primordial sea teeming with giant marine lizards, veritable titans constantly engaged in bloody and mortal battles. The early days of Creation now appeared as a terrible blood-bath, and talk of 'Satan's brood' or ancient mythological figures which had in fact been real spread quickly. Then, in December 1828, Mary Anning found the fragile remains of a ghostly being – half vampire bat and half reptile – with a wingspan of 1.2 metres. Now, the public mind stood completely under the spell of an antediluvian world of hellish nightmares. Not just the depths of the sea but the airy heights of the heavens had once been inhabited by unearthly giants; their lives had amounted to one vast slaughter. Nor had these creatures been like Frankenstein's monster or the wicked spirits of Milton's *Paradise Lost* (1667); they were not incorporeal phantoms from the underworld but as massive and undeniable as the rock where they were found. How could one face up to anything so incredible, how could one understand what no human being had ever seen before?

News from the Primordial Sea: Scenes of the Incommensurable

The new fossils opened the gates to an alien landscape, revealing a history of the earth wholly unknown until this point. To the greatest amazement of the public, an age had existed when a multiplicity of wondrous and terrifying reptiles had ruled the land. But if this was so, what had the land looked like? And above all: what had happened to these life-forms, bizarre and terrible in equal measure, to make them vanish from the face of the earth?

In 1830, the scenery of the prehistoric landscape appeared for the first time. De la Beche's *Duria antiquior* painted the picture of a primeval Dorset, where both familiar creatures and ones newly discovered ate their fellows, only to be devoured in turn. This book, accompanied by lithographs by George Scharf (1788–1860), sold extraordinarily well and gave rise to a wave of efforts to visualize the sunken world. The first point of reference was certainly provided by bones that had been discovered and the reconstructive method that Cuvier pioneered. In addition, pictorial representations of the prehistoric age borrowed from the tradition of Bible illustrations and romantic conceptions of early human history. For example, Scheuchzer's *Physica Sacra* (1731–3) features the Garden of Eden and the Flood; the focus of John Martin's (1789–1854) *The Deluge* (1828) is plain enough.[92] At the same time, the imagery connected to the pictorial traditions of sixteenth-century world landscapes, the scenic genre of great hunts, the moody and atmospheric countrysides depicted from the seventeenth century onwards, late-eighteenth- and early-nineteenth-century representations of fighting animals and contemporary devices for portraying natural history.[93]

The particular sensation of De la Beche's illustrations was the submarine perspective they presented. It is unclear where he got this idea, but he is known occasionally to have gone on dives off the coast of Dorset,[94] it is therefore probable that he was familiar with contemporary images illustrating the underwater vistas that diving revealed. His innovations are particularly clear if one considers, as a point of contrast, the standard way that views of marine biology were depicted: as a rule, the sea features as an impenetrable surface, and its occupants are 'draped' in positions along the beach or in the picture frame.

De la Beche's revolutionary 'diving perspective' would not become the norm in visual culture until the end of the nineteenth century. This merits notice inasmuch as it offered a solution to a problem that was not inconsiderable: namely the fact that the great majority of fossils came from marine organisms that had lived entirely underwater. Accordingly, every effort to represent the vanished world faced the challenge of lending the submarine world visual form. It was not just a matter of imagining a *pre*-human world, but also one that was *non*-human. The fact, then, that De la Beche's approach gained acceptance only with great delay indicates how difficult it must have been for his contemporaries to conceive of an underwater perspective at all. Evidently, it was easier for them to travel billions of years in time than to picture themselves two or three metres beneath the sea's surface. The ocean depths seem to have counted as more alien than the depths of the past.

For all that, the scenic genre that De la Beche developed was copied right away. The first 'successor' was, in all likelihood, 'Jura Formation' (1831) by August Goldfuss (1782–1848), which appeared in *Petrefacta Germaniae* (1826–44); it was followed by 'Organic Remains Restored' by John Philips (1800–74) in *Penny Magazine* (1833). In 1834, the *Dictionaire pittoresque d'histoire naturelle* displayed *animaux perdus*, and Thomas Hawkins's (1810–89) *Memoirs of Ichthyosauri and Plesiosauri* offered readers 'Extinct Monsters of the Ancient Earth'. Rampaging primeval beings were also to be seen in *Beschreibung und Abbildung von dem in Rheinhessen aufgefundenen colossalen Schädel des Dinotherii gigantei* (Description and Reproduction of the Colossal Skull of Dinotherium giganteum Found in Rheinhessen) (1836) by Johann Jakob Kaup (1803–73) and August Wilhelm von Klipstein (1801–94), Peter Parley's (1793–1860) *Wonders of Earth Sea and Sky* (1837), and George Fleming Richardson's (1796–1848) *Sketches in Prose and Verse* (1838). Clashing titans also appeared in Mantell's (1790–1852) *Wonders of Geology* (1838), in Thomas Hawkin's (1810–89) *Book of the Great Sea-Dragons* (1840), and Richardson's *Geology for Beginners* (1842/43).

By the mid-nineteenth century, an imaginative world of pictures had evolved, which granted the prehistoric past a high degree of drama, grandeur and mystery. Biblical, mythical and fantastic motifs played into the mise-en-scène; Romantic atmospheres and elements borrowed from the gothic genre gave vivid expression to the complete otherness of the vanished creatures. A hint of the miraculous – which every imaginative journey brings along even under the strictest epistemological control – clung to even the most precise reconstructions. That said, many researchers did nothing to work against such connotations. Emile Boblaye (1792–1843), for example, used a caption to describe the organisms that were pictured as expressions of a sick fantasy,[95] while Goldfuss considered *pterodactylus* the product of a Chinese artist's unbridled fancy.

Significantly, the sequence of different periods of the earth's history, which geologists worked out painstakingly, hardly left a trace in public consciousness at first. In the main, a static perspective predominated: 'the primitive world' – *die Urwelt, l'ancien monde* – had once existed, and its singularity derived from its otherness and the absence of human beings, above all.

The first work to present a dynamic picture of prehistory to a broader public was Franz Xaver Unger's (1800–70) *Die Urwelt in ihren verschiedenen Bildungsperioden. 14 landschaftliche Darstellungen mit erläuterndem Text* (The Primeval

World in Its Various Periods of Formation. Fourteen Landscape Illustrations with Textual Commentary) (1851). The illustrations were magisterially composed by the Graz painter Josef Kuwasseg (1799–1859); they afford a history of life and, at the same time, a history of death. The final plate presents 'man'. Although nature still provides the focus, the accompanying text celebrates the arrival of the human race as the central event. For Unger, as for so many of his contemporaries, the evolution of life was ultimately a story of the world's gradual preparation for the coming of mankind. This is evident, not least of all, in the Arcadian natural landscape in which Kuwasseg places his representative of humanity, as well as the Biblical allusions the image offers. It seems that this white-skinned Central European wants for nothing in the Garden of Eden; he stands naked without sweat on his brow. But all the same, the Tree of Knowledge features prominently, and the image is superscribed, 'Present World'.

The work that ultimately determined the visual style of prehistorical scenes was Guillaume Louis Figuier's (1819–94) *La terre avant le Déluge* (1863). Figuier enlisted the aid of Edouard Riou (1833–1900), a Parisian landscape artist and illustrator, who prepared more than two dozen pictures for the book – while, incidentally, also working for Jules Verne. Many of the scenes are arranged conventionally; once again, the sea stands as an expanse of water, and the ocean animals lie in loose distribution on the shore. Figuier's book was a thoroughgoing success; in little time, it went through multiple editions and was translated into several languages. Riou's illustrations established a gripping, new type of image; it is employed to frame prehistoric matters to this very day.

The 'wonders of science' – above all, the achievements of geology and paleontology – and the 'wonders of nature' intersected in the discovery and reconstruction of a primeval world. The latter were evident in the overwhelming dimensions and bizarre forms of primal beings, which had surfaced as mysteriously as they then disappeared; here, it seemed that the incommensurability of the earth's history was reflected in its entirety. The inhabitants of the land-before-time were perceived as monstrous jokes of playful nature: nightmarish, on the one hand, and riddling, on the other. Why had God created these creatures? How had they vanished? Did a connection exist to the world of humans today?

Prehistorical scenes developed into a conventional feature of popular books on the 'wonders of geology'. Small toy models of saurians were manufactured, and gigantic illustrations hung on classroom walls. In *Penny Magazine, Illustrated London News, La Nature, Pfennig-Magazin, Gartenlaube, Die Natur, Kosmos* and *Über Land und Meer*, a visual rhetoric emerged for the unbelievable 'revelations' of natural history. In this context, antediluvian monsters embodied the ungraspable, pure and simple. They were the stars of museums of natural history, and 'true-to-life' reproductions – for example, in the Crystal Palace at Sydenham – were mass attractions.[96] The creatures stormed through the novels of Jules

Verne and made it into Giorgio de Chirico's manifesto *Sull'arte metafisica* and the collages of Max Ernst. The fact that they were extinct separated them from the monsters of sixteenth- and seventeenth-century broadsheets and the deformities displayed in cabinets of wonder; at the same time, however, their horrible form – the combination of nightmare and natural-historical enigma – connected them to these phenomena. Their replicas displayed a hazy border between imaginary traits and features admitting scientific justification; pictorial elements inherited from tradition appeared in the scenery and were combined with Darwinian erudition. In striking fashion, the primordial beings epitomized the modern figure of wonder in both vertical and horizontal terms: the abyss of inconceivable time from which incomprehensible entities emerge. When these creatures moved into the general cultural consciousness of the nineteenth century, the modern history of wonder experienced one of its most decisive transformations: contemporary shocks and commotions became evolutionary catastrophes and yielded a pandaemonium of riotous fantasy; here, the 'living fossils' were at home, so to speak. The abyss of prehistorical time corresponded to the phantasm of an oceanic underworld leading to the mysterious origins of life itself.

2 WONDER AND MYSTERY

In the nineteenth century, the underwater world transformed into a fertile space for dreamlike fantasies. Beneath the swell and surge of mighty waves, the unfathomable depths hosted a dazzling array of images and ideas: epistemological models and artistic motifs, narratives, legends, poems, seascapes and accounts of catastrophe.

Invariably, a look beneath the water's surface prompted reflection on the radiant heavens above:

> [T]he contemplative mariner, as in mid-ocean he looked down upon its gentle bosom, continued to experience sentiments akin to those which fill the mind of the devout astronomer when, in the stillness of the night, he looks out upon the stars, and wonders. Nevertheless, the depths of the sea still remained as fathomless and as mysterious as the firmament above.[1]

While the skies could, in principle, be observed through a telescope, the depths of the sea remained hidden to the eye – our preferred organ of perception. It proved a matter of great vexation that two-thirds of the earth's surface would not yield to human curiosity. If not for mankind, Matthias Jakob Schleiden (1804–81) wondered (as did many of his contemporaries),

> für wen [sei] denn jener Reichthum an Glanz und Schönheit ausgebreitet, welchen jene blaue Decke verhüllt, deren spiegelnde Fläche den Lichtstrahl zurückwirft und meist dem neugierigen Lauscher fast wie spottend nur das eigene Bild zeigt?'

> then for whom did it all exist, this wealth of expansive brilliance and beauty wrapped in a blue cloak, whose reflective surface throws back rays of light and for the most part, almost mockingly, offers the inquisitive only their own image? (translation by Erik Butler, hereafter EB.)[2]

It was lamented that possibilities for gaining insight into the 'mysteries' of the sea proved highly unsatisfying[3] – a regret that still haunts marine research today.[4] Soon, the underwater world came to count as the epitome of all that is mysterious and concealed. Henceforth, the sea presented the arena in which baroque cabinets of natural wonders transformed into modern marvels of life and the mysteries of nature.

The Poetics of Not Knowing

To be sure, the nineteenth century did not invent the idea of mystery. The Western history of ideas already knew the topos of Nature and 'her' secrets, and the iconography of the 'Veil of Isis' reached back millennia.[5] Alongside efforts to uncover the mysteries of nature in the strong sense, another view existed which did not see a curtain to be lifted so much as a sign of hidden depths. This fundamental ambivalence concerning the mysteries of nature marked nineteenth-century views of unexplained phenomena in general. Wherever one looked, it seemed that scientific progress was celebrating victory: mysteries solved and secrets revealed. Voices boldly declared that what was unknown today would count as certain truth tomorrow. At the same time, and almost to the point of triviality, the age subscribed to the dictum, 'the more you know, the more you know that you don't know'. As revelations proceeded, the nineteenth century unfolded as an age of enigma. Discussions of hidden mysteries flourished as natural science charted success after success – each of which generated new questions that called for further exploration. Likewise, ambivalence attended discussions about the outer limits of knowledge: the matter of what can or cannot ever be known. In addition to these two main senses of mystery, the one expansive and the other restrictive, the Veil of Isis represented uncertainty in another way, too. For a discovery to *qualify* as a discovery, it must appear as an instance of disclosure; in other words, and by definition, it occurs through the performative act of unveiling. The veil that suggests profundity also stands for inability to decide whether what is being uncovered is actually the same as what has been concealed until now. It goes without saying that the invisible, once it has been made visible, differs from the invisible that remains so. It follows that each instance of unveiling also represents a media trick played *with* the mystery – which, when disclosed, either vanishes or lends a riddling character to the revelation. In either case, this hypothesis accounts for a common observation to be made apropos of nineteenth-century natural science: the fact that discoveries do not lose their aura of mystery, even for their discoverers. The operative assumption here is that items of investigation are perceived as aesthetic objects; alternately, the aesthetic aspect of discovery does not destroy mystery so much as it heightens it (or, indeed, 'releases' it in the first place). Two brief examples may serve to illustrate this point.

Mysterious and Peculiar Beings in the Works of Johannes Müller and Ernst Haeckel

Reading the account that the physiologist, anatomist and marine scientist Johannes Peter Müller (1801–58) offers of the discovery of plankton, one is struck by the latter's insistence on the mysteriousness of his microscopic specimens.[6] Again and again, Müller identifies the materials under examination as '(r)ätselhafte Unbekannte des Meeres' (mysterious unknown beings of the sea);

even after they have been identified as floating life forms, they remain 'durchaus dunkel und rätselhaft' (dark and mysterious indeed (translation Erik Butler, hereafter EB)).[7] Müller's first maritime journey, in 1831, had started from Scheveningen in Holland. He wrote to his wife about his encounter with the North Sea: 'Fast lief ich von unendlicher Sehnsucht nach dem großen Schauspiel des heiligen Meeres, das kein Sterblicher nicht gesehen haben sollte, ehe er die Erde verlässt'. (I almost ran, full of infinite longing for the great spectacle of the holy ocean that no mortal should miss before departing from earth (translation EB)).[8] Afterward, Müller devoted himself to marine biology. With a lady's stocking stretched over a wire ring, he scooped plankton out of the ocean – bizarre crustaceans, transparent worms and medusas (jellyfish) suffused by light. His reports to the Prussian Academy describe a mysterious animal' whose 'strangest feature (is) a starfish attached at the part of the body from which the arms spring'. He also writes of the 'bizarre form' of the 'Rococo' larva[9] and the 'very mysterious' tentacles and 'curious arms' of certain barbed echinoderms resembling urchins.[10] Müller's *Bericht über einige neue Thierformen der Nordsee* (Report on Several New Animal Forms of the North Sea) (1846) introduces the 'mysterious *Tomopteris onisciformis Eschscholtz*' as well as 'several new ... very mysterious sea animals'.[11] These animals (*vexillaria flabellum*), often only a few millimetres long, display 'such a curious form and structure that, at the present moment, it is impossible to guess to what class they may belong'. Their inner organs 'are so convoluted and so oddly shaped that explaining their connection by analogy to other animals (is) impossible'.[12] The metamorphosis of several 'curious'[13] creatures barely a millimetre in size especially excited Müller:

> Wunderbar ist die Verwandlung des mit einer Staffelei verglichenen Thierchens, des *Pluteus paradoxus*; es wird daraus ein Seestern und zwar eine *Ophiura*. Als ich die ersten Anzeigen von dieser Verwandlung wahrgenommen hatte, fühlte ich mich aufgefordert, ihr die ganze Zeit meines Aufenthaltes am Meere zu widmen und sie bis zu ihrem definitiven Ziel zu verfolgen.

> The *Pluteus paradoxus*, a small animal that resembles a scaffold performs a wondrous transformation; it turns into a starfish, an *Ophiura*. When I noticed the first signs of this metamorphosis, I felt the call to devote the entirety of my stay at the sea to them, following them to their ultimate goal. (translation EB)[14]

Müller contributed significantly to the sea's emergence as a field of study in the German-speaking world. In a petition to the Reichstag supporting the foundation of a marine biology research centre in Naples, Italy, Hermann von Helmholtz (1821–94), Rudolf Virchow (1821–1902) and du Bois-Reymond wrote:

> Johannes Müller, dessen bloßer Name Ehrfurcht gebietet, hat während der letzten fünfzehn Jahre seines Lebens fast in jedem Frühling und Herbst sich nach der See aufgemacht, um mit Schleppnetz und Mikroskop *den Wundern der Tiefe* nachzuspüren.

> Johannes Müller, whose very name commands respect, has spent every spring and
> autumn at the ocean for the last fifteen years of his life, studying *the wonders of the
> deep* with a dragnet and a microscope. (translation EB)[15]

Mysterious and odd objects also occupy a prominent position in the works of
Müller's student, Ernst Haeckel (1834–1919).[16] Irenäus Eibl-Eibesfeld, endors-
ing Haeckel's own view, offers a characterization of the 'scientist as an artist':

> (J)e mehr Einblick der Naturwissenschaftler und hier wohl insbesondere der Biologe in
> die Details organismischer Schöpfung gewinnt, desto mehr bestrickt ihn das Wunder
> dieses Seins, die Schönheit selbst im kleinsten. Entgegen mancher Behauptung entzau-
> bert Wissen nicht. Es erschließt vielmehr vor unseren Augen immer neue Wunder.[17]

> The more the natural scientist (especially the biologist) learns about the details of
> creation the more he is entrapped by wonder and the beauty of the minute. Knowl-
> edge does not lead to disenchantment, as some people might think, but discovers ever
> new wonders instead. (translation EB)

In August 1854, Haeckel accompanied his teacher to Helgoland for the first
time. The few weeks on the rocky island convinced the young man that the
North Sea offered a divine vision. The natural diversity of the tiniest of sea
creatures, hitherto unknown to most biologists, lent nature the appearance of
overwhelming expressive frenzy, an 'unerschöpfliche Wunderwelt des Seethier-
Lebens' (inexhaustible world of wonders, full of living sea creatures, EB).[18]

In the nineteenth century, the living world still presented many empty spaces
that needed to be filled. Haeckel, an avid traveller, authored one standard work
after the other. His books describe guancha sponges, medusas and especially radiolarians, whose mineral skeletons struck him as the hidden jewels of the ocean.[19]
Haeckel became an internationally celebrated authority in the field. When the
specimens gathered on the famed expedition of the British *Challenger* needed to be
analyzed, he numbered among the few foreigners whose expertise was requested.[20]

As is generally known, Haeckel held a Darwinist position, which he sought
to popularize with the 'Zauberwort' (magic word) of 'Entwicklung' (development).[21] In his *Natürliche Schöpfungsgeschichte* (1868) (*The History of Creation*,
1880), he wrote:

> (T)he mystic veil of the miraculous and supernatural, which has hitherto been
> allowed to hide the complicated phenomena of this branch of natural knowledge,
> is removed ... The dimming mirage of mythological fiction can no longer exist in the
> clear sunlight of scientific knowledge.[22]

Needless to say, many things are 'still enveloped in the greatest obscurity'.[23] Indeed,
much will prove forever obscure, because '(t)he *ultimate causes*, it is true, remain
... concealed' and 'mysterious to us'.[24] One should not feel disappointed, however.
Everything that *has* been discovered – the 'wonder of the phenomena of life' – is
so great and offers such beauty.[25] The 'clear sunlight of scientific knowledge' shone
bright when Haeckel wrote of the enigmatic design of the Whip-swimmer,[26] the

'wonderful organism' of the monera,[27] 'the wonderful class of the Rhizopoda, or Ray-streamers',[28] the 'astonishing variety of forms' displayed by fresh water snails,[29] the 'unknown function' and 'wonderful richness'[30] of radiolaria, the 'curious ... Bathybius' dwelling in the 'immense depths of the ocean'[31] and the 'puzzling nature' of protista.[32] As he saw it, the natural world abounds in marvels that fully reveal their wondrous mystery only in the bright light of progressive reason:

> There can be no doubt as to the purely mechanical material nature of this process. But here we stand full of wonder and astonishment before the infinite and inconceivable delicacy of this albuminous matter. We are amazed at the undeniable fact that the simple egg-cell of the maternal organism, and a single paternal sperm-thread, transfer the molecular individual vital motion of these two individuals to the child so accurately, that afterwards the minutest bodily and mental peculiarities of both parents reappear in it.[33]

In this context, the last, 'invaluable secret' of the theory of descent is 'that we may see this incredible wonder at any moment, and follow it with our own eyes'.[34]

The Wonder and Mystery of Nature in Popular Literature

Publications on marine science in the nineteenth century teemed with mysteries wondrous and strange. Many writers held that biology represented 'das höchste Problem, welche(m) wir uns überhaupt stellen können' (the greatest problem that we can address); whoever wished to understand the universe 'musste an das Meer, die Wiege allen Lebendigen gehen' (must go to the sea, the cradle of all life).[35] 'There, in fact, may be discovered the principles of life',[36] 'the mystery (deepest of mysteries) of its first appearance'.[37] The goal was to lift 'die Schleier des Meeres' (the veil of the sea) and enter the deep to bring what it conceals to light.[38]

By the same token, it was widely believed that the animal poses the greatest and most difficult mystery for thinking human beings.[39] Water animals, in particular, were thought to possess 'Seelenleben (spiritual life) hidden behind 'several veils'.[40] As soon as one takes one's first step on the shore, one is 'confronted with a mystery'.[41]

> We can scarcely poke and pry for an hour among the rocks at low-water mark, or walk with an observant downcast eye along the beach after a gale, without finding some oddly-fashioned, suspicious-looking being, unlike any form of life that we have seen before. The dark, concealed interior of the sea becomes thus invested with a fresh mystery; its vast recesses appear to be stored with all imaginable forms, and we are tempted to think there must be multitudes of living creatures whose very figure and structure have never yet been suspected.[42]

A great number of scientists considered echinoderms:

> das größte Räthsel der Thierwelt ... Nicht bloß ihre Gestalt, ihr Körperbau, ihre Lebensverrichtungen, auch ihre Entwicklungsgeschichte ist so räthselhaft und so eigenthümlich, so ganz von allem Ueblichen entfernt, dass man Mühe hat, auch nur die gewöhnlichen Benennungen auf sie anzuwenden.

the greatest mystery of the animal kingdom ... Not only their shape, their body structure, their way of life, but also their genesis is so mysterious and odd, so far removed from the usual, that one has trouble applying even the most common designations to them. (translation EB)[43]

Others, in turn, regretted

wie wenig ... man eigentlich von dem speciellen Treiben der bereits bekannten Seefische (weiß); von ihren Verhältnissen zu den anderen Meeresgeschöpfen; von den Gesetzen, welche ihre eigenthümliche Existenz bestimmen? Ist ja sogar der Lebenslauf des Härings noch immer in Dunkel und Geheimniß gehüllt; die Frage, woher er kommt, wohin er geht, noch immer unbeantwortet geblieben.

how little ... one actually (knows) about the specific activities even of familiar sea fish – their relationship to other sea animals (and) the laws that govern their curious existence. Even the life of the herring is still concealed by darkness and mystery. The question of where it comes from and where it is going has still not been answered. (translation EB)[44]

At any rate, the matter of primary importance, and the primary source of wonder, was the abundance of marine life.[45] Practically every nineteenth-century work on the sea begins with a phrase to the effect of, the 'bosom of the ocean is full of mysteries'.[46] The sea brimmed with organisms of all orders, genera, species and classes; countless animals and plants constituted 'a whole world of curiously-shaped animals, which the naturalist only superficially knows, and may, perhaps, never be able to fathom.'[47]

The untold profusion of the undersea world roused wonderment, curiosity and admiration. The interplay of form and function, harmony and purpose, produced delight. Observers were thrilled by the improbable metamorphoses on display, and they extolled the elegance, subtlety and artfulness of creative nature – when they did not fall mute before the dazzling colours and indescribable intricacy of marine life. Even the lowliest of creatures could inspire rapture. For example, the following account of humble annelids:

(T)he marine annelides may well be reckoned among the handsomest of creatures ... We find an astonishing regurlarity and art; these elegant little tubes, which we may often pick up on the strand ... consist of particles of almost equal size, so artistically glued together, that the delicate walls have everywhere an equal thickness ... From each aperture stretches forth a neck ornamented with concentric rings of golden hair, and terminating in a head embellished with a tiara of delicately feathered, rainbow-tinted tentacula. The whole looks like a garden-bed enamelled with gay flowers of elegant form and variegated colours.[48]

Over the course of the century, marine science whipped its prose into a lather of effusive description. All that language had to offer for the wonderful, delightful, astonishing and magnificent was brought to bear on these grotesque, bizarre and enchanting creatures. Today, the aesthetic intensity seems exaggerated. Even the commonest beach invertebrates were often described as exquisite wonders; a writer might enthuse, for instance, about a mussel's 'smaragdgrün oder rubin-roth glänzenden mohnkerngroßen Aeuglein' (little shining emerald-green or ruby-red eyes, the size of poppy seeds, EB) hidden on the edge of the mantle, shimmering like embedded jewels.[49] Likewise, it seems difficult now to regard the common jellyfish (*aurelia aurita*) that every beach visitor has encountered as a 'Brillianten des Meeres' (brilliant gem of the sea).[50] However, the issue is not so much whether these descriptions are based on authentic experiences; what is important are the motifs of scientific passion that are articulated when knowledge is aestheticized. Like their forebears in the late-seventeenth and eighteenth centuries, who described ants, moths and bees as works of heavenly splendour and precious jewels, nineteenth-century explorers pursued their research with the aesthetic charge a new beginning affords. Their study of nature did not proceed by classification – asking *what* a given being is – but by examining *how* it looks. Thus, knowledge was born of an aesthetic process, whereby a field of extraordinary phenomena emerged from the interplay between media of perception and media of representation. The rhetorical exuberance, far-flung figures and effusive metaphors bear witness to sublime pleasure fusing *thaumazein* (striving for knowledge) and *hedone* (enjoyment of virtuosity). Eagerly awaiting encounters with further marvels, scientists and amateur researchers focused their attention on organisms without knowing what questions to pose. They greeted them time and again with appreciative exclamations: *Wunderbar!* Wonderful! *Merveilleux!* Such outpourings hardly conform to the gravity and seriousness convention dictates today. However, this early stage of marine research and the tremendous mass of discoveries that it afforded present another image, too: a science that was both driven by passion and productive. A cold bearing does not belong to the practice of science, but to its ideology.[51] What is more, the creativity and excited rhetoric evident in scientific accounts may be read as exercises in virtuosity; as the historian of science Hans-Jörg Rheinberger has observed, this is precisely what is necessary when mapping out new fields.[52] As such, the generation of new ideas through description formed an integral part of the media practice of marine research.

Early marine research literature is distinguished by complex modes of appropriating the semantics of wonder and mystery, and by subtle differentiation between its many aspects. In the history of science, the conventional view derives such 'poetisation' from marketing strategies, religious thinking or simple ignorance. However, the texts themselves point to a different background. Counter to general assumption, references to wonder and mystery do not occur only in popular works of non-fiction; they are also present in scientific writings. As Andreas W. Daum has shown in a path-breaking study on the popularization of science in the nineteenth century, popular works were not necessarily written by semi-educated, business-minded lay-people; authors included renowned natural scientists

such as Christian Gottfried Ehrenberg (1795–1876), Schleiden, Müller, Haeckel, Matthew Fontaine Maury (1806–76), and Louis Agassiz (1807–73).[53] In
keeping with Lorraine Daston's history of rationality, writers aestheticized texts
and emotionalized research topics for strategic reasons, seeking to convince the
public of the urgency and significance of their research – that is, its worthiness of
support. In this light, the discursive presentation of underwater worlds of wonder follows from an 'economy of affects and values'[54] of new branches of science,
whose purpose is to convey and popularize new data. For all that, such a perspective overlooks the fact that, in the nineteenth century, authoring popular works of
natural science qualified both as a technical task and as an artistic pursuit; accordingly, it held literary merit.[55] Alexander von Humboldt (1769–1859) was a true
pioneer in this regard; his programme involved '(t)he combination of a literary
and a purely scientific aim', whereby he sought 'to engage the imagination, and
at the same time to enrich life with new ideas by the increase of knowledge'. His
works provided a model for many whose goal was to popularize natural science.[56]

In this context, the marine sciences brought forth a spectrum of voices employing metaphors and analogies like those formerly applied to cabinets of wonder. At
the same time, poetic works of science also featured religious motifs. In the nineteenth century, many popular authors sought evidence for the divine creation of
nature; scientists often adhered to a Christian worldview, too. Researchers would
wander the beaches as if strolling through a gigantic open-air museum displaying the
masterpieces of God's craftsmanship. The broad category of 'wonder' provided an
answer to all questions. Additionally, in the sense of the ancient world's *thaumazein*,
it offered the starting point for studying the incomparable multitude and richness of
life born of nature's creativity. Like the natural philosophers of the seventeenth and
eighteenth centuries, nineteenth-century scientists viewed the spectacle as nature's
playfulness at work – *ludi naturae* that competed with works of art. Julia Voss, a historian of art and science, observes that the connection remained valid long into the
nineteenth century, when Charles Darwin endeavoured to introduce the principle
of chance in natural history with his theory of evolution. The complicated architecture of a blossom, or of an organ, evoked lasting admiration – an astonishment that
had its religious side and was embedded in the so-called 'argument from design',
which states that nature's complex structure refers to a creator, as Reverend William
Paley wrote in his influential book Natural Theology, in 1802.[57]

Nineteenth-century discussions of nature was abound in religious motifs. The
works of Emil Adolf Roßmäßler (1806–67), a well-known popular writer, intone
hymns to nature as a temple and a mirror of divine revelation. Conversely, the
monistic natural philosophy of the internationally renowned Haeckel celebrated
and revered evolutionary nature. Both perspectives met with broad approval,
but they were criticized in certain quarters, too. For the purposes at hand, the
decisive matter is the qualitative relationship between religiosity and the scientific study of nature. Conventionally, these orientations are held to be mutually

exclusive; the semantics of wonder and mystery usually counts as the pre-modern residue of an overly sentimental discourse of nature, overcome in our own day. Such oversimplification, however, repeats the ideological conception of science in the eighteenth- and nineteenth centuries; instead of focusing on the diversity of discourses and conditions that in fact prevailed. Matters of historical import remain silenced or invisible so long as the 'culture war' continues, which deems science incompatible with religion (which is supposed to be 'hostile to objectivity'). Neither in the Enlightenment nor in the nineteenth century was such a view evident in concrete research practices. It proves much more rewarding to address contemporary debates between the state, the church and scientific institutions – which were certainly intense and far-ranging – than to write yet another account of conflict between 'backward' believers and forward-looking, 'real' scientists.

It is equally untenable to equate wonder and ignorance. Remarking 'wonders of nature' does not mean that nature has not been understood. Indeed, a well organized human mind, as William Marshall observed, can close itself to enchanted amazement only with difficulty.[58]

> The naturalists of yore esteemed the ocean to be a treasury of wonders, and sought therein monstrosities and organisms contrary to the law of nature, such as they interpreted it. The naturalists of our own time hold equal faith in the wonders of the sea, but they seek the links in nature's chain instead of apparent exceptions.[59]

The more knowledge of evolutionary biology increased, for example, the more one marvelled at the wondrous works of nature: the overflowing creativity that had the anarchic power to produce unimaginable multiplicity. Alongside discussions of purposefulness, functionality and the like, one finds aesthetic impressions that exemplify the long-standing tradition of nature's playfulness in fashioning all her wonderful works of art. Thus, Darwin's *Origin of Species* (1859) includes passages invoking 'the beautiful and harmonious diversity of nature' – and many similar phrases, which are far from empty.[60] After roughly five hundred pages, the author concludes the work that shaped an entire age as follows:

> There is grandeur in this view of life, with its several powers, having been originally breathed into a few forms or into one; and that, whilst this planet has gone cycling on according to the fixed law of gravity, from so simple a beginning *endless forms most beautiful and most wonderful* have been, and are being, evolved.[61]

In sum, words of wonder, the wonderful, and the wondrously beautiful did not serve a religious or a strategic function; nor do they attest to ignorance. Instead, such language stands at the heart of the culture of knowledge. It cannot be evaluated in terms of natural theology or efforts to secure a broader audience and public acknowledgement. One must not reduce it to linguistic functions of communication or reception; nor does it represent 'what the natural sciences cast aside on their race to the top'.[62] These are valid considerations, but they uniformly omit the possibility that such expressions of wonder directly concerned objects of admiration, too. One should not overlook the fascination, astonishment, shock and sensation of being overwhelmed that was provoked by the discovery of terrain that had previously resisted the human search for knowledge. The shadow cast by such resistance made the sea even wider and deeper, even more fantastic and exciting – a chaos of forms and rush of colours constituted by incomprehensible species and shapes. In this light, the discourse of wonder represents an aesthetically structured focus that permeated the nineteenth-century culture of knowledge. The sea offered a perfect – and perfectly absorbent – medium, offering explorers and nature-lovers alike a realm of phenomena that ultimately refused to be 'demystified'. Accordingly, the semantic range of wonder in the context of observation, description, and the production and dissemination of knowledge often struck sentimental notes – but subtle ones, too. After touring the Bay of Kiel in July 1862, Karl August Möbius (1822–89) wrote:

> An Bord der Yacht ,Marie': Hier empfinde ich froh die schöne Einsamkeit des Meeres, die der Waldeinsamkeit den rang streitig macht. Alles ist ruhig rund herum. Das Ohr hört nur dann und wann leises Plätschern, als erzählte das Wasser dem Schiff von den Wundern, die es verschleiert hält.

> On board the yacht *Marie*: Here I happily experience the beautiful loneliness of the sea, which can compete with the loneliness of the forest. All around me is silence. Only now and then does the ear hear a quiet splashing, as if the water were telling the boat of the wonders it keeps hidden. (translation EB)[63]

Midnight Dreams and the Visual Culture of the Underwater World

In the nineteenth century, the sea epitomized all that is mysterious about nature; revealing its secrets counted as a paradigmatic gesture. A hybrid world of imagination characterized by the dramatic play of light and darkness emerged; here dwelled paradisiacal constellations, explosions of colour and figures of the nighttime sky. The theme of 'observing the underwater world' that entered into circulation presented the plunge in terms of the quasi-divine perpective enjoyed by those on the ship watching the landscape pass below. In significant ways, accounts provided by the first balloonists and early marine researchers overlapped. Aeronautical associations merged with marine metaphors to describe a symmetrical universe of wonder that reached up to the stars and down into the depths of the sea.

'We Plunge into the Crystal Liquid of the Indian Ocean'

> we shall see realized therein the marvellous appearances of the fairy tales of our infancy. Fantastic shrubs are decked with living flowers; massive meandrina and astræa contrast with the tufted explanariœ, which expand in the form of goblets, with madrepores of elegant structure and varied ramifications. Everywhere with brown and yellow; rich purple tints subside into the liveliest red or intensest blue. Rosy, yellow,

or peach-coloured nullipores cover the decayed plants, and are themselves enveloped in the black tissue of the rétipores, which resemble the most delicate carved ivory. By their side waver to and fro the yellow and blue fans of the gorgons, richly wrought like jewels of filigree. The sand is besprinkled with sea-hedgehogs and sea-starts, of fantastic forms and varied colours. The flustra and escara cling to the coral branches like mosses and lichens, and patellæ, striped with yellow and purple, adhere to them like great cochineals ... The humming-birds of ocean – small gleaming fishes, some bright with a metallic spendour of azure or vermilion, some with a gilded green or dazzling silver lustre – play around the coral bushes. Light as spirits of the abyss the white or blue bells of the medusa float through this enchanted world. Here the violet isabelle, and the golden green, and the coquettish flame of fire, black and streaked with vermilion, pursue one another. There, among the ocean groves, wind to and fro, like long silver ribands, shot with rosy and azure lights, the nemerta, the sepia, glittering with all the colours of the rainbow, which alternately cross one another, shine forth, or swiftly wane ... When the day declines, and the darkness of night sinks into the depth, this radiant garden kindles with new splendours. Microscopic medusas and crustaceans sparkle in the obscurity like glow-worms. The pannatula, which wars by day a garb of cinnabar red, floats in a phosphorescent light. Every angle beams and shines ... And to complete the marvels of this enchanted night, the large silver disc of the sea-moon (orgathoriscus mola, the sun or moon fish gently moves through the whirl of tiny stars.[64]

Fairy tales, magic, glimmering sparks, wonder, dreams, spirits, stars: this excerpt from *Die Pflanze und Ihr Leben* (*The Plant and Its Life*, 1870) (1858) by Matthias Jakob Schleiden (1808–81), numbers among the most famous and frequently quoted passages of nineteenth-century science. The author takes his readers on a virtual dive, leading them through a metaphorical landscape that, from the hummingbird to the silvery sunfish, offers all the charms of a magic garden. It is worth remarking that when Schleiden wrote these words, most people had no opportunity to immerse themselves in the 'crystal liquid' themselves. All the same, many other descriptions of the sea performed the same vertical movement:

As soon as we descend a little below the surface, what interesting species and elegant forms conceal themselves, so to speak, in organisms of the simplest character, because adapted to the uniform existence which they all lead! What richness may be found in that relative poverty! – what profusion of life in those abysses to which we have not even access![65]

The classic perspective for the underwater spectacle, however, was the view from a travelling boat above. Johann David Schöpf (1752–1800) – whose works represent a treasure trove of motifs for later writings on the sea – provided one of the earliest descriptions of the 'heavenly' vision:

By going out in a small boat among the little islands the rarest and most splendid sight is to be had. The boat swims on a substance of crystalline fluidity in which, as in the air, it seems to hang. Those unaccustomed are like to grow giddy at the sight. Below, on the pure white sand covering the bottom, one can make out every detail, reptilia of a thousand forms, sea-urchins and sea-stars, slugs, shell-fish and particolored fishes; one floats above whole forests of stately sea-plants, gorgoniae, corals, alcyoniae, flabellae, and many sorts of shrubby spongious growths, their colours not less delighting the eye, and as softly moved by the waves back and forth as a flower-strewn field of the earth.[66]

The Wondrous Art of Swimming in the Air

The view of the researcher or traveller through the clear water to the bottom of the sea was often described as a heavenly journey into the sky. Such a godlike view from above, however, presupposed a complex cultural construction that recognizes no vertical restriction of perspective. Accordingly, Schöpf's image of a boat floating in the air, beneath which a submarine landscape passes by, is no coincidence. In 1783, the year his ship launched, the first *mongolfières* and *charlières* ascended. In 1788, when his book was published, *aéropetomanie* – also known as *fiévre aérostatique* – stood at its height. It is unlikely that Schöpf himself ever reached the skies by balloon, but he lived in a culture that, for more than two hundred years already, had professed enthusiasm for 'Icarian views'.[67] Under the spell of the great voyages of discovery, sixteenth-century cartographers had devised maps of land and sea, city plans and globes, which made the 'aerial view' (or, alternately, the 'bird's-eye-view') a general possession. A 'veritable cartographic turn' had occurred, betokening a deep caesura in the history of knowledge.[68] The symbolic and religious maps of the world that had been made in Middle Ages yielded to a new regime of visibility, which adopted the perspective of the heavens and, accordingly, made the observer 'fly'. These maps were just as familiar to Schöpf, the seafarer, as survey maps of the globe had been to travellers during the Renaissance. He had likely read at least one of the fantastic literary narratives about travel to the Moon; at any rate, he was surely familiar with airships, for example, the much-discussed design of Francesco Lana de Terzi (1631–87), a Jesuit priest, or the hoax, notorious throughout Europe, about the trip Bartholomeus Lourenco des Gusmaos (1685–1724) was supposed to have made on the *Passarola* from Portugal to Vienna on 24 June 1709.[69] Thus, the echo in Schöpf's 'aeronautic' vision is less the poetic representation of a concrete, aesthetic experience – that is, a view from a boat through crystal-clear water onto the ocean floor – than an expression of the affinity between the skies and the seas: the idea that both elements are to be travelled in the same way, namely swimming or diving, in small rowboats or elegant sailing vessels.[70]

Airships and Fish Balloons

In his *Vermischte Nachrichten über die aërostatischen Maschinen* (Varied Reports about Aereostatic Machines), Georg Christoph Lichtenberg (1742–99) imagines the immense pleasure of a balloon trip: 'Man bedenke auch nur das Atmen der Alpenluft, das Baden, Plätschern und Schwimmen im Lichtmeer und in

Gesellschaft der Morgensterne, während die Hälfte der Welt unter einem noch im Schlamm der Nacht ruht' (Just think about breathing the air of the Alpes, of bathing, splashing, swimming in the sea of light accompanied by the morning stars, while half of the world still rests in the mud of night).[71] While Schleiden, Schöpf, Adalbert von Chamisso (1781–1838), Tilesius von Tilenau (1769– 1857), Ehrenberg[72] and many others described the seabed's population of flowers, sea pens, glow-worms and little birds, Jean Paul (1762–1825) wrote an entire book detailing the adventures of the aeronaut Giannozzo in a marine Eden and other airy seas.[73] In the sky, aero*nauts* and 'artists of aerial natation' (*Luftschwim-mkünstler*) floated by; baroque gondolas and imposing vessels rowed across *seas* of clouds. More than any other time, the late-eighteenth and early-nineteenth centuries witnessed the verticalization of wonder. The period lent concrete expression to Aristotle's equation of water and air[74] and the ancient view that birds swim through the liquid air.[75] Thus, one reads in *Zedlers Universallexikon* (1735) that although mankind was not made to fly, the practice might be learned:

> (Für ein Leben im Wasser ist) der Mensch nicht erschaffen, gleichwohl könne er durch Kunstgriffe so weit kommen, dass er schwimme. Da nun zwischen dem Wasser und der Lufft ziemliche Gleichheit sey, folglich müsse es so gut in diesem als in jenen angehen

> Man (has) not been created for a life in the water, yet by tricks of art he manages to swim. Since a seeming likeness exists between water and air, in consequence it must work as well here as there.[76]

The view found expression most visibly in the many designs for fish-balloons that were made. August Wilhelm Zachariae (1769–1823) – a teacher of mathematics at the convent school of Roßleben in Thuringia and probably the 'inventor' of the paper airplane[77] – was a particularly prominent and outspoken advocate. He had the idea of enclosing the balloon, like an air bladder, in a vessel shaped like a denizen of the deep in order to create a kind of 'flying fish'.[78] What stands out in the considerations he offers is the ready shift from flying to diving, the move from the highest regions of the sky to the night-dark depths of the sea. In the first chapter, '*Vorbereitende hydrostatische Sätze*' (Preliminary Hydrostatic The-ses), he reflects on how different hollow bodies rise in the water, but by § 22 he has already moved from water in general to the sea in particular. The issue is no longer buoyant spheres, but human beings drawn to the sea floor. From this point on, the author focuses on a diver gliding through the water by means of various 'sails'; consideration then turns to an air-filled globe navigated with the aid of four 'sails' moving in circles. Zachariae speaks of 'feathering', 'luffing', 'maneuvers' and pumping; at the end, experimentation occurs through a diving bell alone (p. 82), which is described as having the 'elongated form of a capsized boat' (p. 84).

Above all, Zachariae discusses different variants of submarine vessels in his *Luftschwimmkunst*. Accordingly, in § 66 he asks where it would lead if one equipped divers with all means necessary for moving at will beneath the surface of the water.[79] His answer attests to the prevalent scientific ideal; it also stands as one of the earliest records of how to claim the deeps for human knowledge:

> Nichts verdient zwecklos genannt zu werden, was darauf abzielt, die Summe der menschlichen Erkenntniß zu vermehren; und so kann auch der Wunsch, den Menschen bis in die Abgründe des Meeres hinabsteigen zu sehn, nicht unter die nutzlosen Phantasien gerechnet werden.

> Nothing deserves to be called useless if it contributes to human knowledge; therefore man's wish to descend to the depth of the sea should not be assigned to idle phantasies.[80]

For Zachariae, the discovery of the ocean and research of the heavens represented adjacent terrains; activity in one sphere seems to lead straight to the other. After sounding the depths of various arguments, the author shares his project for an airship in the form of a fish. If the airfish was never built, this is not because the idea was so far-fetched. As late as 1866, Arthur Mangin – who remained skeptical about the possibility of flight – reported that, now and then, one finds idle engineers who kill time by building flying fishes that they steer in enclosed spaces protected from air currents. That said, Mangin continues, nobody pays these experiments much attention.[81] The statement is only true in part. Even if success proved elusive, fish-balloons abounded in visual culture throughout the nineteenth century-until they were finally replaced by dirigibles and zeppelins (if only as models on paper, as in the *First Aeronautical Exhibition* (1868).

In sum, considerable points of overlap exist between the reports made by the first balloon-travellers and early oceanographers. Intersecting aeronautical imagery and marine metaphors conjure up a symmetrical world of wonder; its vertical axis traverses the imagination of the Enlightenment as a whole. An all-but paradigmatic passage written by the French zoologist Armand de Quatrefages (1810–92) reads:

> This marvellously limpid condition of the water produced another charming illusion. Leaning over the side of the boat, we could see flitting beneath our eyes a vision of plains, valleys, and hills. After gazing intently for a while at the picturesque scene beneath our eyes, we scarcely perceived the intervening liquid element which served for its atmosphere and bore us on its clear surface. We seemed to be suspended in empty space, or, rather, realising one of those dreams in which the imagination often indulges, we appeared to be soaring like a bird, and to contemplate from some aërial heigth the thousand varied features of hills and dale.[82]

The vertical dimension leads both up to the stars and down into the secrets of the sea, connecting the impossible with the unknown: soaring in the clouds means navigating an oceanic element, so to speak. Henceforth, talk of the 'unerforschtem Äthermeer' ('unexplored ethereal sea', translation EB)[83] and the black night of the ocean proved a constant feature of discourse.

A Sea of Stars

> Scarcely had night spread its veil over the surface of the ocean, when it had the appearance
> of being all over on fire. Every wave that broke had a luminous margin or top; wherever
> the sides of the ship came in contact with the sea, there appeared a line of phosporeal
> light. The eye discovered this luminous appearance every where on the ocean; nay, the
> very bosom of this immense element seemed to be pregnant with this shining appearance.

> Johann Reinhold Forster (1729–98), Cape of Good Hope, 30 October 1771[84]

Bioluminescence is one of the most enchanting phenomena of the sea. The
ineffable glow when thousands of tiny sparks illuminate the water holds an established position in scientific writings, but only there.[85] Art and poetry are largely
silent about the gold-green marine light that sparkles and shimmers, but the
great voyages of discovery in the seventeenth century intoned a Circean hymn
to the spectacle. The fact merits notice inasmuch as phosphorescent floods of
light and self-lighting organisms do not represent an exotic phenomenon; they
have always been observable on all the coasts of Europe. What is it, then, that
suddenly prompted writers to seek out images of sprays of light, will-o'-the-wisps
and firestorms to describe their magic? One may venture two conjectures.

For centuries, marine luminescence, like other natural phenomena, had likely not
been perceived as anything mysterious by coastal inhabitants. Living in a world filled
with willful forces and dark powers, they would have considered it more important
to know the spawning grounds of herring than that the reasons why the sea glowed.
Bioluminescence was a matter one simply observed – a riddle, perhaps, but not a
mystery. Only when natural science developed an interest in marine phenomena in
the course of the eighteenth century did it demand explanation.[86] Mysteries emerge
at the height of knowledge, as if to spite it. In other words: Knowing too little always
also means not knowing that there is a mystery.[87] However, as soon as the glow of
the sea was recognized as something mysterious, it became possible to view it in its
complex, theatrical dimension-which consitutes the essence of mystery. Hereby, it
stepped onto a stage defined by Enlightenment thinking – *Aufklärung, Lumières,
Illuminismo* – at the centre of which, needless to say, stood light. Light encountered
light: illumination in the metaphysical perspective of the wholly other.[88]

Glowing marine organisms are nighttime phenomena. Delicately shimmering, they swam up out of the jet-black sea and assumed a place in the symbolic
order of culture. It is difficult to conceive a natural phenomenon occurring
against a background that is more ecstatically, mystically and epistemologically
charged. 'Consciousness begins with light', Hartmut Böhme has noted; 'jede
Äußerung des Lebendigen ist allererst vom Licht abhängig' (every expression of
life depends on light in the first place, EB).[89] Almost universally, light represents
the primordial being of Nature, its *quinta essentia*; for the West, too, essence

appears only through light. Light is the immaterial soul's medium in global darkness: natural substance with an aesthetic mode of being, which coincides with the self-transparency of *logos*.[90] Whatever one pictures as being, appearing and/or possessing life or mind is caught up in the semantics of light. Light is an 'absolute Metapher' (absolute metaphor), in Hans Blumenberg's phrasing: unavoidable, untranslatable into conceptual language and pointing to nothing but itself.[91]

In this context, the enchanted descriptions of an opaque, phospherescent sea acquire a different resonance. Luminous phenomena in the dark of night presented an erotic spectacle of sparkling arcs and serpentine fires pouring into the obscure womb of the fertile depths.[92] One might allude to the Orphic tradition of radiant Eros just as readily as to the age-old battle between the powers of Darkness and Light. Within a few years, a phenomenon long familiar to fishermen and seafarers had transformed into an enigma surrounded by myth, which countless parties endeavoured to explain.[93]

At the beginning of the nineteenth century, the view gained general acceptance that marine phosphorescence stems from tiny organisms, microscopic infusoria. In 1830, Gustav Adolf Michaelis (1798–1848) published the highly regarded study, *Über das Leuchten der Ostsee* (On the Phosphorescence of the Baltic Sea). In 1835, Ehrenberg's *Das Leuchten des Meeres* (Marine Phosphorescence) appeared; it quickly became the standard work.[94] In addition to presenting first-hand observations and interpreting them, Ehrenberg goes through the entire bibliography from Aristotle onwards – 435 authors in all – where mention of luminescence in animals or plants occurs. Still more, he draws up the fullest survey to date of self-luminous organisms, including the time and place of their first sighting, offering a comprehensive cartography of a history of discovery as brief as it was intensive.

But for all that, the magic and mystery did not simply vanish. Matter-of-fact description, however precise, could not dispel the poetic lustre of the topic. Michaelis, for example, speaks of a 'wundervolles Schauspiel' (wonderful spectacle) and 'magische Effekte' (magical effects) which account for an unexpressable experience.[95]

Ehrenberg also spoke of a 'hochpoëtischen' (highly poetic) and 'tief erregenden' (deeply stirring) 'für die Idee des Lebens, wie es scheint, wichtigen Naturerscheinung' (phenomenon of nature that is, it seems, important for the idea of life).[96] In the simple forms of this smallest life, the infusoria, he saw the basis for all organic formations, including human beings; investigation, he believed, would yield the key to the secrets of life. This idea – that one might 'das unter dem Schleier der Kleinheit verborgen wirkende riesenhafte Leben ans Licht … ziehen' (bring to light the enormity of life, which seems to be hidden under the veil of diminutiveness)[97] – made the sea into a place of revelation, a shrine which promised the answer to all questions. Where is the origin of life to be found? How does matter transform into spirit? How do the species arise in their vast multiplicity, and why? In the course of the nineteenth century – and long before

Darwin's *Origin of Species* (1859) – the search began for the primal forms containing the secret of life; the model was a primal ocean, at the bottom of which crept the firstborn of creation. At least for a few decades, this last of all secrets and the greatest of all mysteries had a precise location: the seabed, where all the answers to the beginnings of time and the 'riddles of existence' might be found.

> These organized atoms; these imperceptible zoophytes, are the torches of Ocean; they possess that subtle principle which all religions, all philosophies, all poetries have proclaimed to be the emblem of the Divine mind – Light! And this viscous, fatty matter, the residuum of the decomposition of innumerable beings, plants, and animals, this mucus secreted by the fishes, is yet a source of Light – what do I say? it is a source of *Life* – it is the universal nutriment of the oceanic flora and fauna; it is the *milk* in whose centre are born, and upon which are nourished, all those ephemeral creatures, so weak and so delicate – the infusoriae, molluscs, and radiatas – those infinitesimal atoms, whose power, nevertheless, owing to their numbers, is incalculable, and whose numbers, thanks to their fencundity, are beyond computation, and which play, in the marine world, a far more important part than its gigantic monsters, its sharks and whales. For these living atoms must be counted by myriads and myriads of myriads of legions; and it is these, we must not forget, which convert Ocean into an enormous reservoir of life, a vast organism wherein matter moves, circulates, renews and transforms itself, organizes itself, accomplishes its work, and incessantly recommences the circle of its mysterious evolutions.[98]

As the preceding examples from the first half of the nineteenth century illustrate, the scientific-aesthetic *mise-en-scène* of mystery unfolded in perfect concert with the character of the sea. The delicate glow would not yield to argument or biochemical analysis; instead, it was perceived in terms of gesture and form: impressions of light and dark, soft and hard, growth, decline and recurrence. More than all else, it offered a a blank space symbolizing where language and meaning fail. It called for a form of objectivity that would not simply be the denominator of a number in an equation, but rather express the poetic core of the oceanic: 'livingness' (*das Lebendige*). As the search proceeded at the border of scientific and aesthetic discourses, marine mystery promised improbable and unattainable treasure, stood for overwhelming fertility and, at the same time, presented the image of 'another' creation. The glow of sea sparkles (*Noctiluca miliaris*) and fire algae (*Pyrrhophyta*) seemed to contain a sea of of stars; the depths of the ocean melted into the dark nighttime sky. As the poet and naturalist Adelbert von Chamisso (1781–1838), a world-traveller, put it:

> Ich werde zu den Schönheiten des Himmels ein Schauspiel rechnen, welches ... sich (in südlichen Gefilden) in reicherer Pracht zu entfalten pflegt. Ich meine das Leuchten des Meeres.

> Among the beauties of the heavens I will count a spectacle that tends ... (in southern regions) to unfold its majesty more fully. I mean the phosphorescence of the sea.[99]

To this very day, such parallels between stars in the sky and in the sea, the ocean depths and the night, pervade accounts in untold variation. Marine biolumines-cence, life that sparks itself, counts not just as one of the strangest qualities of the sea, but also as the fullest aesthetic expression of unearthly beauty.

Marine Research in the Nineteenth Century: An Overview

> Natural History of the ocean began modestly, at the shoreline. Beachcombing, to search for beautiful and unusual shells gained popularity in the eighteenth century and remained a central feature of marine natural history.[100]

At mid-century, enthusiasm for sea algae surfaced,[101] and toward the turn of the century a heightened interest in fossils. Between 1775 and 1800, members of the prosperous middle class took to investigating the beaches. They started to exam-ine the broad expanses of the Wadden Sea exposed at ebbtide; before long, they made small nautical excursions, too. Already in 1750, it seems, dragnets were used in Italy to collect marine animals. In the 1770s, the Danish naturalist Otto Frederic Müller (1730–84) spent four summers at the southern coast of Norway, where he dredged the sea at a depth of approximately 50 metres (30 fathoms). For all that, dredging became an established practice only from the 1830s onwards, when amateur scientists began using trawlnets to explore the flora and fauna of the ocean.[102] On average, operations proceeded at a depth of 36 to 180 metres (20–100 fathoms). There was no pressing reason to go deeper; as a rule, a few dives at moderate depth yielded enough results to keep one busy all winter long.

In educated circles, collecting and studying marine animals and plants rep-resented a genteel diversion. It filled the idle hours of amateur researchers and refined vacationers when the ocean entered their homes, conversations, thoughts and dreams. Natural science by the sea offered adventure and enjoyment at the same time. Hereby, one should note:

> Nineteenth-century natural history was not the pastime of pottering gentleman end-lessly classifying things to no apparent purpose, with a few truly scientific giants like Charles Darwin emerging miraculously from their ranks. Marine Zoology provides a particularly good example of an undertaking grounded in traditional natural history yet simultaneously at the forefront of new professional science. Naturalists carried a vivid image of Newtonian astronomy into their belief that, from the facts they observed of similarities between living things, a set of unifying principles would soon emerge.[103]

The greater part of zoological discoveries in the sea originated among the impas-sioned natural studies of these early enthusiasts. They were keen observers, rapt dreamers and meticulous archivists in equal measure. Charles Kingsley (1819–75) sums up the prevailing spirit in *Glaucus; Or, the Wonders of the Shore*:

> Why not discover some of the wonders of the Shore? For wonders are there around you at every step, stranger than ever opium-eater dreamed ... wonders of *classification* ... marvels, which meet us at every step in the *anatomy* and the *reproduction* of these creatures.[104]

In the nineteenth century, the 'wonders of the sea' were much more than – and something altogether different from – curiosities that ladies with nothing better to do collected for the decoration of their boudoirs (as Lynn Barber has claimed).[105] They were both aesthetic and epistemic objects.[106] Their wondrousness was evident in external features that occasioned amazement, to be sure, but above all they proved fascinating for their singular anatomy and the mystery of their reproduction – that is, the difficulties they posed when one sought to understand their place in God's well-ordered creation.

The study of the sea could not have been more 'modern'. It beseemed people of a better sort to take up the multiplicity of lifeforms, their characteristics and particularities, and to be able to distinguish bread-coloured chameleon stars from serpent stars with bluish-black and crimson stripes and rings, amaranth-red feather stars, violet-black urchins, flesh-coloured leeches shaped like worms, black sea cucumbers and orange-coloured starfish. The famous 'sea craze' of the nineteenth century may have been a matter of fashion, among other things. More importantly, it involved a public that, when the HMp *Challenger* went on the first deep-sea expedition (1872–7), did not ask why it is worth retrieving living sea lilies in the first place, and which enthusiastically participated in speculations on the strange nature of the *Bathybius*, which was held to be a kind of 'primordial soup'.[107]

'Discovering' the Deep Sea

The 'discovery' of the deep sea stands among the most important discursive events of the latter half of the nineteenth century. Historical accounts begin, as a rule, by observing that the land masses of the Earth had largely been explored by mid-century – with the exception of the poles – and that coasts had been well mapped for the most part. And yet, even though ships of varied provenance had plowed the Seven Seas for more than two hundred years, hardly anything was known about life in the deep or the nature of the sea floor at a couple hundred or thousand metres' depth.

Mysteries of the Deep

> There flourish immense forests, even more mysterious than the sacred woods of olden time. Fish, molluscs, crabs, are the happy denizens of these shady retreats. But who can flatter himself that he is familiar with these haunts? Do they not rather seem for ever closed against the intrusion of man? Who can presume to fathom the mystery of these immense tracts, denser with foliage than the virgin forests of the New World? ... We imagine all that is mysterious beneath its waters ... The sea and all that therin is appears surrounded with a poetic and miraculous aureole, which is the birth at once of fear and of superstition.[108]

The emotional, or, alternately, aesthetic, ambivalence embodied by the sea was attended by an equally ambivalent projection. On the one hand, the ocean was perceived as an enchanted land: 'its inhabitants are self-luminous; they thunder upon their enemies from a distance; they harden themselves into stone'.[109] On the other hand, it offered a reservoir for imaginative dives into uncertainties as dark as the night: an ideal stage for fear and superstition. What might be easily dismissed today as naïve credulity surely felt different two hundred years ago.

At a time when new events and scientific theories were overturning so many received certainties, keeping a cool head must have proved difficult. For instance, when utterly fantastic fossils of marine dinosaurs were found at the beginning of the nineteenth century, sightings of sea monsters experienced a dramatic upswing. In the 1860s, Darwin's *Origin of Species* encouraged views of the deep as a 'storehouse of "living fossils"',[110] and the creatures that researchers actually managed to draw to the surface seemed to confirm the image of prehistory as an infernal regime of terror.

'By and by', reads a letter that Ehrenberg wrote in 1860, 'we shall learn how to bring up larger samples of the bottom, and bring up some Leviathans'.[111] Scientists stressed time and again that the creatures discovered in the ocean depths could in no way be dragons, giants squids, Leviathans or other fantastic beings. All the same, there was always the possibility that, as the result of scientific progress, ultimately the prophecies of legend would be fulfilled when gigantic animal forms emerged from the deep.[112] Accordingly, the picture of a pandaemonian underworld, with a host of images reaching back centuries, came along with the disclosure of the deep sea and shaped perceptions, expectations and habits of seeing in decisive fashion. The border between the possible and the impossible turned into a dangerous realm where the credulous and agnostic alike could quickly lose their reputations. 'Does the great sea-serpent exist?' asked Georg Hartwig, an author of works of popular science. Given the fossilized remains of long-vanished plesiosaurs, many deemed it possible. Thus, Louis Agassiz is reported to have said: 'I can no longer doubt the existence of some large marine reptile allied to Ichthyosaurus and Plesiosaurus yet unknown to naturalists'.[113] Henry Lee (1826–88), a natural scientist and the director of the Brighton aquarium, observed in *Sea Monsters Unmasked*:

> It appears to me more than probable that many marine animals, unknown to science, and some of them of gigantic size, may have their ordinary habitat in the great depths of the sea, and only occasionally come to the surface; and I think it not impossible that amongst them may be marine snakes of greater dimensions than we are aware of, and even a creature having close affinities with the old sea-reptiles whose fossil skeletons tell of their magnitude and abundance in past ages.[114]

Rozwadowski sums up the matter: 'sea serpent investigation was treated seriously, if cautiously by men of science'.[115] Expeditions set out to capture the sea serpent – to no avail. A famous case occurred in 1842 when, in the hall of the Royal Iron Foundry in Berlin, the skeleton of a monster thirty-five metres in length was exhibited as the sensation of the century. 'Unfortunately', it quickly turned out that the fossilized bones belonged to at least five different ancient whales, which a certain Dr Alfred C. Koch had arranged with as much skill as fantasy.[116]

It is a matter of no small irony that a large 'sea serpent' really does exist: the 'oarfish', which has a coral-red, spiky crest on its head and a red, ridged fin running down the length of its back. To this day, hardly anything is known about how this extraordinarily rare animal lives. Ichthyologists suppose, however, that this 'monster with a horse's head' and a flaming mane probably grows to more than fifteen metres and may occasionally swim in undulating movements on the sea's surface.[117] Needless to say, it is another matter altogether whether this retiring creature was responsible for the mass-sightings of sea serpents that occurred between 1816 and 1847,[118] when they were seen swimming for days in harbours, posing in front of eager observers and accompanying ships out onto the high seas until every last sailor could provide a personal account.

Thalassiotrocus Telegraphicus – Living or Lifeless Deep?

Until as late as the second half of the nineteenth century, it was commonly assumed that only death lurked in the abysses of the deep sea. In the introductory passage of Maury's *Sailing Directions* (1854) – a milestone of oceanographic research – one encounters not just sober tables of figures, but also Clarence's dream-speech from *Richard III*, which presents the newly measured basin of the Atlantic as a fearsome crypt:

> Could the waters of the Atlantic be drawn off, so as to expose to view this great sea-gash ... from the Arctic to the Antarctic, it would present a scene the most rugged, grand, and imposing. The very ribs of the solid earth, with the foundations of the sea, would be brought to light, and we should have presented to us at one view, in the empty cradle of the ocean, a thousand fearful wrecks, with that dreadful array of dead men's skulls, great anchors, heaps of pearl and inestimable stones ... making it hideous with sights of ugly death.[119]

For centuries, the ocean depths counted as the most inhospitable and uncanny place imaginable. The idea was widespread among sailors that capsized ships and people or things flung overboard would sink only as far as the water density still matched their weight.[120] In 1842, the botanist, zoologist, and geologist Edward Forbes (1815–54) sounded the depths of the Aegean Sea and found no life below 500 meters.[121] This reinforced the view that the depths were an 'azoic' zone where the cold, lack of light and water pressure allowed nothing to exist. Independent of Forbes, the same conclusion was reached by the Swedish zoologist Sven Lovén (1809–95), even though he left it open where the border of life might actually lie.[122] It was above all Forbes and Lovén's readers who made theoretical conjecture into doctrine; for decades afterward, voices to the contrary found no echo.[123] The field of hydrography, for example, had known for some time that warmer layers also exist in the ocean. Time and again, deep-sea fishers and whalers had offered proof that survival at great depths was possible, for example, when harpooned sperm whales dove and dragged the lines hundreds of metres beneath the surface. In 1819, at a depth of 1,828 metres (1000 fathoms), the naval officer Sir James Clark Ross (1800–62) had encountered different kinds of worm and a starfish.[124] In the 1840s, while measuring the depths of the Algerian coast, Georges Aimé (1813–46), a physicist and pioneer of the study of waves, retrieved various animals from a depth of 1,800 metres.[125] All this information circulated in shipping-industry publications, accounts of voyages, newspapers and journals. Its failure to command due attention speaks for the fact that natural scientists and seafarers – even when the latter were officers – lived in completely separate 'epistemological worlds'. The language and form that professional seafarers used to report their discoveries did not correspond to what was recognized as 'scientific'. What is more, neither Ross nor Aimé presented any specimens gathered from the deep. In the early days, marine biology – which already faced the problems posed by abundant sightings of sea serpents and mermaids – relied on empirical currency.

Finally, on 15 October 1860, George C. Wallich (1815–99) managed to bring thirteen living starfish from a depth of 682 meters on board the HMp *Bulldog*. He proudly noted: 'the deep has sent forth the long coveted message'.[126] But to Wallich's great disappointment, his discovery was not accepted as a decisive demonstration of life in the depths; as critics observed, the starfish might have fastened on the measuring line somewhere on the long journey from the ocean floor to the surface. Ultimately, an accident provided conclusive proof of life in the deep. Shortly after Wallich's find, a defective telegraph cable on the Algerian coast between Sardinia and Bona was brought up from a depth of 3,600 meters.[127] It was encrusted with sea creatures – *Grantia, Plumularia, Gorgonia, Caryophyllia, Alcyonium, Cellepora, Retepora, Eschara, Salicornaria, Ascidia, Lima, Serpula* – as well as a new kind of coral that, in honor of the occasion, was christened *Thalassiotrocus telegraphicus*. Undoubtedly, the cable had spent a considerable time underwater, and the depth had been measured numerous times – an absolute sensation! Only after this point did the notion of a lifeless deep count as refuted 'officially' and definitively.

The First Expeditions: The Deep Sea Comes to the Surface.

> Between 1840 and 1880, the ocean ceased being a wasteland and highway and was transformed into a destination, a frontier, an uncivilized place ripe for conquest and exploitation.[128]

Around 1860, the exploration of the ocean exercised an irresistible pull for many scientists. Their growing fascination for bizarre forms of life coincided with technological enthusiasm for measuring ever-greater depths. Accompanied by economic interests – laying cables between the continents had encouraged general acceptance that exact knowledge of the sea floor would yield gains – the exploration of the depths began in earnest. The general public hailed the development. The colonialistic resonances are unmistakable:

> [B]eneath the waves, there are many dominions yet to be visited, and kingdoms to be discovered; and he who venturously brings up from the abyss enough of their inhabitants to display the physiognomy of the country, will taste that cup of delight, the sweetness of whose draught those only who have made a discovery know.[129]

In the late 1860s and early 1870s, the voyages of the HMS *Lightning*, HMS *Porcupine*, and HMS *Challenger* established the ocean depths as a significant research focus in the biological and physical sciences. In the summer of 1868, on board the the HMp *Lightning*, the Scottish zoologist Charles Wyville Thomson (1830–82) and the English physiologist William B. Carpenter (1813–85) brought up a rich array of creatures never seen before, from a depth of 958 metres (530 fathoms). They found sponges, rhizopods, enchinoderms, crustaceans and molluscs that yielded a new picture of the seabed as a space densely inhabited by the most improbable beings.[130] The following year, the crew of the HMp *Porcupine* dragged up specimens from a depth of 4,450 metres (2,435 fathoms).[131] In due course, samples from 80 to 100 fathoms counted as negligible, and those from 300 to 800 fathoms as only moderate.

Finally, the journey of the British corvette *Challenger* (1872–6), directed by Thomson, achieved the status of legend.[132] Never before had such a large-scale expedition been sent out specifically to investigate the physical, chemical, geological and (above all) biological conditions of the deep. The combined costs reached a sum corresponding to approximately ten million pounds today; until then, a comparable amount had never been awarded to a single scientific project. Besides Thomson and Carpenter, the scientific crew consisted of the Scottish chemist John Young Buchanan (1844–1925), the zoologist Henry Nottidge Mosely (1844–91), John Murray (1841–1914) and Rudolf von Willemoes-Suhm (1847–75), as well as the Swiss draftsman Jean-Jacques Wild. Over the course of 713 days at sea, the *Challenger* made 362 stops to measure depth and temperature, take samples of sediment and water and collect animals and plants with trawls and plankton nets. The methods employed were not new – dredging, plumbing, chemical analysis, observation through microscope, temperature measurement and lab experiments – but the intensity was; research was pursued across the globe, with a focus on deep sea regions. In the Mariana Trench, an abyss of almost 8,200 metres was sounded, and researchers were able, for the first time, to sweep the ocean floor with a net. Even the exceptionally sober chief engineer, William James Joseph Spry (1835–1906), got caught up in the breath-taking discoveries:

Doubtless we shall be told of wondrous facts which will read like fairy tales; for previously no sounding-line had ever traversed the great oceans, mapped out their figure. We now know ... of the myriads of curious creatures, organised with delicacy and beauty, existing in these previously unsounded depths; creatures with numberless eyes, and others without any; starfish, growing on long and slender stalks; of beautiful phosphorescent avenues of vegetation; fish of all hues, blue and gold, striped and banded, in all colours and sizes, from the tiniest infusoria to the huge whale.[133]

When the *Challenger* finally dropped anchor at Sheerness in May 1876, it unloaded 563 crates containing 2,270 glasses with objects preserved in alcohol, 1,749 small, corked bottles, 1,860 glass tubes, 176 tin cans, 180 tin boxes and twenty-two barrels with objects in brine awaiting scientific analysis. Four years later, in autumn 1880, the first volume of the *Challenger Report* appeared. Huxley greeted the publication with euphoric words: it promised 'at last [to] reveal some of the secrets of the busy life which, contrary to all the beliefs of the naturalists of a past generation, blindly toils and moils in the darkness and cold of the marine abysses'.[134] Fifteen years later, the final installment appeared. In total, the results of the *Challenger* expedition comprise forty large-format volumes that, ever since, have represented a vital input to the beginnings of oceanographic research in almost all the world's libraries. The report is based on international cooperation and contains thousands of illustrations from scientific specialists all over the world, to whom specimens of individual animal-groups had been sent.[135] 'Thereafter', positivistic historical discourse affirms, 'the ocean – including its greatest depths – remained the province of scientists from a wide spectrum: zoology, geology, chemistry, and physics'.[136]

The voyage of the *Challenger* represented a milestone of marine exploration. It was followed by numerous research expeditions on both sides of the Atlantic. They included the travels of the pp *Vøringen* (1876–8), with Georg Ossian Sars (1837–1927) and Henrik Mohn (1838–1921) on board, to investigate the North Atlantic seabed, and, in the 1880s, those of the French vessels *Travailleur* and *Talisman*, which, under the direction of Alphonse Milne-Edwards (1835–1900), sought out the Bay of Biscay and the West African coastal region as far as the Cape Verde Islands. From 1883 to 1886 the North American *Enterprise* sailed the seas. In 1889, the German plankton-expedition of the pp *National*, with Victor Hensen (1835–1924) on board, was launched. In the early 1890s, the Austrian research ship *Pola* explored the Mediterranean; on 28 July 1891, fifty sea miles southwest of Cape Matapan, at 35° 44' 48' N., 21° 45' 48' E., it determined the deepest site of the Mediterranean – 4,404 metres – which has been known as the 'Pola Deep' ever since.[137]

Discovery of the deep proved a defining social event in the second half of the nineteenth century. The public followed the first expeditions keenly; lay people and professionals greeted exhibitions and publications enthusiastically. Expert discourse could count on a solid number of well-informed amateurs to disseminate information through public lectures, popular books, and journal articles. The dramaturgy could hardly have been more enthralling. With each expedition, new sensations came to light. If, previously, above all tiny organisms had been brought up from the depths, now there were true 'giants', too; in this light, the ocean depths seemed to present a dreamscape of absurd proportions: sea firs as tall as a man, mammoth crabs, crayfish with tentacles extending one and a half meter, sea squirts which measured more than two meters, huge mantis crabs and numerous most peculiar cuttlefish.[138]

Abyssal gigantism offered rich stuff to feed theories that held the depths of the sea to be a refuge for mythical and prehistorical beings. Outsized animals simply fit too perfectly into the portrait gallery of beasts and monsters to be assigned an unproblematic place in established nomenclature. At the same time, discoveries surpassed all that had ever been seen before. Accordingly, the organisms – which, most often, were shrivelled and squashed – had a precarious status as objects. On the one hand, they were epistemic entities pointing to a Creation that followed completely different rules and was wholly incomprehensible; on the other, these phantasmagoric enigmas radiated immense aesthetic fascination. In museums of natural history and at the great world exhibitions, enthusiastic visitors leaned over mauve-coloured sea cucumbers and crustaceans that seemed to have been painted firecracker red; they marvelled at fish bearing their own 'lantern' and snakelike entities whose heads looked like a pelican with terrifying teeth. Who could understand it all? Instead of looking for answers, responses were mainly aesthetic; the encounter produced masterworks that have not been equalled to this day. The creatures were drawn and painted, tinted and retouched, photographed and scupted, until – in a thousand and one forms – they had permeated the visual culture of the *fin de siècle*. The deep night had arrived in human fantasy in ways that proved eloquent and visually arresting in equal measure.

Carl Chun and the Journey of the Valdivia – 31 July 1898 to 1 May 1899

Among the grand nineteenth-century voyages, the German deep-sea expedition of the *Valdivia*, which occurred from 1898 to 1899, was especially significant. It yielded far-ranging material – discoveries included 180 new kinds of deep-sea fish alone! – finally confirming that vast populations teem not just on the sea floor, but at all layers of the ocean. That said, it was perhaps the *Valdivia's* contribution to the visual culture of the *fin de siècle* that proved most important of all. The two reports – a scientific account and a version for the general public – present an *oeuvre* that shows, with unprecedented clarity, how certain forms of reflection are accessible only in aesthetic terms: a kind of authentically scientific-aesthetic knowledge, or, alternately, ability is called for; this mode of experience and cognition cannot be disqualified as compensatory and irrational or dismissed as an unscientific eruption of sentiment. The credit goes chiefly to the leader of the expedition, the German zoologist and deep-sea researcher Carl Chun (1852–1914). Chun's work represents a singular engagement with the aesthetics of ocean animals in words and images. His scientific-artistic method was prized both by the public and among scientists, and his colleagues wrote laudatory articles; they esteemed his work on invertebrate marine animals, especially his 'wunderbare Ästhetik dieser duftigen Gebilde, die nur gelegentlich und unerwartet aus magischem Dunkel der salzigen Wasserflut nächtlich emporsteigen, um vorübergehend einem glücklichen Auge sichtbar zu werden' (wonderful aesthetics of those filmy figures which only occasionally ascend from the magical darkness of a deep salty night as transient apparitions for a fortunate eye).[139] Chun's unimpeachable reputation as an accomplished man of science is also attested by the fact that Kaiser Wilhelm II saw fit to supply him with the handsome sum of 300,000 marks for his expedition.

The expedition was a thoroughgoing success. The public was happy, research-colleagues enthusiastic, and the Kaiser altogether content. With the twenty-four-volume report and its popular summary, *Aus den Tiefen der Weltmeere* (From the Depths of the Ocean), Chun gave the sea-mad world of the turn of the century a gift that was a scientific work and an aesthetic-poetic dream in equal measure.

'Delightfully Illustrated': On the 'Magic' of Life in the Deep Sea

The twenty-four illustrated volumes of *Wissenschaftliche Ergebnisse der deutschen Tiefsee-Expedition auf dem Dampfer 'Valdivia' 1898–99* (Scientific Results of the German Deep-Sea Expedition aboard the Steamer *Valdivia*, 1898–9) (1902–40), as well as the account for the general public, *Aus den Tiefen der Weltmeere* (From the Depths of the World Ocean), present 'treasures lifted from the sea' as an exquisite panorama of 'marvellous forms'. Both in photographs and in numerous scientific illustrations, their foreignness and extravagance are presented in radiant detail. The shades of colour reflect Chun's enthusiastic attention to the finest nuances. The spatial arrangement of the beings on the paper is varied, dramatically elegant, confounding and moving. The filigreed limbs do not simply fill out the framework of space in orderly and decorative fashion, as is the case in Haeckel's *Kunstformen der Natur*, for example; instead, they describe circles over the page with vivid gracefulness. Fishes swim out at the viewer with murderous jaws or rush 'busily' over the pages and the centre fold, as if the book were a paper aquarium. Presenting the organisms in this way – that is, not subordinating them to flat, graphic logic but rather taking the page as space in which they can 'move' – is a unique occurrence in oceanographic research literature. Moreover, the individual depictions of the life forms prove 'so frappierend und seltsam, dabei aber auch wieder so genau, dass man sie in vielerlei Umzeichnungen noch heute in den Lehrbüchern zur Meereskunde wiederfindet' (so striking and strange, but then again also so exact, that one finds reworkings of them still today in the textbooks of oceanography, EB).[140]

Both the *Valdivia* report and in the popular account of the voyage offer the viewer a language of images that immediately rouse aesthetic curiosity in the borderland between scientific and aesthetic discourse.[141] On one page, a polyp shoot leans gently to the right and seems to push the text into an asymmetrical half-oval (*Aus den Tiefen der Weltmeere*, p. 482); elsewhere, the enormous gullet of a pelican eel bends the words into an arc (*Aus den Tiefen der Weltmeere*, p. 521); in another section, a gigantic starfish swims across the page (*Aus den Tiefen der Weltmeere*, p. 197). These and other unconventional layout features heighten the effect of the pictorial matter and dynamize the relation between text and image; what is more, they evidence engagement with the space of knowledge constituted by scientific images. It would certainly be going too far to presume that the two famed zoologists Willy Kükenthal (1861–1922) and August Brauer (1863–1917) – the artists responsible for the polyp and pelican eel, respectively – engaged in theoretical reflection *avant la lettre* about the way that images shape knowledge. Even so, in the second half of the nineteenth century, natural scientists readily acknowledged that scientific illustrations played a key role in the research process, offering an important medial praxis for dividing scientific attention and, in consequence, constructing theory. Illustration was a solid component of natural-scientific training and promoted familiarity with the objects of study; thereby, one came to understand their physiognomy and defining traits. In this context, one can also understand the unconventional arrangement of space and the versatile staging of newly discovered species to articulate a mode of self-reflection that did not view scientific illustrations as foreign bodies framed by an autonomous text, but as an element that contributes to the genesis of knowledge by organizing texts and changing them, or, indeed, making them possible in the first place. Thus, the texts represent even more than the height of professional and technical mastery; they provide a rare instance, in scientific literature, of the medial conditions of epistemic insight and its connection to aesthetics.

die uns berechtigen, ſie zu einer neuen Gattung zu er=
heben, ſo dürfte doch die Erwartung, daß ſie ungeahnte
Aufſchlüſſe über die Beziehungen der Wirbeltiere zu nie=
deren Formen gebe, nicht in Erfüllung gehen. Sie erweiſt
ſich in jeder Hinſicht als echte Appendicularie, und keines
ihrer Organſyſteme fällt auffällig aus dem gewohnten Rah=
men heraus.

Was die pelagiſch lebenden Tiefſeefiſche anbelangt, ſo
glauben wir uns wohl kaum einer Übertreibung ſchuldig zu
machen, wenn wir ſagen, daß eine ganze Welt von neuen
Formen durch die Anwendung der Vertikalnetze entdeckt wor=
den iſt. Der Bearbeiter der Fiſche, Dr. Brauer, teilt mir
mit, daß dieſelben nicht weniger als 180 Arten angehören,
unter denen ein auffällig großer Teil mit bisher bekannt ge=
wordenen nicht zu identifizieren iſt. Sie gehören meiſt den Fa=
milien der Scopeliden (Myktophiiden), Stomiatiden, Lophiiden
und Muräniden an. Es iſt indeſſen weniger die große Zahl
von neuen Arten, Gattungen und ſelbſt Familien, die hier
überraſcht, denn die wunderbare, oft monſtröſe Geſtalt und
die höchſt eigenartigen Anpaſſungen an den Aufenthalt in
unbelichteten Tiefen, welche dieſelben erkennen laſſen. Meiſt
ſind ſie ſchwarz und faſt ſtets mit Leuchtorganen ausge=
ſtattet; in ſeltenen Fällen ſind ſie ſilberglänzend oder bunt
gefärbt. Da uns die merkwürdigen Anpaſſungen der
ganzen Körpergeſtalt an eine räuberiſche Lebensweiſe
in der Tiefſee noch in einem anderen Zuſammen=
hange beſchäftigen werden, ſo ſei hier nur darauf
hingewieſen, daß unſere Kenntniſſe über die Bio=
logie dieſer Organismen inſofern eine wichtige
Bereicherung erfahren haben, als wir mit aller
Schärfe den Nachweis führen konnten, daß viele
bisher für typiſche Grundbewohner gehaltene
Formen pelagiſche Lebensweiſe führen. Dies gilt
namentlich für eine Anzahl von Tiefſee=Aalen
und Lophiiden, deren wir einige im Bilde vor=
führen. Die Phantaſie eines genialen Teniers ver=
möchte kaum bizarrere Monſtra auf die Leinwand
zu zaubern, als ſie hier unter den Lophiiden uns

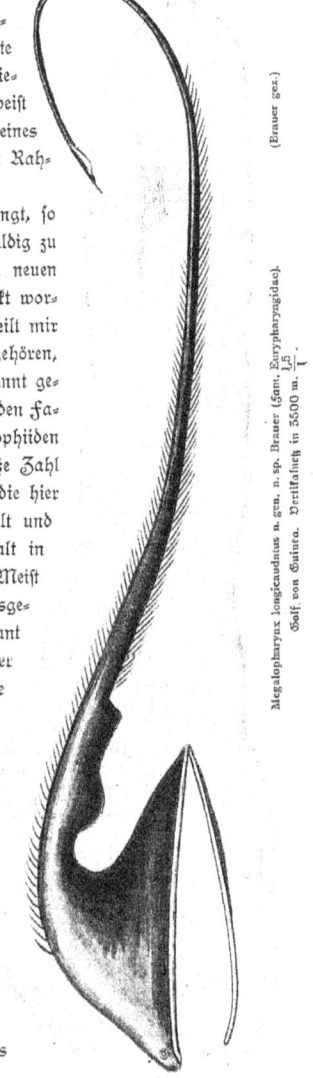

(Brauer gez.)

Megalopharynx longicaudatus n. gen. n. sp. Brauer (Fam. Eurypharyngidae).
Golf von Guinea. Vertikalnetz in 5500 m. $\frac{1.5}{1}$

Figure 2.1: *Pelikanaal*, C. Chun, *Aus den Tiefen des Weltmeeres. Schilderungen von der
Deutschen Tiefsee-Expedition* (Jena: Fischer, 1900), p. 557.

Media Practices of Visualization: Drawing and Taking Photographs

From at least the seventeenth century on, visual experience of nature's multiplicity of colour and form has been an established feature of scientific publications. Every expedition included painters and illustrators, and the results of their painstaking labours eventually wound up in private and public collections. This graphic praxis emerged as an important medium of exchange between scientific and artistic methods of thinking and recording, and it established an aesthetic space open to experiments, *mise-en-scène* and explorations of every kind.[142] In the tradition that emerged, illustration was a medial praxis for understanding the 'Leben und Weben aller Natur' ('organic construction' and the 'interwoven life of all nature', EB).[143] 'There must be observance of the ruling organic law', the critic, artist, and historian John Ruskin (1819–1900) wrote in *Elements of Drawing*: 'only in this way can one see how life, past and present, finds expression'.[144]

The nineteenth century experimented with many new media and modes of visualization – cyanotype, photogram, photogenic drawing, amphitype, chrysotype, *dessin fumée*, nature printing and other techniques; in the latter half, photography emerged as a further method of scientific visualization.[145] The voyage of the *Challenger* was the first research expedition to include photographers, even though their possibilities were still very limited at the time.[146] The equipment was heavy and unwieldy, and the need to develop every plate as quickly as possible restricted movement. What is more, emulsions were so slow that only immobile objects could be recorded. As is well known, matters changed soon enough; already on the *Valdivia*, the darkroom proved so popular that Chun spoke of a veritable *furor photographicus* taking hold of all the crewmembers.

The conventional view holds that photography displaced scientific illustration as higher standards of objectivity emerged. Lorraine Daston, Peter Galison and others have remarked a fundamental paradigm shift: the emergence of the ideal of 'mechanical objectivity'.[147] According to this view, the aesthetics of the eye and the hand yielded to the dominance of mechanical recording instruments and methods of taking protocol that avoided intervention; investigation by artistic means was replaced by a sharp separation between scientists and artists. The general validity of these findings has been sufficiently critiqued and relativized (or, alternately, refuted) in numerous studies.[148] One cannot say that photography *per se* was employed as a general antidote to fallible, individual human observation; nor may it be said that it replaced scientific illustration (or, alternately, that the latter lost its status as a means of proof and insight). In this context, Leopold Dippel (1827–1914), a Darmstadt botanist emphasizes:

> Eine jede Beobachtung ist, wenn sie überhaupt der Wissenschaft zu Gute kommen soll, durch eine deutliche und naturgetreue Zeichnung festzuhalten. Letztere kann durch kein anderes Mittel unnöthig gemacht oder ersetzt werden.

> Each observation must be captured by a distinct true-to-life drawing if it's meant to
> promote science. There are no other means which can substitute or reduntantise the
> drawing practice.[149]

Scientific illustrations were more than, and different from, mere copies of
sensory impressions; to this day, they represent an important part of all *observational sciences* involving objects that have not been epistemologically secured.
This also holds for marine research; as Barbara Maria Stafford demonstrates, to
great effect, in *Artful Science*, its praxis stood in a tradition with a strong visual
orientation that valued close exchange between artistic and scientific methods.
Although scientific images did not count as works of art, they certainly incorporated aesthetic procedures, models and heuristic images with artistic, mythical
and symbolic origins. Accordingly – as images with manifold cultural contexts –
scientific illustrations were both read and contemplated; indeed, the exceptional
chromatic vividness and multiplicity of forms in pictures of the sea often made it
impossible to avoid an aesthetic encounter.

Just as there is no such thing as a non-aesthetic image,[150] zoologists and botanists could not do without scientific illustration. Far into the twentieth century,
artists travelled on expeditions and were responsible for preparing illustrations
for scientific publications. Still today,

> [n]otwithstanding the amazing developments in image technology, including holog-
> raphy, digital photography and the wonders of computer enhancement and virtual
> reality, no-one has been able to improve on the hand and the eye of a superb natural
> history artist when a particular morphological feature or the subtlest nuance of col-
> our needs to be shown to maximum effect.[151]

Sketches made after nature prove significant in proportion to the unfamiliarity
and unusualness of an object's appearance. Especially for the explorers of the
ocean in the late nineteenth century – who encountered objects that did not
connect with known patterns of form or colour and could not be assigned a place
in a classificatory structure through analogy or metaphor – artistic investigation represented *the* decisive medial praxis for approaching unfamiliar, foreign,
strange and incomprehensible forms of life. The illustrated volumes of the *Valdivia* report are exemplary in this regard. They articulate a search for recurrent
patterns, underlying structures and points of agreement, as well as exceptions,
variable elements and points of uniqueness. More still, they document that, in
the early phase of 'media diversity', photography and illustration did not stand at
odds with each other; instead, they entertained a complementary relation.

The illustrations of the *Valdivia* report confirm the media-aesthetic insight
that observing research objects cannot be separated from the techniques that
make them visible. Whether examining crabs, urchins, serpent stars or sea
cucumbers, both media showed different aspects of the creatures. Whereas the

illustrations afford nuances and arrangements that a photograph can hardly provide, the black-and-white pictures unfold a very particular, realistic-yet-fantastic
dimension that is difficult to define. They far exceed what one commonly considers the task of scientific photography: securing an object optically. Instead,
they provide evidence for a moment of intensive engagement with the possibilities and appropriateness of different media and modes of representation.
As much becomes clear especially in pictures of glass sponges (*Hexactinellida*),
which can only have been made after intensive experimentation with graphic
and photographic approaches. The results are surprising: drawings are full of
architectonic precision, presenting a blueprint and a model at once; every con-
vexity and concavity can be located exactly. On the photographs, in contrast,
the same creatures display a presence that is difficult to determine and almost
'magical'; taken together with the blackness of the background (as the 'deep of
night'), they yield a dream-like constellation. The uncanny and riddling darkness
is what gives the object its artful and auratic effect in the first place. It isolates
the phenomenon, enshrouds it, and communicates the impression of isolation
and concealment – only to have the object emerge unexpectedly, so to speak.
For all that, the 'where' and 'why' of these curious, milky-white beings remains
a complete enigma. One beholds them in black and white, yet the primal matter formed of light still seems unfathomable. On this score, contemporary ideas
about photography as a replica of the real amount to nothing. If anything, the
cloudy shapes leave it uncertain what, exactly, they are offering for contemplation. This superficial 'objectivity' has too much about it that is fantastic: the
absence of any points of reference in known forms of life, the murky transitions
between elements, fuzzy borders (even though, at the same time, dimensions
seem monumental). Some of the photographs can be unfolded to a metre in
length; here, the eye, shocked and fascinated by the undeniable evidence of what
has been seen, loses itself in the thousands and thousands of chambers and cavi-
ties, pores, ramifications and vesicles. Thus, the newest technological medium
leads to a marvel-inducing art of presentation that dates from the beginnings of
modern science: the fold-out drawing of a flea in Robert Hookes's *Micrographia*
(1664) – the symbol of the general aesthetic curiosity about wonders of nature,
which the photographs of glass sponges continue in virtuosic fashion.

What comes out in the pictures, in addition to an overwhelming, breathtaking component, is the insight that seeing also means showing. In the act of
photographing, aesthetic decisions must be made that induce excitement, even
if they seem counterintuitive. There is a clear preference for ambiguity, unclear
transitions and unfocused elements, as well as for renouncing the conventions of
evidence. Instead of prescribing an instrumental-positivistic gaze, photography
here opens a space of possibility in which new ways of seeing and representing
can be explored, and standing dispositives treated in new ways. Ironically, the

medium of illustration presents the glass sponges in perfect exactness; after all, every single line serves to define concrete forms and distances, and even what cannot be determined absolutely calls for a concrete decision. The photographs, in contrast, leave the murky habitations of *Aphrocallistes beatrix J. E. Gray* (pl. XII) or the frayed, disappearing edges of *Pheronema carpenteri Wyv. Th.* (pl. XV) as they appear: indeterminate, transitional forms between dark nothingness and white structuredness that seem to dissolve in myriad intermediate shades.

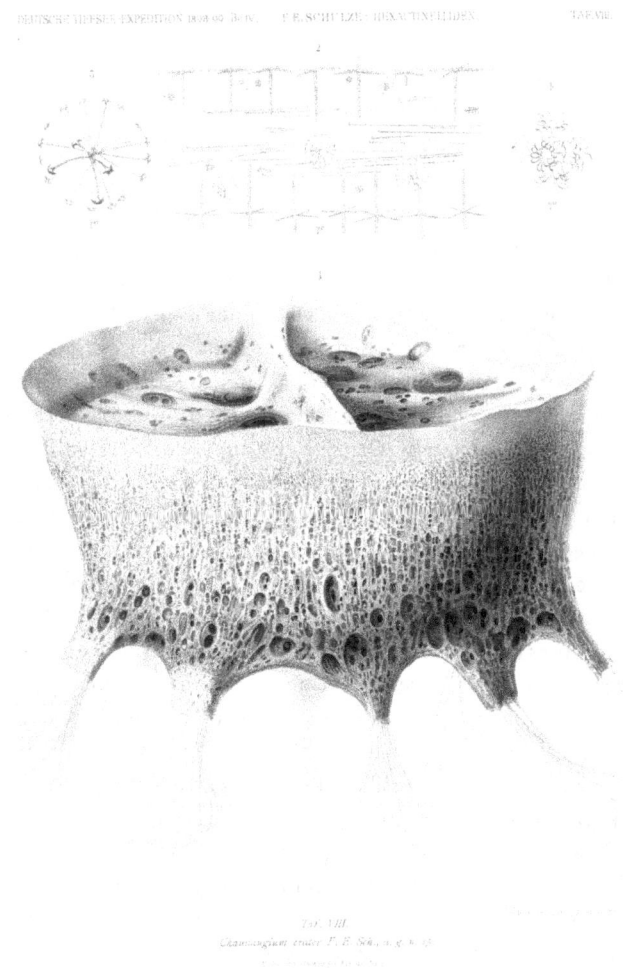

Figure 2.2: Chaunangium Crater, **C. Chun** (ed.), *Wissenschaftliche Ergebnisse der Deutschen Tiefsee Expedition auf dem Dampfer 'Valdivia' 1898–1899'*, **24 vols** (**Jena: Gustav Fischer, 1902–40**), **vol. 4**: *Hexactinellida*, **ed. F. E. Schulze** (**Jena, 1904**), **Tafel VIII.**

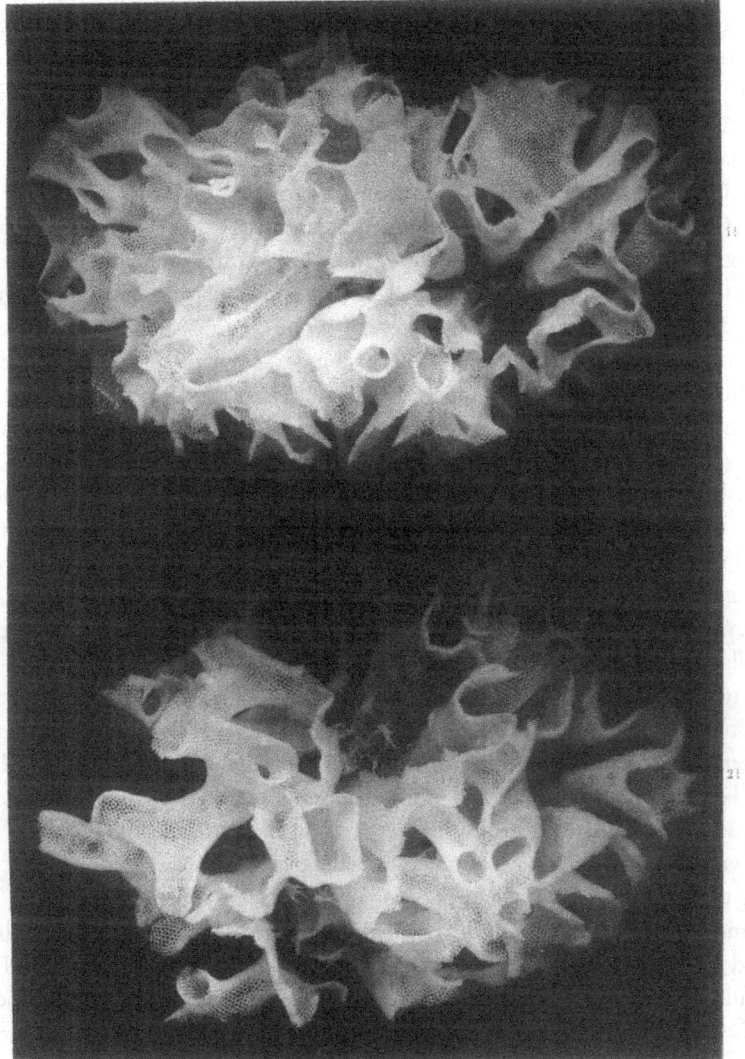

Taf. XIII.
Aphrocallistes beatrix J. E. Gray

Figure 2.3: *Aphrocallistes beatrix J. E. Gray*, C. Chun (ed.), *Wissenschaftliche Ergebnisse der Deutschen Tiefsee Expedition auf dem Dampfer 'Valdivia' 1898–1899'*, 24 vols (Jena: Gustav Fischer, 1902–40), vol. 4: *Hexactinellida*, ed. F. E. Schulze (Jena, 1904), Tafel XIII.

Mystery in Sight and Wonders within Reach: The Zoological Illustrations of Carl Chun

Carl Chun was known for the 'glänzende künstlerische Ausführung seiner bildlichen Darstellungen' (brilliant artistic performance of his visualizations).[152] His drawings inhabit the space where art and science intersect, and they thematize a doubt that any such thing as objectivable nature exists at all. Chun's collaborator for many years, Otto Steche (1879–1945), wrote that his work was distinguished above all by 'love for the object treated'.

Chun numbered, it follows, among those scientists who also possess great artistic gifts, for whom the process of illustration and research coincide. The matter is all the more remarkable for the fact that marine life, in particular, offers the greatest of challenges to scientific illustrators.[153] If one consider's Chun's graphic work – how the 'delicate colours and forms of these fragrant marine blossoms' or 'the contours of these organisms often emerge as delicate as a breath yet with plastic definition'[154] – the naïve question arises how these creatures 'really' looked when they first appeared aboard the *Valdivia* or another research vessel. What did Chun and his contemporaries see when they studied these confusingly foreign organisms and tried to give them form on paper? The question proves problematic and interesting in equal measure. For one, it is clear that these illustrations – like all scientific images – construct something that does not exist as such empirically;[155] that is, they produce the very object they claim simply to depict. None of Chun's creatures will ever swim past as they are drawn. On the other hand, Chun's pictures – like all non-digital scientific images – display a certain 'materiality and solidity of their own', which determines their epistemic function in essential manner; after all, they refer to materially existing objects, organisms dragged up from the deep, which are still preserved today, in solution or dried out, in the collections of universities and museums of natural history. Whatever relation the living creatures and the preserved specimens may have assumed over the decades, it plays a constitutive role for the ways the drawings and photographs function as scientific images.[156] This brings us back to the deck of the *Valdivia* and the question of how, exactly, muddy and wet objects were converted into reproducible pictorial representations. How could deep-sea cephalopods, which evolved to inhabit a completely opposite sphere of life – and whose appearance is considerably changed by differences of air, light and water-pressure (to say nothing of the torture of being hauled up in an overfilled net) – turn into a magic lantern, *Liocranchia valdiviae*, angel octopus, *Velodona togata* or a vampire squid, *Vampyroteuthis infernalis*?

An insight of cultural history and the history of science holds that the production of scientific images is accompanied by complex processes of transformation. The 'visibility of things', Peter Geimer observes, is 'not simply a given quality ... but must be shaped and investigated through experimentation in ateliers and laboratories'.[157] This applies in equal measure to Chun's pictures, which, like many zoological illustrations, resulted from manifold reworkings and reinscriptions that incorporated methods of investigation employing the most varied technologies and media. If they were still alive, the sea animals and plants caught with the dragnets of the *Valdivia* were immediately placed in seawater cooled with ice – the ice-machine, which provided at least five kilos daily, proved priceless – and they were observed for a few hours more: their *habitus* was captured by snapshots, and their colouration by aquarelle. Either in laboratories onboard or at a later point in time, they were measured, weighed, dissected, placed under a micoscope, chemically analyzed and palpated on every side; then they were assigned a place in relation to species that were already known (i.e., alongside reproductions, models and preserved specimens). In this way, each individual object was woven into a network of different activities and thereby fragmented step by step, until it was transformed into an abstract 'container' for the most varied forms of data. Thus, in the course of observation, description, experimentation and analysis, Chun's desk brought together highly divergent types of images that, despite their specific semiotic qualities and prehistories, related to each other and yielded connections between the various epistemological elements that they represented. Against the solid background of a primary object, from which they derived, they constituted a shared scientific field and turned into instruments of the research process. This field, however, hardly presented antecedents in any given case – the deep sea had just surfaced in the research landscape – and so Chun and his colleagues hardly had visual reference points to connect with. The decisive question, which arises in every scientific process of visualization, but especially here, is how a disparate structure of reference between 'uncertain' phenomena can yield a (unified) image that does justice to the precarious character of the thing represented and, at the same time, fulfills its scientific function as an object of research.

The medial gesture of drafting permitted Chun to bring together all these different elements and considerations and transfer them into an 'order of showing'.[158] The accompanying texts provide an initial point of orientation that is strongly marked by haptic language. Time and again, they mention surface impressions and textures; an issue is grasping material qualities and physical properties. Chun's descriptions develop an elaborate metaphorical field of life and the living. Colour exists only in perceptions of movement such as shimmering, glowing and change. At the same time, such properties are the most conspicuous features of his drawings, in which crabs, medusas and cephalopods seem to be floating by – as if they were making a brief stop on the page. Accordingly, the mediating movement of drawing proceeds as the interplay of the eye and the hand; it identifies Chun as an observer guided by the sense of touch.

Chun's work engages with the riddling aesthetics of marine animals. When he has a pale-blue Venus girdle (*Cestus veneris*) float over a black background, for example, it represents a break with a lack of epistemological definition. One must, as it were, 'side' with objects of investigation if one wants to learn anything about them. This amounts to making an aesthetic decision, which precedes epistemological appraisal and opens a space where insight may occur. Normally, scientific procedure smoothes over preliminary aesthetic determinations, so that one only encounters 'polished', rational objects. Chun's works, in contrast, foreground precisely these aspects of mediated *mise-en-scène* – for example, when a crab's feelers extend beyond the frame of the picture or a coral expands onto the lower part of the page. In such cases, Chun undermines the ontological logic of representational correspondence and, at the same time, foregrounds the status of what is shown as a specific form of pictorial knowledge. The latter is based on the 'delightful', 'delicate' and 'bizarre' materiality of 'his' deep-sea organisms, which hybrid 'Strömen von Spuren' (streams of traces, EB)[159] make evident. Instead of smoothing them over, Chun's drawings leave objects in their unfathomable foreignness; instead of dissolving their confusing form into schematic glosses, so to speak, they embed them in an aesthetics of mystery. In the process, the illustrations connect to the early-modern tradition of representing natural wonders – or, alternately, transform them into a modern form of rapt amazement; either way, they follow a gesture of unveiling that always holds new and even more breathtaking, realities at the ready.

Chun's scientific work, many sources attest, was marked by 'deep admiration for the enormous power of nature to give shape'.[160] His visual work confirms as much: the elegant drawings rouse aesthetic curiosity, and the spectacular arrangements are meant not just to be understood, but also to be enjoyed; not least of all, the modest yet commanding use of lines and colours brings together the delicacy and subtlety of the organisms and the joy in their discovery. Many of Chun's contemporaries shared his view of the wonder-inducing, marvellous and incommensurable as an important component of research work; however, he united aesthetic praxis and scientific insight at a level that others seem to have failed to attain even to this day.

At the turn of the century, when the *Valdivia* made its voyage, the oceanic expanse still stood for the unknown, the extreme and the inconceivable. This expedition proved the thoroughgoing inhabitation of all water layers; for all that, measuring and naming, inventorying and classifying, dissecting, disarticulating and preserving specimens did not make the underwater beings any more familiar. With each new 'revelation', another wave of peculiar and confusing forms entered the visual culture of the *fin de siècle*, which preserved its incomprehension as aesthetic amazement. From today's perspective, it might seem that the history of marine and deep-sea research happened somewhere 'out there', in damp subterranean realms, or at the margins of scientific endeavour; in fact, it occurred at the heart of modern culture.

3 *TWENTY THOUSAND LEAGUES UNDER THE SEA*: THE OVERTURE TO A PASSION

The year 1866 was signalised by a remarkable incident, a mysterious and inexplicable phenomenon, which doubtless no one has yet forgotten ...[1] I was perfectly up in the subject which was the question of the day. How could I be otherwise? I had read and re-read all the American and European papers without being any nearer a conclusion. This mystery puzzled me. Under the impossibility of forming an opinion, I jumped from one extreme to the other. That there really was something could not be doubted, and the incredulous were invited to put their finger on the wound of the *Scotia*.

Upon my arrival in New York several persons did me the honour of consulting me on the phenomenon in question. I had published in France a work in quarto, in two volumes, entitled 'Mysteries of the Great Submarine Grounds'. This book, highly approved of in the learned world, gained for me a special reputation in this rather obscure branch of Natural History. My advice was asked. As long as I could deny the reality of the fact, I confined myself to a decided negative. But soon finding myself driven into a corner, I was obliged to explain myself categorically. And even the Honourable Pierre Aronnax, Professor in the Museum of Paris, 'was called upon by the New York Herald to express a definite opinion of some sort. I did something ... ' After examining one by one the different hypotheses, rejecting all other suggestions, it becomes necessary to admit the existence of a marine animal of enormous power.

The great depths of the ocean are entirely unknown to us. Soundings cannot reach them. What passes in those remote depths – what beings live, or can live, twelve or fifteen miles beneath the surface of the waters – what is the organisation of these animals, we can scarcely conjecture. However, the solution of the problem submitted to me may modify the form of the dilemma. Either we do know all the varieties of beings which people our planet, or we do not. I we do *not* know them all – if Nature has still secrets in ichthyology for us, nothing is more conformable to reason than to admit the existence of fishes, or cetaceans of other kinds, or even of new species, of an organisation formed to inhabit the strata inaccessible to soundings, and which an accident of some sort, either fantastical or capricious, has brought at long intervals to the upper level of the ocean.

My article was warmly discussed, which procured it a high reputation. It railed round it a certain number of partisans. The solution it proposed gave, at least, full liberty to the imagination. The human mind delights in grand conceptions of supernatural beings. And the sea is precisely their best vehicle, the only medium through which these giants (against which terrestrial animals, such as elephants or rhinoceroses, are as nothing), can be produced or developed.[2]

So begins one of the most adventurous stories of the nineteenth century, the overture to a passion: *Twenty Thousand Leagues under the Sea* (1870), by Jules Verne (1828–1905). The novel concerns a strange event, an unexplained and inexplicable phenomenon of nature: an enormous creature infinitely larger and infinitely faster than a whale. In equal measure, the book tells of a region of darkness for the natural sciences: the vast, oceanic depths concealing completely unknown and utterly remote reaches which refuse to surrender their secrets. The main character of the novel is in fact the Sea, an *imaginarium* of inconceivable abysses and extraordinary, hidden realms.

Underwater Sites and Sights: The Sea as a Discursive Concert

Figure 3.1: Frontispiece, J. Verne, *Vingt Mille Lieues Sous Les Mers.* Voyages extraordi-
naires (Paris: Hetzel, 1869), dessinateur, Édouard Riou.

Remarkable visual curiosity pervades *Twenty Thousand Leagues under the Sea*. Hardly a scene occurs that does not display a 'demonstrative character' by presenting the voyagers with a '*curieux spectacle*'.[3] Magical vistas alternate with hunts through the deep; contemplative scenes of reflection at the glass windows of the *Nautilus* follow upon excursions to find giant oysters and battles with enormous sea spiders and giant sharks; tranquil ichthyological studies break off when gripping spectacles erupt – for instance, an orgiastic massacre of whales or the spectacle of a foundering warship. The formerly impenetrable veil of the ocean is lifted by means of scientistic optimism and fantastic technology, revealing the planet's ultimate secrets as a pageant for the rapt crew, a marvel of natural history and aesthetics alike.[4]

More still, the novel is shaped by images of verticality, depth and penetration into an inner *sanctum*. Like mythological leitmotifs, the element of water and swimming motion point to profound depths inhabited by complementary desires: yearning for security, on the one hand, and longing for the dissolution of boundaries, on the other.[5] For Captian Nemo, the sea is synonymous with absolute peace and absolute freedom:

> 'Yes; I love it! The sea is everything ... The sea is only the embodiment of a super-natural and wonderful existence. It is nothing but love and emotion; it is the "Living Infinite" ... The globe began with sea, so to speak; and who knows if it will not end with it? In it is supreme tranquillity ... Ah! sir, live – live in the bosom of the waters! There only is independence! There I recognise no masters! There I am free!'[6]

Here, the sea appears as the blessed sphere where the ego returns to its home inside itself. *Twenty Thousand Leagues under the Sea* may be Verne's greatest novel, according to Friedrich Wolfenzettel, because both the plot and the symbolism as a whole express this profound, dreamlike dimension.[7] As Verne describes him, Nemo is a mysterious visionary who inducts the first-person narrator, Pierre Aronnax, into the secrets of the deep; thereby, he makes him more and more familiar with the dreams he harbours – practically infecting him and drawing him down with him. On their many excursions into the marine world, down through underwater forests, among enormous ocean animals, over submarine volcanoes and, finally, to the ruins of the legendary city of Atlantis, Nemo never explains anything; rather, he presses onward, mutely pointing to 'the wonders of this hidden world'.

> Where was I? Whither had Captain Nemo's fancy hurried me? I would fain have asked him; not being able to, I stopped him – I seized his arm. But shaking his head, and pointing to the highest point of the mountain, he seemed to say –
> 'Come, come along. Come higher!'[8]

Nemo's silence expresses an 'ungraspable mystery'[9] into which the captain draws his guest against his will. After Nemo has conducted Aronnax to the ruins of Atlantis, both men sink into a wordless reverie:

> Whilst I was trying to fix in my mind every detail of this great landscape, Captain
> Nemo remained motionless, as if petrified in mute ecstasy, leaning on a mossy stone.
> Was he dreaming of those generations long since disappeared? Was he asking them
> the secret of human destiny?[10]

In the shimmering glow of an underwater volcanic eruption, both travellers witness a world that is not just a spectacle, but also a revelation, not just *féerie*, but mystery, too.[11] Like the host of other events on the ocean floor, it strikes them as a quasi-supernatural phenomenon; ultimately, it can be 'described' only by stock phrases about its ineffability. The ocean depths and its secrets mark a site where history pregnant with meaning has vanished. As such, they turn into a symbolic space for everything else that has disappeared and gone missing, too. Here, in fantastic images of an overwhelming universe beneath the waves, Verne formulates a motif that recurs throughout the nineteenth century: diving into prehistory to draw forth one's own identity from the depths.

At the very beginning of the nineteenth century, cultural attention had already moved inward and steered its gaze down into the deep. Sunken layers of the earth and the unfathomable seas became reservoirs of symbols that would feed into and structure the discovery of the unconscious.[12] In the course of the following decades, the ocean circled inward, downward and outward until the imaginative potential of nature filled the whole of culture. As much is attested not just by Verne's bestseller, but also by influential works such as Victor Hugo's *Toilers of the Sea* (1866), Gustave Flaubert's *Temptation of Saint Anthony* (1874) and, above all, Jules Michelet's gospel of the ocean, *La Mer* (1861). These books present the sea as the materialized form of infinite abundance, the site where all life originates, as well as a totality of the wholly Other, inaccessible to human understanding. In the process, they portray underwater spaces of yearning and experience; their aesthetic and media-technical conditions of possibility and representation are closely interwoven with contemporary marine research.

*Figure 3.2: **Aquarium**,* J. Verne, *Vingt Mille Lieues Sous Les Mers.* Voyages extraordi-
naires (Paris: Hetzel, 1869), dessinateur, Alphonse de Neuville.

We remained mute, not stirring, and not knowing what surprise awaited us, whether
agreeable or disagreeable. A sliding noise was heard: one would have said that panels
were working at the sides of the *Nautilus* ... Suddenly light broke at each side of the
saloon, through two oblong openings. The liquid mass appeared vividly lit up by the
electric gleam. Two crystal plates separated us from the sea ... What a spectacle! ... But
in the middle fluid travelled over by the *Nautilus*, the electric brightness was produced
even in the bosom of the waves. It was no longer luminous water, but liquid light.

On each side a window opened into this unexplored abyss. The obscurity of the
saloon showed to advantage the brightness outside, and we looked out as if this pure
crystal had been the glass of an immense aquarium ... For two whole hours an aquatic
army escorted the *Nautilus*. During their games, their bounds, while rivalling each
other in beauty, brightness, and velocity, I distinguished the green labre; the banded
mullet, marked by a double line of black; the round-tailed goby, of a white colour,
with violet spots on the back; the Japanese scombrus, a beautiful mackerel of these
seas ... the brilliant azuros, whose name alone defies description ... Our imagination
was kept at its height, interjections followed quickly on each other. Ned named the
fish, and Conseil classed them. I was in ecstasies with the vivacity of their movements
and the beauty of their forms ... Suddenly there was daylight in the saloon, the iron
panel closed again, and the enchanting vision disappeared. But for a long time I
dreamt on till my eyes fell on the instruments hanging on the partition.[13]

This passage brings together everything we have been discussing: mysterious darkness and wondrous illumination, unlimited fullness and the infinitely unknown, unbounded amazement and the classic scientific reaction: classification. On the one hand, *Twenty Thousand Leagues under the Sea* reads like a vast biological textbook, exploring marine life in its subtlest nuances of colour and taxonomical traits. The researchers whose authority Pierre Aronnax invokes are not fictitious; they are estimable contemporaries – even if foreign names are occasionally garbled.[14] The Erhemberg in question is none other than Christian Gottfried Ehrenberg – a researcher who, as was noted previously, had written in 1835 that marine luminescence may stem from tiny infusoria. Deploying scientific language with precision, Verne engineers rich resonances with other texts and literary genres. The discourse of the characters covers the sea with orders of language and knowledge that incorporate professional jargon from handbooks, lexicons and the findings presented in scientific journals. Their accounts of their journeys yield labels for what they find on the unknown continent; accordingly, it is hardly by chance that a library occupies the centre of the *Nautilus*, from where Captain Nemo organizes the expeditions.[15]

On the other hand – and at the same time – Verne juxtaposes the enthusiastic Aronnax, ever prone to reverie, with his servant Conseil, who thinks in schematic terms and tries to classify everything. There is such a thing as a mania for naming, which necessarily fails when seeking to extract discrete unities from the vast indeterminacy of the sea.[16] Verne has no interest in condensing the multiplicity of knowledge into a harmonious, symphonic 'portrait'; instead, he emphasizes the movement displayed by objects of knowledge. After all, the motto of the *Nautilus* is *mobilis in mobili*: the ocean itself symbolizes motion. The uncountability, imponderability and incalculability it exhibits echoes in the register of oceanographic literature that, as we have seen, chose to strike a tone of poetic wonder. With a sweeping gesture of intoxication, Verne presents the sea as a vast riddle continuously bringing forth new life, subjecting its inhabitants to unceasing transformation and opening fluid boundaries between the living and the inanimate, and flora and fauna. The ocean refuses to be frozen into inert epistemological terrain; instead, it gives rise to a superabundance of acts of description, enumeration and cataloguing. The paradox of such a paradisiacal landscape, which also admits scientific description, fuses positive knowledge and imaginative depth into one.[17]

Jules Verne, the *Nautilus* and Media Technology

Through the glass window of the *Nautilus* and the goggles of their diving suits, Captain Nemo's voyagers enjoy a view of an undersea 'land of wonder'.[18] In the process, the *Nautilus* alternates between different functions, serving by turns as a theatrical, wondrous and magical mobile, a floating museum of natural history, a cabinet of curiosities and a *féerie scientifique* – a scene of science fiction, as well as a private dream made technological reality.[19] Through the round portholes of this 'technological wonder', the characters behold the 'wonders of the sea' as if they were attending one of the universal exhibitions that were fashionable at the time. This impression is no coincidence. Verne wrote the first part of his novel aboard a boat in the Atlantic; in spring 1867, he returned to Paris.[20] There, at the world's fair which had just opened, he found all the spectacular 'wonder' – machines he needed for his imaginary voyage to the ocean depths: electric lights, dynamo-electric motors, a diving apparatus for which the designers, Benoît Rouquayrol (1826–75) and Auguste Denayrouze (1837–83), had received a gold medal and, most importantly, a pageant of natural history combining the marvels of technology and the wonders of nature as if in a dream. In particular, this meant the three great aquariums made by Caumes and Bétancourt in the *jardin réservé*. Two of them had the form of grottoes – one was shaped like the interior of a body, and the other presented a limestone cave with columns and stalagtites wreathed with seaweed. The third and largest aquarium was built like a gigantic crystal chamber with walls and ceiling made of glass.[21] Surrounded by water suffused with light and life, visitors felt they were wandering in a 'magical castle on the ocean floor' and looking upon the 'secrets of the frigid deep'. Strange fish swam overhead and delicate jellyfish floated by. It seemed to be a world inhabited by mermaids, Nereids and Undinas.[22] Verne beheld his first underwater visions here, and his illustrator, Édouard Riou (1833–1900), received inspiration on how to lend the aquarium's swaying kelp forests and blooming coral cliffs into oceanic dimensions. Now, Verne yielded to his contemporaries' passion for the sea. He read all the 'cult books' of the day – *La Mer, Toilers of the Sea*, Henry Milne-Edward's *Mystères des grands fonds sous-marins* (1867), Mangin's *Mystères de l'Ocean* and Alfred Fredols' *Le monde de la mer* (1865) – and he applied himself to studying the latest reports from the *HMS Lightning*. In 1868, Verne travelled to Le Havre to visit the nautical exhibition and view the celebrated aquarium there, which had been modelled on Fingal's Cave on the basalt coast of the Staffa, an island in the Inner Hebrides.

'A World Behind Glass'

The *Nautilus* combines two of the most important media-technical strategies for aestheticizing conditions of perception and representing marine life as that which, as a matter of principle, remains inaccessible: the aquarium and diving equipment. By these means, the human eye was able to peer into a 'night' that, until now, had proven impenetrable; at very least, a simulation came into view. The huge panes of the salon make the sea look like an aquarium. Likewise, the *Nautilus* does not just enter alien worlds in the manner of a spaceship; it also provides a submarine habitat from which diving excursions onto foreign terrain can be made. In this milieu of hybrid mediality and aesthetic-somatic ambivalence-after all, aquariums may be described as simulations of immersion, whereas diving requires that the observer be cocooned in water-resistant equipment and therefore remain 'outside' the aqueous element – the *Nautilus* emerges as the symbolic site where models are formed, or, alternately, where the sea is conceptualized as a precarious space of knowledge and experience. Verne demonstrates that all accounts and representations of the sea and its inhabitants are influenced by the media and instruments that shape observation and research, which inscribe dives, virtual and real, with both a surplus and a deficit of sensory depth. What is more, the studies conducted by Nemo and Arronax correspond to a paradigm shift in the sciences, which made the *Nautilus* and the 'aquarium of the ocean' possible in the first place by shifting attention from the classification of dead specimens to the study of living ones in their own environment. A novel praxis of experimentation was inaugurated to investigate a broad array of ways that creatures live and develop; in the process, the ocean world entered contemporary discourses of knowledge, technology and art in the most varied forms. The key difference between aquariums and diving apparatuses is that the latter became omnipresent in the world of entertainment in no time at all; in contrast, physical immersion underwater was reserved for only a few. The fourth and fifth chapters will explore these two theatres in detail. What follows now, however, explores the media-aesthetic ambivalence of the *Nautilus* – or, more precisely, its circular observation window – which left many traces in the visual culture of the turn of the century.

The View from an Underwater Window

When the unwilling visitors aboard the *Nautilus* first see the submarine vessel's iron plates open and reveal an enormous window – 'as if this pure crystal had been the glass of an immense aquarium'[23] – the dreamlike scene appears as a 'world behind glass'. At the same time, the three figures seen from behind stand in for the observer of this superb pageant. The combination of the glass window and the illuminated ocean night not only opens a new order of visibility; it also provokes media-ontological irritation, as it were: henceforth, gazing *into* the aquarium and looking *out of* the submarine vessel into the unknown expanses of the ocean serve as aesthetic and epistemological equivalents.

Alphonse de Neuville's famous illustration of the scene represents one of the points of origin for views of an underwater realm abounding in riddles and dreams; as a site for both the production and the transmission of knowledge, it fulfilled contemporary fantasies of technological invention. When Stuart Paton (1883–1944) adapted Verne's novel for the screen in 1916, this setting immediately advanced to a leading position in the repertory of images used to portray submarine research expeditions. Ever since, views through the windows of diving vessels have formed a solid component of the visual rhetoric that shows the conquest of the deep in documentary and fictional cinema alike. The visionary project of marine research, still in its infancy, took up established patterns of perception: on the one hand, technological media that made it possible to look upon, and give objective form to, alien worlds largely lying beyond the realm of human experience; on the other, the repertoire of images associated with the window as a metaphor symbolizing separation, insight, perspective and passage from one sphere to another.[24]

The gaze through the window concentrates the boundlessness of oceanic space optically and transfers it into a circumscribed zone of perception and experience.[25] Neuville makes use of a motif employed by European painters from the early Renaissance onwards: views from windows, entryways and open arcades that reach into the landscape beyond. Already toward the end of the Middle Ages – and especially north of the Alps – the conquest of visible and representable space had occurred through the window perspective, which eventually came to occupy a central position in visual experience.

Neuville transferred the thrust of this historical progression – expansion into nature and the conquest of nature from within a building – to the interior of the *Nautilus*; in the context of modern 'offensives' to explain the world, this amounted to a positivistic claim to power. What is new about Verne/Neuville's work is the mise-en-scène of the underwater window as the largest possible 'lens' that science and technology can provide; now, the probing gaze has access to realities that have been unavailable until this point. By framing a picture, focusing vision and assuring transparency in equal measure, the window emerges as a site of multiple meanings and prompts critical reflection. From the outset, Verne's primal scene reveals four ideal aspects: a screen, a window space, a frame and a showcase.[26] With this panoply of meanings, it provides a visual analogy well suited to the ambiguities that arise from the surplus and paucity of the alien, underwater world; at the same time, it offers a preview of conditions under which submarine reality may be adequately assessed (whatever ambivalence may then persist). Neuville still presents an enormous window that discloses fantastic life-forms – a view for observers who are 'illuminated' in both a literal and a figural sense – illustrating a totalizing claim of perception. In contrast, in later variations on the same motif, the underwater window only offers a slice of the field of vision. That is, the window serves as a means of distantiation: reduced to the size of the human head, it affords only a hint of the ocean – say, a glimpse of luminous blue or the glow of underwater lamps. All there is to see is left to the observer's fantasy.

The window of the *Nautilus* presents the depths of the sea as a framed worldview. The sea appears as an aesthetic event, where scientifically arranged observational space and immeasurable foreignness come into close contact. Even though the silhouettes of the four characters also have the function of signalling the relationship between these two worlds, the positions they occupy remain firmly situated within the discursive arrangement of observational protocol. In fact, they do not have access to the outside; they stand at a partition, on the threshold between worlds: only their gaze and their thoughts – which are hidden to us – can pass through. Under these circumstances, the window all but becomes the symbol of a mystical fusion of an earthly interior and vast, cosmic expanses accessible only to visionary contemplation. Now, the underwater window is both a paradigm of the insuperable separation between humankind and the sea, and a way to subvert this same border by means of aesthetic experience: a site of yearning and a medium of insight alike. Inasmuch as it makes it possible to observe nature from a water-resistant cocoon in the first place, it stands for the empirical fact that an otherwise inaccessible world may now be viewed by human beings; at the same time, however, it signals the impossibility of intervening – or, at any rate, the fact that such experience is strictly limited to the optical sphere. This meeting of transparency and prohibition greatly intensifies all that is seen and heightens the enigmatic power of suggestion. Not only do scientific processes of observation obtain a new claim to objectivity; the subjectivity of the gaze becomes an issue, as well – that is, the possibility that looking through the window is more than a matter of 'revealing' and involves 'staging' which works by stimulating the forces of imagination. Accordingly, the gaze extends from inside the *Nautilus* – as from all later submarine vessels – into a sea of creatures that can be identified biologically and, at the same time, into a metaphysically-charged and subjectively transfigured reality.[27]

It is a fortunate circumstance that windows in diving vessels are usually circular. In architecture, a round or oval window is called an *oculus* – a name pointing to the reciprocal symbolism that connects portholes, eyes and the faculty of sight.[28] Against this backdrop, Mangin's observation (quoted above) that the human eye had penetrated a gruesome night finds a corroborating image. It is also worth noting a change in window-perspective that occurred when films began to be made underwater. The works of Verne/Neuville and Paton depict observers in a protected, interior space; they symbolize collective interest in the scientific object and offer themselves to readers/filmgoers as observers by proxy. However, in the course of the twentieth century, the outside view into the diving vessel emerged as another option. Looking out the window was joined by its counterpart, looking into the window. Now, the spectator beheld the bewildered and melancholy eyes of the deep-sea traveller. The latter changed from being the stand-in observer to playing the part of the mirror image; the former drifted off into the unanchored freedom of the submarine world. In the course of this transfer, the window that connected and divided them went from being a means of sounding the 'gruesome night' to representing a means of self-knowledge.

Alien, underwater visions, then, result from the medial and aesthetic regime of the perspectives described above. More still, they are themselves windows in other optical arrangements: segments of pictures, photographs and films that mark the ocean view as an ambigous see-through form that is able to merge the two different spheres as well as reality and illusion.²⁹ Fittingly, Captain Nemo calls his vantage point, which reaches deep into the oceanic expanse, a 'magic window'. His words gesture to the doubling that occurs in the process of seeing: the observer's projection duplicates the gaze of the marine researcher and conqueror. At the same time, the words point back to the first setting of all, which stands at the origin of all underwater window-rhapsodies: the 'observatorium' onboard the *Nautilus,* which promises to show the sea as a unified aesthetic phenomenon. As the paradoxical site of intersection between nature and artifice, domestic interior and oceanic outside, the observatorium affords sensory representations of worlds hitherto unknown, which now yield the model for a mediological *episteme.* Thereby, the *Nautilus* represents a diving vessel in the 'aquarium of the ocean' – an extraordinarily productive assembly of media-experiments that, as the condition of possibility for fabulous wonders, has travelled the seas of art and science ever since.

Many-Tentacled Dreams and Nightmares

Below the thunders of the upper deep;
Far, far beneath in the abysmal sea,
His ancient, dreamless, uninvaded sleep
The Kraken sleepeth: faintest sunlights flee
About his shadowy sides: above him swell
Huge sponges of millennial growth and height;
And far away into the sickly light,
From many a wondrous grot and secret cell
Unnumbered and enormous polypi
Winnow with giant arms the slumbering green.
There hath he lain for ages and will lie
Battening upon huge sea-worms in his sleep,
Until the latter fire shall heat the deep;
Then once by man and angels to be seen,
In roaring he shall rise and on the surface die.

Alfred Tennyson, *The Kraken* (1830)³⁰

The sea in Verne's novel does not just offer wondrous, dreamlike visions; it is also a realm of terror and nightmares. Densely inhabited by dark figures and fearsome apparitions, it stands for the last geographical abyss, one of the darkest 'unknowns' the world knew in 1870. We have already discussed its obscure and awe-inducing aspect: monstrous entities – in particular, a marine creature of extraordinary might, a beast infinitely faster than a whale. At the beginning of his work, Verne provides a commanding description of the unsteady alternation between legends of horror and scientific enlightenment that characterized the mid-nineteenth century. Yielding to the spirit of progress, the public no longer viewed mammoth bones as the remains of giants, and con artists had stopped fabricating fake birds of paradise or fastening narwhal horns to horses. All the same, the known world was suddenly rent asunder and everything – or at least a much of it – that had been taken for granted disappeared into the gulf that emerged. In one fell swoop, the earth had become millions of years old; giant lizards had roamed its surface before suddenly going extinct. It seemed that horrific catastrophes had occurred, which, all the same, had somehow led from single-cell organisms to human beings. If inconceivable life had emerged from the depths of time, what might still be creeping and crawling, sucking and slobbering, in the eternal darkness of the ocean deep, the preserve of protoplasm and abiogenesis – indeed, the beginning of everything at all – lurking and waiting for anyone or anything unlucky enough to approach?

What animals, Verne has Aronnax wonder, reside in regions no plumbline can reach? Perhaps giant shellfish, huge lobsters or enormous shrimp? Why, a popular line of thinking held, wouldn't the unknown depths have preserved enormous specimens from another age, the last exemplars of titanic species? After all, more recent organisms included giants – the horrifying, murderous entities that had unsettled sailors for centuries. Still in the nineteenth century, reports and eyewitness accounts confirming the biological possibility of such creatures were presented to the public. The octopus represented one of the most significant figures around which the sea's primordial, improbable and fantastic dimensions crystallized. In the 1870s, the creature became one of the most sought-after guests in aquariums; the eager public wanted nothing more than to get to the bottom of rumours surrounding this animal steeped in myth. 'An aquarium without an octopus is like a plum-pudding without plums', wrote Henry Lee (1826–88), the director of the Brighton Aquarium, the first such institution to put one of these animals on display.[31] The octopus owed its great popularity as the 'chief of curiosities'[32] to the thoroughgoing success of Hugo's *Toilers of the Sea* and Verne's *Twenty Thousand Leagues under the Sea*, which eternalized the kraken – or, alternately, giant squid – as the epitome of monstrosity in literature.

Battle with the Sea Monster: Viscous Hatred and Zoological Sensation in the Works of Victor Hugo and Jules Verne

The battle with the dreadful squid forms a particularly impressive chapter of Verne's *Twenty Thousand Leagues under the Sea*. It stands in parallel to one of the most admired scenes in Hugo's *Toilers of the Sea*. Herman Melville also devoted a chapter to the legendary monster. In *Moby Dick; Or, The Whale* (1851), the crew of the Pequod spots something ghostly northeast of Java. It is not the white whale they have been pursuing but one of the

> most wondrous [phenomena] which the secret seas have hitherto revealed to mankind. A vast pulpy mass, furlongs in length and breadth, of a glancing cream-colour,

lay floating on the water, innumerable long arms radiating from its centre, and curl-
ing and twisting like a nest of anacondas, as if blindly to clutch at any hapless object
within reach ... [It] undulated there on the billows, an unearthly, formless, chance-
like apparition of life ... [It was t]he great live squid, which, they say, few whale-ships
ever beheld, and returned to their ports to tell of it.[33]

Accounts of various nautical expeditions echo in Melville's description. For
instance, in *Voyage de découvertes aux terres australes* the French scientist François
Péron (1775–1810) had reported an enormous cuttlefish, as thick as a barrel, floating off the coast of Van Diemen's Land (Tasmania), with colossal arms that moved
like fearsome snakes on the water's surface.[34] As is well known, the novels of Verne
and Hugo also combine contemporary works of natural science with a grammar of
monstrosity and wonder. Hugo is said to have hatched the idea for his sea creature
while pursuing his own studies of nature; they began after a trip to Sark in 1859,
one of the Channel Islands, where he saw an octopus for the first time and was
fascinated by the names locals had for it: 'devilfish' or, alternately, 'bloodsucker'.[35]
Hugo reports that seafarers called the monster *polyp*, whereas the scientific name is
cephalopod (literally, 'head-foot').[36] When he describes the creature's anatomy, musculature, the way its suckers are arranged and function, and its hunting and mating
behaviour, he casts it in eerie language that conjures up unspeakable terror. In so
doing, he adopts the tone of the famous chapter in Michelet's *La Mer* (1861) which
depicts the animal as a pirate of the deep, a 'sea rover'.[37] Even though the book for
the most part offerns an homage to the ocean world as a realm of love and fertility,
when Michelet comes to the octopus, he leads his readers into a gloomy empire
of death, 'a world of war [and] slaughter'.[38] Here, the principal inhabitant is cast
as having mutated from primeval invertebrates that puffed up beyond all measure,
becoming 'absorbing bladder[s] ... ever craving and ever consuming'.[39] The resulting
portrait presents an animal of abyssal malignancy that does not kill for food but
simply in order to destroy: the horror of indiscriminate violence made flesh.

In *La Mer,* Michelet expresses an aversion to the octopus of relatively recent
date, which first surfaced toward the end of the eighteenth century. It seems that
this modern myth started with Pierre Denys de Montfort's (1766–1820) *Histoire naturelle, générale et particulière, des Mollusques* (1802–6).[40] Until this point,
works of natural history had described the octopus as possessing strength, intelligence and dignity.[41] Denys de Montfort, on the other hand, presented it as a
malicious monster, in the moral sense of the word. For the first time, mentioned
was made of an irresistible urge to destroy and gruesome pleasure in killing.[42]
Now as then, it is unclear what provided the basis for this savage description (or,
alternately, the motifs underlying it). In the Christian cosmos, the octopus had
occasionally represented the Devil or sinful womanhood; there are also instances
where it provided the image of seduction, betrayal, mendacity or avarice. For all
that, however, it had never counted as the symbol of moral ruin *per se*. Alcide

Dessalines d'Orbigny (1802–57) speculated that his predecessor consciously invented a fantastically revolting animal to take revenge on his colleagues for their general credulousness and stupidity. Shortly after his work went to press, Denys de Montfort is supposed to have said: 'I claim that the tentacles of my kraken reach across the streets of Gibraltar'. More still: 'If my giant squid is well received I will claim in the second edition that he overran a whole squadron'.[43] At any rate, his intended revenge – or irony – did not work. Instead, his study made his name synonymous with gullibility, and for decades afterward it proved difficult for scientists even to think about the existence of gigantic cephalopods. Notwithstanding attitudes among researchers and scholars in the natural sciences, his portrait of the octopus as a criminal (so to speak) passed into modern legend when Michelet took it up and Hugo then lent it unforgettable literary form.[44]

Hugo pictures the octopus not as a dangerous cephalopod but as a sucking, soulless nightmare. It is a representative of hatred, the embodiment of hypocrisy and disgust, a 'riddle ... of Evil' in the framework of divine creation.[45] Under the heading 'The Monster', he writes:

> It is difficult for those who have not seen it to believe in the existence of the devil-fish. Compared to this creature, the ancient hydras are insignificant. At times we are tempted to imagine that the vague forms which float in our dreams may encounter in the realm of the Possible attractive forces, having power to fix their lineaments, and shape living beings, out of these creatures of our slumbers. The Unknown has power over these strange visions, and out of them composes monsters. Orpheus, Homer, and Hesiod imagined only the Chimera: Providence has created this terrible creature of the sea. Creation abounds in monstrous forms of life. The wherefore of this perplexes and affrights the religious thinker.[46]

For Hugo, the octopus is the numinous-made-ooze, the nightmare of belief, God's masterwork of the disgusting. It is not just a biological entity, but the incarnation of an aesthetics of evil: a phantom, a demonic abyss, a ghost. By endowing it with all conceivable attributes of the slimy, dark and devious, Hugo wrests the animal, once and for all, from conceptions that had existed ever since antiquity and moves it into the grey zone of the nightmarish. Here, it couples with the bloodsucking spider and vampire, Medusa's crown of serpents and the nine-headed Hydra.[47] The interplay of imagination and reality yields a figure of horror that defies Creation:

> These animals are indeed phantoms as much as monsters. They are proved and yet improbable. Their fate is to exist in spite of à priori reasonings. They are the amphibia of the shore which separates life from death. Their unreality makes their existence puzzling. They touch the frontier of man's domain and people the region of chimeras. We deny the possibility of the vampire, and the cephaloptera appears. Their swarming is a certainty which disconcerts our confidence. Optimism, which is nevertheless in the right, becomes silenced in their presence. They form the visible extremity of the dark circles. They mark the transition of our reality into another. They seem to belong to that commencement of terrible life which the dreamer sees confusedly through the loophole of the night.[48]

In fashioning an incarnation of incommensurable dread, Hugo clearly struck a nerve. Following the famous battle he staged between the novel's protagonist, Gilliat, and the hateful 'bloodsucker', the struggle between man and *Kraken* became a kind of 'popular mythology' – a fixed feature of popular culture in adventure novels and films.[49] In the process, an 'innocent' organism familiar throughout Europe metamorphosed into a hellish monster of Biblical proportions surpassing human language and understanding. Henceforth, encountering the octopus meant meeting an enemy that had to be killed. In contrast, Verne's giant octopus – in fact, a squid – took the underwater stage as a hybrid being combining legend and scientific sensation. The author presents it as a true monstrosity, a fantastic incidence of nature – *nature's play*, as one would have said in early modern times.[50] The first meeting with the animal at the 'aquarium-porthole' of the *Nautilus* frames it as a rare exemplar in the sea's cabinet of curiosities; accordingly, it elicits amazement – the classical European reaction to miracles. Killing comes only later in Verne's book: in the famous battle scene where the crew must defend itself with axes and harpoons against a horde of aggressive giant squid; in the process, they lose one of their comrades.

It might seem remarkable, today, that Verne refers to his monster both as a squid and as an octopus – which are known to be two rather different creatures. When the voyagers study the animal through the windows of the 'aquarium', the word 'squid' is always used. On the other hand, 'octopus' dominates in the famous battle scene. Verne might have made this 'mistake' because he was thinking about Hugo's striking scene of combat. When Monsieur Aronnax seeks to describe the terrible fight, words fail him and he explicitly refers to Hugo: 'To paint such pictures, one must have the pen of the most illustrious of our poets, the author of "The Toilers of the Deep".'[51] Roger Caillois takes this as an indication that Verne did not mean to describe a squid at all, but rather an octopus especially since there is some inconsistency about the number of tentacles.[52] He claims that only the *Kraken* would inspire fear and terror, whereas the squid would neither move nor fascinate readers.[53] This cannot be, however. For one, the giant squid exercises fascination to this very day in literature and film.[54] Moreover, in Verne's time it was not unusual for the words to be used interchangeably, even in scientific discussions.[55] Although the text consistently specifies eight arms – which points to the octopus – the illustrations by Neuville and Riou feature an animal that has, by turns, eight and ten limbs; at the same time, the creature's overall shape identifies it as a squid.[56] As Aronnax reports, it has a 'spindle-like body'[57] – a morphological definition that bars any confusion. Finally, the legends that Verne's protagonist recounts as a preview to the battle clearly refer to the myth of the giant squid.[58] Pierre Aronnax represents a man of unimpeachable scientific repute, a cultural type of the nineteenth century; he may have read *Toilers of the Sea,* yet he is interested in the zoological description of any monsters that might exist – not in Hugo's ghostly suns or slime bubbling over with hatred. Notwithstanding confusion about the number of arms that different cephalopods have, Verne presents his readers with a creature that stands on the threshold between myth and scientific discovery, belief in miracles and enlightenment. Monsieur Aronnax describes a giant creature attested by a long history of eyewitness reports and even by actual remains; all the same, it seems to belong to the world of legend more than it fits in a reference work of natural history such as *Brehms Tierleben*. At any rate, the real payoff of Verne's literary 'discovery' is that even now, almost one hundred and fifty years later, the wondrous creature still numbers among the most mysterious beings on the planet. Most importantly: it really exists; it is called *Architeuthis*, the giant squid.

Today, giant squid are known to us as great athletes of natation.[59] Their bodies contain an ammonium chloride solution that is lighter than ocean water and provides buoyancy – in addition to giving the animals a strong smell. For this reason, giant squid prove inedible to human beings; the sperm whale (*Physeter macrocephalus*), on the other hand, savours their taste, as the stomachs of dead specimens have revealed. Whalers have long marvelled at the squid's beak, which can be as large as a human palm, and at the huge scars left on sperm whales, which come, it seems, from suckers equipped with teeth and as large as soup plates. Efforts are ongoing, on the basis of these scars, to determine the size of the squid that made them in self-defense; calculations prove difficult, however, because they expand as the whales grow over the years. All imponderables aside, we may be sure of one thing: giant squid are enormous. Animals up to eighteen metres in length have been recorded, and researchers deem it possible that much larger specimens inhabit the ocean depths. The individual organs of the giant squid also exceed all measure: its penis, for example, is supposed to be about a metre long, and the eyes, which can have a diameter of up to twenty-five centimetres, are among the largest in the whole animal kingdom. Much about the giant squid's biology remains unknown and must be inferred from the anatomical material that is available – and from the habits of smaller, better known relatives. Until a few years ago, no human being had ever seen a giant squid that was still alive. Finally, in September 2004, a Japanese research team headed by Tsunemi Kubodera and Kyoichi Mori succeed in observing one at a depth of 900 metres.[60] All the same, now as then, giant squid number among the largest – and, at the same time, least studied – organisms in our biosphere.

Given its enigmatic past, one might ask how the squid made it into Verne'sbestselling novel. Despite uncertainty in matters of taxonomy, the author appears to have been relatively well informed about this rare animal. One can answer the question in two ways: first, by adopting the perspective of epistemological history, and second, by exploring the contexts of cultural and media aesthetics.

As we have noted, the question about the giant squid's existence was an important theme in nineteenth-century science.[61] It was scarcely possible for any work of natural history on the sea and its inhabitants to avoid the creature. The state of knowledge in Verne's day was roughly as follows. If one omits ancient reports of sea monsters, the first comprehensive account should be considered *Historia de Gentibus Septentrionalibus* (1555) by Olaus Magnus (1490–1557), the archbishop of Uppsala in Sweden. In the course of his travels, he collected accounts made by nordic fishermen who reported enormous, multi-limbed marine creatures reaching up to sixteen metres in length. A similar picture emerges from the collection made by the bishop of Bergen, Erik Pontoppidan (1698–1764); *Kurzgefaßte Nachrichten, die Naturhistorie in Dännemark betreffend* (1765) includes matters that convinced the author of the unusual animals' existence. In the meanwhile, squids's remains in various states of decay had washed ashore all over the globe. The works of Belon, Rondelet and Gesner describe these finds as 'sea monks'[62] – at any rate, cryptozoologists assume that the name refers to (misrecognized) giant squid. Reports made over the course of the seventeenth-, eighteenth- and early-nineteenth centuries eventually concluded that a strange fish or invertebrate – an especially large squid – was at issue.[63]

Then, in the mid-nineteenth century, the Danish researcher Japetus Steenstrup (1818–97) presented evidence for the giant squid's existence. He examined the beaks of two specimens that had washed up on the Danish shore near Aalbækstrand in 1853 and 1855; on this basis, he concluded that they had belonged to a creature he christened *Architheutis*, the 'first' or 'greatest' squid, in 1856. Finally, five years later – and this is presumably the report that impressed Verne the most, since he has Pierre Aronnax recount the matter in a key passage – the sailors of the French clipper *Alecton*, sailing off the coast of Tenerife, found themselves in the presence of a giant squid. A huge animal with a brick-red body was sighted swimming at the surface. When the warship approached, the creature tried to get out of the way, but it did not dive. Lieutenant Bouyer, the ship's captain, gave orders to harpoon it, but the hooks penetrated the soft flesh without catching on anything solid. Bullets also passed through its body. Finally, however, jets of blood and foam shot up and a strong, musky smell filled the air. The sailors had spent three hours trying to kill the beast. When they cast a cable into the water to heave it onboard, it cut right through; the head and tentacles fell into the sea and vanished in the waves. What was left – a forty-pound 'tail' – was taken to Tenerife.

Naturally, Lieutenant Bouyer's report proved a sensation; members of the French Academy of Sciences paid due notice. Experts of renown – for example, Henry Milne-Edwards (1800–85) (who, incidentally, provided the model for Pierre Aronnax) – declared that the ocean floor was likely home to a vast array of other giant cephalopods, and he pointed to the findings in Scandinavia as well as observations made by Péron, Jean René Constant Quoy (1790–1869), Joseph Paul Gaimard (1796–1858) and Rang.[64] In so doing, Milne-Edwards gave voice to a broadly held view – especially in the realm of popular science –

that, especially in light of recent discoveries in paleontology and marine biology, wondrous creatures and improbable beings were a real possibility.

But for all that, most natural scientists were not convinced by the report of the *Alecton*'s crew. Prominent researchers such as Richard Owen (1804–92), the head of the British Museum of Natural History, held that giant squid were a fiction tainted by superstition – surely the captain and crew of the *Alecton* had been 'deceived'. What might strike us today as 'hardheaded' – did the thirty sailors on the ship experience a collective hallucination? – is not easy to dismiss as ideology or dogma in the context of the times. Then again, might it not be a sign of healthy common sense to ask – like Mangin, a famous French natural scientist – whether it was reasonable to affirm the existence of an invertebrate

> Whose body alone is from sixteen to eighteen feet long, 4400 pound in weight, and whose flesh has so little consistency that the harpoons cannot bite in it, and the slip-knot with which they attempt to seize it cuts it in two lika a mass of jelly; this animal which, hunted, shot at, harpooned, benevolently remains for three hours within range of its antagonists, instead f regaining the depths of the abyss ... and whose upper portion dives and disappears only after it has been separated from the lower![65]

As this controversy illustrates, a central problem of the age was making claims seem credible. Within a few decades, a series of scientific 'revolutions' had made it necessary to revise, from the bottom up, the conceptual coordinates inherited from the past; in the process, the standard distinctions between the possible and the impossible had loosened. Yet even in this context, the scientific establishment's resistance to the giant squid represents a source of irritation, especially since the evidence was constantly growing. The same year that the *Alecton* encountered the giant squid, the Dutch natural scientist Pieter Harting published *Description de deux céphalopodes gigantesques*.[66] In 1871, a creature wound up in the nets of American fishermen in the Grand Banks; Steenstrup examined a photograph of the animal and identified it as *Architeuthis monachus*. A year later, a specimen 15.8 metres long washed ashore at Thimble Tickle Bay in Newfoundland. A year after that, a giant squid attacked a small boat at Portugal Cove in Conception Bay.[67] Two of the tentacles were hacked off and brought ashore. Eventually, they landed on the dissecting table of Addison Emery Verrill (1839–1926), a professor of zoology at Yale. Verrill duly incorporated them into his research on giant squid in eastern North America; between 1874 and 1882, he published approximately thirty studies about *Architeuthis*, the *Colossal Cephalopod of the North Atlantic*.[68]

How, then, can it be explained that the giant squid's existence was doubted until well into the twentieth century? What does it take besides witness accounts, mandible joints, anatomical specimens, photographs and drawings to prove that an organism is real?[69] Even if the history of science tells of many painful instances of falsification and fraud, and even if the experienced natural scientist does well to proceed with caution – instead of marvelling at any anomaly that comes along – the question still holds: How must facts be constituted so that we, who are so modern and enlightened, will actually believe them? Mangin provided an answer exemplary of the times:

> There is no caprices, as I have already said, in creation. Nature is subject to certain fixed laws, and to believe that all animals may indifferently exhibit the most arbitrary and irregular dimensions, is an opinion which could only gain credence among persons completely ignorant of natural philosophy. The evidence is indisputable that there exists ... a necessary co-relation, in virtue of which it is as impossible to place any rational belief in the existence of an infusoria six feet long as of a microscopical elephant, in that of a spider as large as a horse, as in that of a rhinoceros no bigger than a fly. And it is in virtue of this same law that the existence of a sepia or a calamry of the size of a whale must appear a priori inadmissible.[70]

In other words: a 'fact' that overturns the notions and rules which, to our minds, account for the way organisms are organized is not a fact at all. Instead, we call it a 'miracle'.

The second answer to the question about how Verne managed to 'discover' a marine giant *avant la lettre* takes us where scientific explanation hits a limit. At the same time, however, the very question misses the point inasmuch as it suggests that Verne was somehow endowed with the gift of second sight. Such suspicions surface especially quickly in the case of this particular author; popular histories of technology credit him with having 'predicted' submarines, airplanes and even rockets to the moon. None of these putative prognostications stand at issue here. Instead, our task involves contextualizing the author's 'level of knowledge' in cultural and aesthetic terms. *Architeuthis*, the giant squid, is 'the quintessential sea monster', after all; it is 'probably responsible for more myths, fables, fantasies, and fictions than all other marine monsters combined'.[71] It is impossible to isolate a creature like this in a strictly defined epistemological sphere; instead, it wanders in a twilight realm between wish and doubt. Even if one fails to notice as much, every encounter with this being – an entity that rises both from the depths of time and from the depths of the sea – introduces images and elements from older traditions. According to Hans Holländer, this is why the plesiosaur, the ichthyosaur, the Midgard Serpent, Leviathan, the sea serpent and 'the Great Kraken' seem to have so much in common.[72] The fluid transformations of these twisting and turning creatures can be grasped only as they come into view; the 'novel encounter' must give rise to a 'novel vision'. In this context, it is unimportant – and, at any rate, impossible – to verify whether the giant squid outside the porthole actually is *the* giant squid; the point, instead, is to see *what* the creature peering in looks like. This 'disinterested' mode of viewing – which does not seek to assign phenomena places in a system right away, but tries to appreciate all the facets and details of their morphology – is a method the visual arts employ. It is no accident that Riou's illustration does not present the giant squid as a work of natural history might do; instead, the animal dances elegantly in a curved frame. It appears in the oval of the underwater window, which the novel stages as a passageway for gazes, emotions and thoughts; the creature represents an aesthetic event, a phenomenon between visionary transport and the objectivating focus of technological instruments of observation.

Riou's octopus is decorative. Especially the way its arms pour out and extend playfully calls to mind the stylized depictions of anemones, sea stars and jellyfish in contemporary art, which prized the aesthetics of the marvellous. In Verne's account, the big-eyed animal also becomes a wondrous *monstrum*: 'Yet, what monsters these poulps are! what vitality the Creator has given them! what vigour in their movements'![73] Such language attests to awe and admiration, the classical emotions associated with the miraculous. As the squid slowly sucks itself fast on the *Nautilus,* Arronax describes the absolute strangeness of the creature in a fascinated and excited tone. He calls the giant squid a marvellous invention of nature[74] – a formulation that points back to the heyday of connections between art and science, when one spoke of *ludi naturae* and exhibited them in cabinets of art and wonder. As the 'polyps' circle the ship in disquieting number, Verne stages an authentically aesthetic situation: surrounded by his companions and in a state of absolute spiritual tranquillity, Monsieur Arronax makes a drawing of the animal! In the process, an extraordinary encounter transforms into a situation where the monster represents an aesthetic object. The drawing does not serve to secure evidence so much as it affords a medium for achieving insight: 'um zeichnend und malend tiefer in das Geheimnis [der] Schönheit' dieser fremden Lebensform 'einzudringen' ('drawing and painting' offer means of 'penetrating deeper into the mystery [of the] beauty', translation by Erik Butler, hereafter EB) of the alien creature, as Ernst Haeckel once observed when describing the impetus behind his morphological studies.[75] Twenty years later, Adolf Portmann – himself a cephalopod-lover – declared that one has really seen and understood only what one has drawn.[76]

Even though he admired Mangin and read his books attentively, Verne steered the opposite course. Instead of the laws of nature, he was interested in its material elements; instead of affirming the impossibility of the giant squid's existence, he sought the form in which it might prove possible, after all. The point is not to declare that Verne was 'right'. Rather, the point is that Verne took a wide array of contemporary technological and scientific discourses and made them a means of aesthetic engagement. What emerged was an object of fascination that has accompanied modern history and still manages to command our attention almost one and a half centuries later.

'... *Like a Silken Scarf Floating* ... ' [77] – *Love Letters to the Octopus*

Even though the octopus of nightmare occupies a prominent position in the modern bestiary, a sensuous cephalopod exists too, and it has inspired myths that echo the world over. Caillois argues that the innocent, ornamental, sometimes wise, sometimes sensual Kraken of Greco-Roman antiquity has now been displaced once and for all, by the blood-curdling monster Hugo described.[78] But if one takes even a cursory look at the art and science of the nineteenth and twentieth centuries, this claim seems highly dubious. The fascination exercised by the octopus, which wandered through the writings and illustrations of natural history for centuries, did not break off in modernity. The hymns of praise, lyrical outpourings and flowing colour gradients in the works of Aldrovandi, d'Orbigny and Leopold Brunner (1788–1866) still echo in Chun's reports of his expeditions and, more recently, in the popular accounts of Jacques-Yves Cousteau, Norman John Berrill and Alfred Klingel. Here, the octopus is no simple pile of 'ashes'; rather, it is an 'enchanting', 'lovely', 'graceful' inhabitant of the sea with a 'majestic colouration'.[79] Cephalopods are animals of 'curious strangeness', 'the greatest peculiarity' and 'wondrous variety' displaying 'inexhaustible diversity'.[80] Finally, visitors to aquariums, speechless and amazed, have consistently demonstrated not just incredible interest in, but also great fondness for, the incomparable octopus. Observing its amorous sport, Wilhelm Bölsche declared his love for this 'baroque and crazy' animal,[81] and Gustav Schubert wrote that one does not get tired of watching the wonderful movements and lively expressions of the animal; he was speaking for many others, too.[82]

As the taste for the 'ornamental' octopus flourished in Art Nouveau at the turn of the century, the 'clever cephalopod' became a favourite object of study for marine researchers. In the nineteenth century, the brightest minds had already gathered – for example, at the celebrated *Stazione Zoologica* in Naples – to evaluate Darwin's evolutionary theories experimentally; from the beginning, tests with octopuses played a key role. Researchers determined that they were capable of feats of learning and memory, like vertebrates; indeed, it was observed that they even 'train' each other.[83] When the same experiments were repeated in 1992 and the animals' astonishing capacity for learning was confirmed, it generated worldwide attention yet again.

A recurrent theme of scientific enthusiasm for octopuses involves their elegant, seemingly perfect, way of movement – as well as the strange fact that they swim backwards, 'like that legendary, backwards flying bird, which did not care, where he went, but liked to see where he had been'.[84]

A dozen sprawling, lace-like shapes would suddenly gather themselves into streamlines and shoot upwards, het-propelled by the marvellous siphon in their heads, like a display of fairy water-rockets. At the top of their flight, they seemed to explode;

their tails of trailed tentacles burst outwards into shimmering points around their tiny bodies, and they sank like drifting gossamer stars back to the sea-floor again.[85]

The changes of skin colour to reflect emotion offers a neverending source of amazement.[86] 'It is impossible to look at the beautiful play of the chromatophores of the cephalopods ... without being filled with wonder and the complication and the perfection of the play itself',[87] writes Enrico Sereni. Gilbert Klingel observes:

> It always seemed irritated at my presence. Its nervousness may have been caused by fear, for it certainly made no pretence of belligerency, and it constantly underwent a series of pigment alterations that were little short of marvellous. Blushing was its specialty. No schoolgirl with her first love was ever subjected to a more rapid or recurring course of excited flushes than this particular octopus. The most common colours were creamy white, mottled vandyke brown, maroon, bluish grey, and finally light ultramarine nearly the colour of the water. When most agitated it turned livid white, which is I believe, the reaction of fear. During some of the changes it became streaked, at times in wide bands of maroon and cream, and once or twice in wavy lines of lavender and deep rose. Even red spots and irregular purplish polka dots were included in its repertoire, though these gaudy variations seldom lasted for long.[88]

The eyes of the octopus pose an enduring riddle, too:

> I think if you asked any zoologist to select the single most startling feature in the whole animal kingdom, the chances are he would say, not the human eye, which by any account is an organ amazing beyond belief, nor the squid-octopus eye, but the fact that these two eyes, man's and squid's, are alike in almost every detail.[89]

Finally, octopods delight observers with their cunning, curiosity and the legendary ability to be 'Houdinis' – a talent that borders on the anatomically impossible. 'It [the octopus] is the very acme of suppleness', writes James W. Atz, the curator at the New York Aquarium.[90] Playful interactions and affectionate touches count as the most precious moments: all texts reporting such encounters echo with the writers' quickened heartbeats and startled disbelief at receiving the gift of such an extraordinary experience.[91]

> Of one thing we can be certain, however: there are more wonderful things ... still to be found out about the octopus.[92]

La Pieuvre –Jean Painlevé

One of the most interesting encounters with the octopus at the borders between art and science occurs in the eerie yet beautiful films of of the French scientist and director Jean Painlevé (1902–89): *La Pieuvre* (1928) and *Les Amours de la Pieuvre* (1965).[93] Painlevé was not only a biologist, but also an active member of the surrealist movement; in almost two hundred short cinematic works, he combined the ideas of the avant-garde and scientific documentation. He was one of the first to present filmgoers with the wonders of the underwater world, often magnifying its inhabitants, sometimes speeding up their activities or showing them in slow motion, but always casting them in an extraordinarily artful light.

Painlevé did some of his shooting in the sea – with his friend Yves de Prieur (1885–1963), he pioneered the underwater film – but most of his work was conducted in the studio. His *Institut de cinématographie scientifique,* located on the Rue Armand-Moisant in Paris, contained an aquarium that also served as a giant screen; it offered views of sea horses, spider crabs and hermit crabs swimming, with all their strange forms and movements, through images of liquid light. His entertaining and whimsical revolutions of vision went far beyond the possibilities afforded by strict scientific observation and created a 'surrealism of natural phenomena'[94] blurring the borders between science and fiction.

Painlevé was fascinated by octopuses. In an interview, he stated that this affinity stemmed from his childhood and later had grown during time he spent at the Institute for Marine Biology in Roscoff. Above all, he was impressed by the animals' intelligence, emotionality and ability to remember; indeed, his love of the octopus is what prompted him to make films of popular science in the first place.[95] To date, there is no evidence whether he drew inspiration from artistic or literary models – for example, from Lautréamont's *Les Chants de Maldoror* (1869). However, given the enormous admiration that André Breton (1896–1966) professed for this poet, who died at a young age,[96] it is highly unlikely that Lautréamont's view on the octopus did not overlap with ideas Painlevé entertained on his own. As much seems to follow both from the unnervingly surreal introductory sequence in *La Pieuvre* and from the fact that the rest of the film makes a point of not presenting the animal as an incarnation of the abyss. Instead, Painlevé's highly idiosyncratic cinematic language presents an animal whose contours change over and over; gentle cross-fades show it as a zoological specimen and an unknown dream-creature by turns. Painlevé wrote of his methods:

> Does the complete understanding of a natural phenomenon strip away its miraculous qualities? It is certainly a risk. But it should at least maintain all of its poetry, for poetry subverts reason and is never dulled by repetition. Besides, a few gaps in our knowledge will always allow for a joyous confusion of the mysterious, the unknown, and the miraculous.[97]

Both *La Pieuvre* and *Les Amours de la Pieuvre* begin by presenting the octopus as a culturally coded figure of uncanniness. Walking on land, the animal embodies menacing and bizarre otherness. In *La Pieuvre,* it glides off a window ledge, oozes over a doll that lies prone and then slithers its way down a leafless tree. Finally, the creature appears in dark, black water, its limbs encircling a bright, glowing skull. *Les Amours de la Pieuvre*, in turn, opens with an octopus dragging itself across what seems to be an infinite plane of black-grey seaweed; the accompanying music – by Pierre Henry, a pioneer of *musique concrète* – heightens the grotesque atmosphere. A creepy, exaggerated voice that seems to come from beyond the grave identifies the writhing creature as an 'animal horrifique'.[98] Such an 'introduction' stands as an exception in Painlevé's cinematic oeuvre. Otherwise, his films of marine environments bring out the historical constructedness of human perception and the cultural contingency of scientific observation. Painlevé toys with the medial conditions underlying the realm of knowledge and makes operative aesthetic assumptions perform a 'dance' with the marine beings that occupy the spotlight; research activity makes allowances for the role that imagination and fiction play in processes of scientific representation.

Where the octopus is concerned, however, it seems that Painlevé must take a special approach. To account for the animal's biological and aesthetic substance, he lets the images oscillate between reality and representation; ultimately, neither one predominates. The octopus's 'shore leave' – which evidently symbolizes the sphere of culture – stages a territory to which disgust, hate and aversion can be restricted. Here, in the open air, the octopus is a dark, slimy mass and moves like a creeping coil of snakes. But as soon as it reaches the shore, it transforms into a retiring animal – part of the coastal fauna that stands at the mercy of human beings. Finally, when the octopus reaches the water and dives in, the film casts off the affectively charged mass of texts and images – two thousand years' worth of material culled from the history of human imagination. Now, the animal appears as soft as silk and satin.

In *La Pieuvre*, we encounter the octopus as it lies sleeping underwater. A close-up reveals that its eye is closed; the animal's body rises and falls in a powerful rhythm. Slowly, the octopus peers out – an intertitle informs the viewer that its eye is like a human being's – takes a look, and then goes back to sleep. This moment, between waking and drowsiness, is strangely moving. But even though the animal is resting, an array of forces is at work. Its breathing seems to follow a convulsive choreography: delicate and elastic swells billow, draw back together, tense up, relax and blow outward again, as if the body were the mighty instrument of an unknown musician. Despite the absence of sound in the film, one's ears seem to ring from the humming and buzzing that issue from the living bellows – an orchestral spectacle lends the pumping musculature a volume worthy of Poseidon himself.

The sun shines through the water, onto the octopus. Under the falling light, its iridescent skin gleams and brings forth a sparkling spectacle; at a hundred points, an impressionistic pattern emerges. Then comes death.

The octopus dies in a painful, protracted scene, mustering energy only to collapse again. Like a ruffled garland, the snow-white tentacles coil in the half-light; they seem to blossom into a sea star before falling down to the black ocean floor for the last time. The slowness of the underwater movements heightens the dramatic atmosphere; a tormented, final struggle occurs before our eyes that seems to present the image of the inevitable, our own mortality. We behold a dying creature that merits our compassion, the symbol of all life headed for extinction. This is also what the film shows at the end of the scene: the dead octopus is washed ashore. The camera draws close to its eye staring blind and empty. For the ten long seconds that dwell on the lifeless, grey body, one recalls the tender gaze of the animal that only now was still brimming with vitality.

Forty years later, the octopus stood before Painlevé's camera again. Many elements of the first film recur in the second – for example, the trek on land in the opening sequence, the attention paid to the creature's eye and its breathing and the interest taken in the play of colour on the octopus's skin; this time, the images are not in black and white, and a mosaic of grey, blue and yellow tones appears. What stands out, however, is the way that love for the octopus, in the modern age, is coupled with curiosity about its exceptional biology.

As the title of indicates, the film focuses on the remarkable courtship of two octopods. As if onstage in a laboratory, the drama unfolds: the initial approach, a *pas de deux*, and then insemination, as if performed by two figures that have accidentally stumbled into the story. The beautifully reddish-brown animals flit and scurry through the scenery, touch each other timidly and uncoil their tentacles, hesitantly reaching out – before getting scared and fleeing the other's presence. The whole while, it seems that both of them are looking into the camera, which only heightens the ballet-like quality. As captured on film, their movements present a rhythmical series of changing poses that strike viewers as having been choreographed for their entertainment.

Finally comes the act of procreation, and the male inserts his special third arm into the female's mantle cavity to transmit packets of sperm. Now, the couple shifts from dancing to expression in the style of musical theater. The female moves constantly; the accompanying music hints at a piano concert performed by jazzy tentacles. As she drags her partner behind her, a manly singing voice fades in, making the slightly disadvantageous situation of the male into a stage gag. The narrator informs viewers that they are witnessing a process that last hours or even days in nature; on the screen, however, the male extracts his arm from the female's mantle after thirty seconds, then he exits. The following sequence consists of the microscopic close-ups that assured Painlevé's fame by enabling audiences to peek at the tiniest details of submarine life. The embryo's maturation is shown in abbreviated segments until the young creature finally slips out of the egg casing. As the process unfolds, we see the tireless mother animal perform delicately playful motions as she, without any concern for herself, assures that the eggs, stretched out like strings of pearls, remain properly aerated – a vision of the greatest devotion and tenderness. In light of the final sequence of images, it appears all but certain that this affective colouration belonged to the associations that Painlevé intended. Surrounded by a silver shower of thousands upon thousands of shimmering baby octopuses, the mother animal appears again – the poetic symbol of fertility and the abundance of ocean life. Her elegant tentacles describe a dancing gesture that seems to offer a last message: birth, life, profusion – *fin*.

Film scholars, who are rediscovering Painlevé only now, ackowledge him as the master of revealing the invisible. As is always the case with such honorary titles, the question arises what, exactly, is meant. Does revealing the invisible involve aestheticization, by means of technological media, what has not been visible until now? Does it concern something visible that has never been seen in this way before? Or are we standing before some new creation, which only exists in the framework of the media employed? Painlevé's films seem to experiment with all three possibilities. In *La Pieuvre,* the octopus represents an object of aesthetic contemplation above all. *Les Amours de la Pieuvre*, on the other hand, places the strangeness of the animal's anatomy and habits in the foreground. Needless to say, it is not a matter of making the films compete with each other. Each of them develops a particular aesthetics of the octopus, presenting it as a hybrid entity incorporating aspects of a work of art and biological sensation; moreover, both films are concerned with showing points of ambivalence. For all that, *La Pieuvre* unfolds mainly as optical research; the editing, close-ups, and lighting all serve to make the borders of visibility porous. In contrast, *Les Amours de la Pieuvre* stages the grand narrative of life: meeting, sex and birth. *La Pieuvre* makes the octopus transparent, presenting the marine creature in terms of the interplay of movements, colours, rhythms, morphologies and textures. Watching *Les Amours de la Pieuvre*, one sees a theatrical mise-en-scène where two cephalopodic characters offer a performance of the act of procreation. In the process, the positivistic premise of objectivity that normally attends films of animals – and suggests that only naturally given processes are being documented – is dashed to pieces by the way the spectacular nature of the cinematic medium is exposed. Thus, the octopuses on film achieve a state of floating undecidability: on the one hand, there is no doubt that they exist in nature; at the same time, they inhabit the realm of the unreal and the unknown. The space where their new mode of visuality is disclosed is the aquarium. Only rarely does it become this clear that the aquarium is a medium – that it produces images and creates a new framework for perception. Standing between the laboratory and a dream-landscape, it provides the setting for Painlevé's surreal art of science, which opens up a new, ludic and aesthetic dimension for both known and unknown marine life in its untold variety.

All the same, the octopus remains a mystery. Whether it is in the sea, an aquarium or presented in close-up or in high/low literature, something always remains out of reach, a difference that cannot be compensated for, which goes beyond pure incomprehension. Octopuses and squid seem to have their own mysteriousness, a special form of strangeness that is perhaps most accessible to an aesthetic process. This strangeness generates an ambivalent power of fascination that has given rise to a great number of artistic and scientific subtleties and distinctions. For centuries, this power transformed cephalopods into catalysts of wonder, agents of the most diverse processes of transference between scientific, artistic, technological and popular discourses and visual models of a way of life inaccessible to human beings. That said, an encounter with one of these creatures is always a highly emotional experience. Or, the other way around: It is only through the process of amazement that octopuses and squids become constituted as epistemic and aesthetic objects in the first place – through an oscillating affective triangle of the three classic reactions that manifest themselves when one is faced with a wonder: terror, astonishment (*thaumazein* as well as *hedone*) and curiosity. Suspended between horror and admiration, between bafflement and the thirst for knowledge, we encounter an elegantly dancing animal, a powerful predator, a spectacular choreographer and the most terrifying marauder the sea has ever brought forth.

4 *MISE-EN-SCÈNE:* INVENTED REALITIES, OR THE MEDIALITY OF WONDERS – THE SEA IN THE AQUARIUM

The fantastic underwater world cast a deep spell on early modern culture. The dramatic rise of natural science, especially Darwin's theory of evolution, prompted widespread public interest, and efforts to communicate its findings by aesthetic means met with a receptive audience among professionals and laymen alike.[1] A culture of knowledge and experience of extraordinary poetic and aesthetic subtlety emerged, spread by an ever-growing number of illustrated books and gazettes; craftsmen, illustrators and architects found a wealth of material largely unknown until now, which offered models for ornamental design. Artists eager for something new welcomed the intellectual and scientific currents and tides of the day; in particular, the world of marine creatures inspired them to explore the origin of existence, the forces animating organic growth and the rhythms of life in nature. The bizarre diversity of forms of underwater flora and fauna became a key aspect of artistic experience, particularly when the aim was to give form to the incommensurable and the incomprehensible. The fairytale moods inspired simply by seaweed and algae acted as a surreal element giving rise to new forms of representation that opened the way for new flights of abstraction. Odilon Redon (1840–1916) painted abstruse creatures in the dark light of the underwater domain, and Alfred Kubin (1877–1959) transformed the ocean depths into shady realms of evolutionary mystery. Gustav Klimt (1862–1918) created hybrid patterns and collage-like fragments of microscopic marine life forms; plants would metamorphose into fish, and reptiles turned into woman emerging from the water. Hermann Obrist (1862–1927) developed a morphology, or theory of forms, of marine organisms, and Paul Klee (1879–1940) studied fish, which, in his eyes, embodied nature's tendency toward variation. The surfaces of *art nouveau* vases 'phosphoresced' with fish-scales, snake-skin, seaweed, algae and decorative jellyfish. Tiffany lamps cast their soft light on high-end furniture made of ray skin. Mysterious, iridescent water creatures graced the expensive glasswork designed by Emile Gallé (1846–1904), Henri Bergé (1870–1937), Eugène

Feuillatre (1870–1916) and Francois Rousseau (1827–90).[2] In 1900, at the entrance to the world's fair in Paris, one could walk through the impressive *Porte Monumentale* by René Binet (1866–1911) – a magnificent adaptation of Ernst Haeckel's skeleton of a *Nassellaria* (which he had drawn in Plate 65 of the *Report on the Radiolaria, collected by HMS Challenger during the years 1873–76*).[3]

The key question when looking at the storied career of aesthetic underwater impressions is: What real and imaginary foundations were they based on? How could tiny fish preserved in alcohol, microscopic pictures of *radiolaria* or luminous colour prints of candy-coloured tube worms provide an endless source of inspiration? Without a doubt, the highpoints of scientific observations of nature – such as Haeckel's famous *Art Forms in Nature* (1899–1904), which presented wondrous *radiolaria*, *anthomedusae* and *hexactinellid* sponges masterfully rendered by Adolf Giltsch (1852–1911), a lithographer from Jena – influenced many generations of artists. The spectacular meeting between humans and the sea hinged on experiences that promised an encounter with the living inhabitants of the world beneath the waves. The visual culture of the day mirrored a longing for immersion: delving into the strange universe of the ocean and viewing what floated and flowed there. Such an experience (even if it proved virtual) was provided by a singular artefact and media apparatus that emerged in the nineteenth century and represented an exemplary location for forming theories and models of ocean life – a catalyst for forming links across a broad spectrum of discourses: the aquarium.

Representatives of the Sea: Aquariums as Epistemic Objects and a Medium of Speculation

La Vie en Miniature: A New Research Paradigm

Today, in the twenty-first century, it may prove difficult to appreciate the electrifying effect that aquariums had in the second half of the nineteenth century. Now, household aquariums are less likely to be associated with aesthetic or intellectual refinement; if anything, they evoke whimsy. In terms of cultural history, they are commonly viewed alongside dollhouses, train sets and stamp collections. But aquariums offered far more than instances of the fetishism that marked the age as a whole. They were epistemic objects of the first order and, as such, epitomized the medial ambivalence of the aesthetics of knowledge. Media – and this is undoubtedly what aquariums are – raise the question of the scope of possible insight. On the one hand, they concretize, objectivate and lend material form; at the same time, they virtualize, subjectivize and simulate. The combination of authenticity and illusion presents nature in all its wonderful suppleness and mutability: what is otherwise distant appears in close-up; the fantastic becomes accessible. The incommensurable dimension of media involves the paradoxical experience of phenomena being as they seem, while also being something wholly different. They have the magical effect of making present what is absent or lies beyond one's grasp. Needless to say, 'magic' does not refer to a non-rational or esoteric process in this context; rather, it means kind of performativity which combines and separates entities, as well as the ways they achieve representation in determinate situations. Media coordinate differences – for example, the fictive and the real – at the moment of experience. They create alluring spectacles that reveal reality as porous and subject to modification and are themselves plastic.

Aquariums served as models that both reflected possibilities of knowledge and permitted the study of the mediated nature of the epistemological process. Constituting a paradigmatic 'blank space', the dominant 'methods of power and knowledge assumed responsibility for the life processes and undertook to control and modify them'. Thereby, they manifested the 'entry of life into history … into the order of knowledge and power, into the sphere of political techniques' which Michel Foucault has emphasized.[4] Historically speaking, aquariums marked a decisive shift within practices of research from a static conception of nature, in which 'corpses' were dissected and classification tables drawn up, to the study of living beings subject to 'experimentalisation'.[5] At major centres of research, such as Banyuls-sur-mer (founded in 1863), Roscoff (1872), Naples (1872) and Villefranche-sur-mer (1880), aquariums are where the discourses of biology, the natural sciences and engineering converged. Aquariums instituted a new 'order of visibility',[6] a new constellation of visuality and new practices of observation that facilitated the systematic research of all living things – especially their developmental cycles, metamorphoses and modes of behaviour. As sites of an unprecedented form of knowledge-production, aquariums represented a new paradigm for researching life processes, which entailed new modes of producing evidence.[7] They brought together the aesthetic and epistemic domains in a unique way – producing an 'inner aesthetic of research practice and the presentation of knowledge' with numerous points of overlap and interference between science and aesthetics.[8]

Aquariums enabled the observation and study of living organisms' movements, behaviour and metamorphoses as spatial structures. More still, they were symbolic manifestations of the modern fascination with the vertical dimension. This media-technological apparatus promised a view beyond standing borders, into the deep – that is, it promised a glimpse into a part of reality that had proven impenetrable until now. Finally, one might behold one of the last great realms of mystery on earth: the oceanic depths and all they concealed. Behind the glass surface of the aquarium, what had been as dark as night opened up as a habitat to be surveyed by the naked eye. The microcosmic 'world behind glass' seemed to permit the immeasurable ocean to be fathomed – to provide an apparatus for operationalizing underwater infinity. To their great astonishment, however, people soon realized that the mysteries did not diminish in size or number, all

advances in knowledge and scientific method notwithstanding. Indeed, they multiplied. Inconceivable wonders of metamorphosis came to light: 'animal-plants'
and 'stone-animals'[9] fusing three dimensions of matter paraded before rapt viewers, who were left dazzled and full of questions. A particularly famous instance is
the *Eozoon canadense*. For the longest time, this small object was taken to be an
animal – specifically, a petrified foraminifera – and greeted by the public for afford-
ing a glimpse at 'the dawn of life on our planet'. After ten years of debate about its
'true nature', Karl August Möbius (1825–1908) finally succeeded in conducting a
microscopic analysis of ninety specimens; he determined that the *Eozoon* is in fact
a combination of minerals: serpentine, chrysotile and limestone.[10] In the course
of the debate – and other disputes involving 'new forms of life' – one question in
particular became more and more pressing: What, exactly, is 'life' from the standpoint of the natural sciences? What are its biological characteristics?

And so, time and again, the newly disclosed dimension of reality abounding in
fresh insights and knowledge presented as many problems as answers. Nothing, it
seemed, separated what was now revealed from magic and mystery, even – indeed,
especially – when one looked at it, in thousandfold enlargement, through a microscope. The utter foreignness of phenomena proved real and unreal in equal measure.

Within a few years, a veritable craze had gripped Europe and the United States.
Whether in private salons, as a tourist attractions or in scientific laboratories,
aquariums were *the* sensation during the second half of the nineteenth century.
They perfectly suited the widespread enthusiasm for scientific and historical *tableaux* that were staged on a broad scale.[11] Moreover, because they made it possible
to see things that otherwise occupied a field of reference beyond verification, they
functioned as a medium of speculation. Not only could people count on beholding 'the ocean' *en miniature;* it also seemed that the aquarium would resolve the
greatest of all mysteries: the origin of life and the genesis of the diversity of species.

Even though the revolutionary publication of Darwin's *On the Origin of Species
by Means of Natural Selection, or Preservation of favoured Races in the Struggle for
Life* would not occur for some time yet – the book appeared in 1859 – one thing
was already clear for many natural scientists of the 1840s and 1850s: the earth had
emerged from a primordial sea and the earliest life likely followed an aquatic design.
Only on the seafloor, then, were the answers to the beginning of time to be found.

*Suggestive Models of Thought: Underwater Dreamworlds and Their Realistic
Effects*

In short order, aquariums came to represent extremely interesting fields where
the discourses of oceanography, evolutionary biology and geological history
intersected with new experimental practices of amateur science and art. They
provided a significant means by which popular natural science was disseminated,

and they constitued an essential part of image-based entertainment – especially insofar as they connected with the visual culture of the microscope. At the same time, aquariums and their occupants provided figures for the imperialist conquest of the world, extending from the measurement and exploration of the globe to Western culture's triumph over the mysteries of nature.

The principle reason for aquariums' existence was the inaccessibility of their field of reference – the sea. Significantly, however, they were perceived as natural 'biotopes'. With almost touching naïveté, people accepted that they offered a natural, underwater habitat; they expected to be able to study animals now isolated in a glass environment as if they were 'in the wild'. Even though the figural language of literature and the arts had foregrounded the spectacularly artificial and speculative aspects of such simulated 'naturalness' early on, the fact that knowledge derived from aquariums was based on a media aesthetic did not become an issue until much later on, during the 1920s. In purely practical terms, installing and maintaining an aquarium connects with questions of care, nourishment, reproduction and behaviour – i.e., the conditions indispensable for life to flourish. As a matter of course, anyone wishing to study living organisms had to understand their environment, in order to create conditions as 'normal' as possible for them. As such, it was necessary to acquire practical knowledge, too. It follows that aquariums provided important models for thought in the process of identifying ecological habitats.

An exemplary instance of this mode of obtaining knowledge was the cooperation between Möbius and his friend, the Hamburg merchant Heinrich Adolph Meyer (1822–89). Both men shared the same enthusiasm for marine fauna. At mid-century, they started working together to study animals' relationship to their environment in the Baltic Sea and in aquariums which Meyer set up at his house in Hamburg and at his holiday villa in Kiel. These private ocean aquariums, the first of their kind, are what enabled them to conduct their studies of marine biology. Eventually, Meyer donated his home aquarium to the Museum of Natural History in Hamburg; he was also instrumental in the founding of the exemplary and unique seawater aquarium at the Hamburg Zoo.[12] Möbius described 'natural' conditions as a standard for functional aquariums:

> Eine wichtige Aufgabe der zoologischen Gärten besteht darin, die natürlichen Eigenschaften und Thätigkeiten der Thiere vor unsern Augen entfalten zu lassen, und um sie zu lösen, bemühen wir uns, einer jeden Art von Thieren solche Verhältnisse zu bereiten, welche mit den Eigenthümlichkeiten ihres freien Wohnortes möglichst übereinstimmen ... Wir fangen in Flüssen Fische, in Bächen Krebse, in Mooren Schnecken, in Teichen Muscheln und setzen alle in einen und denselben beschränkten Wasserraum. Ist es nicht, als sperrten wir Säugethiere, Vögel, Reptilien und Insekten aus der heissen und kalten Zone, aus trocknen Wüsten und aus feuchten schattigen Wäldern in ein Gemach? Und dennoch erwarten wir, dass sie sich fröhlich bewegen, essen und sich vermehren.

> Wir müssen uns an die Natur anschliessen, wenn wir die Prinzipien der Aquarien
> wissenschaftlich so weit kennen lernen wollen, dass wir fähig werden, sie künstlich in
> möglichster Vollkommenheit einzurichten.
>
> (An important task of zoological gardens is to let the natural properties and activities
> of the animals unfold in front of our eyes. To make that happen, we try to give each
> species of animals those conditions that are most in accordance with the particular
> conditions of their habitat in the wild ... We catch fish in rivers, crayfish in creeks,
> snails in swamps and mussels in ponds and put them all in the same, limited water
> space. Is that not the same as if we were locking mammals, birds, reptiles and insects
> from hot and cold zones – from dry deserts and moist, shady forests in one room
> together? And yet, we expect them to move about happily, eat and procreate.
>   We have to join nature if we want to know the principles of aquariums scientifi-
> cally enough to be able to furnish them artificially to the highest possible perfection).[13]

Möbius viewed the aquarium as an epistemological apparatus; as such, it played
a significant role for understanding relations between organisms and their environment. In 1865, the first volume of *Die Fauna der Kieler Bucht* (The Fauna of
Bay of Kiel), co-authored with Meyer, appeared. The work examined animals in
a full range of vital activity for the first time. It was the first practical ecological
research study to be published – appearing one year ahead of Haeckel's *Generelle Morphologie der Organismen* (General Morphology of Organisms) (1866),
which defined ecology as a term and field of knowledge. Subsequently, in 1877,
Möbius offered the first account of the patterns whereby diverse organisms exist
in relation to environmental conditions: *Die Auster und die Austernwirtschaft*
(The Oyster and Oyster-Culture).

> Science possesses, as yet, no word by which such a community of living beings may be
> designated; no word for a community where the sum of species and individuals, being
> mutually limited and selected under the average external conditions of life, have, by
> means of transmission, continued in possession of a certain definite territory. I pro-
> pose the word *Biocoenosis* for such a community.[14]

To be sure, Möbius would not have been able to develop his model of biocoeno-
sis without intensive study of nature 'on location' in the Bay of Kiel (which he
conducted from Meyer's sailing yacht *Marie*) and on the oyster banks of the
Atlantic (which he visited as a member of the Prussian government's Fisheries
Commission). At the same time, an extremely close connection clearly existed
between his studies and the aquarium as a research laboratory and medium
of observation. In particular, popular articles Möbius wrote for *Hamburger
Nachrichten* and *Zoologischer Garten* described the aquarium as the model for
contemplating the interaction between marine life forms and their surroundings
and at the same time as a modern version of one of the seven Wonders of the
World, the Hanging Gardens of Babylon.[15]

In the mid-nineteenth century, enthusiasm for aquariums unfolded on two levels. For one, almost all the larger cities of the Western world witnessed the construction of aquariums, which often housed important research stations, as well. Secondly, small glass structures were produced on a massive scale, enabling researchers and amateurs alike to observe animals (or merely decorate their homes). In terms of function, it is easy to distinguish between a scientific laboratory for experimentation, a tourist attraction and a stylish domestic appointment. However, in terms of manufacture and use, there are many continuities between scientific aquariums and the market for aesthetic sensation and enjoyment. Before addressing the latter point, let us examine the conditions – in terms of the history of culture, science, aesthetics and technology – that made it all possible.

Aquarium Fever: The Ocean as an Aesthetic Event

The Great Public Aquariums and the Mediatization of the Ocean

> [M]odern science, not less ingenious in its popularizing processes than patient and bold in its investigation of the secrets of nature, has found a means of opening up to us the wondrous scenes of the submarine world. She has created small oceans in miniature, tiny seas for our domenstic chambers, wherein, through walls of crystal, we may watch the fish, the crustaceans, the molluscs, and the zoophytes living their normal life in the bosom of the 'briny wave' among rocks, and coral, and sea-weeds. I refer to those aquaria which, within the last few years, have been established in some of the public museums of natural history, especially at the Zoological Gardens in Regent's Park and the Jardin des Plantes at Paris.[16]

From Showcase to Living Polyrama: London and Paris

The London Aquarium was the first of its kind.[17] The impetus for its construction came from David W. Mitchell (1813–59), Secretary of the Zoological Society, who conceived the 'living museum' on the model of Joseph Paxton's (1803–65) Crystal Palace of 1851. In December 1852, Philip Henry Gosse (1810–88), a well-known natural historian and the author of works of popular science, gave the Zoological Society a small collection of zoophytes and annelids[18] and, in 1853, the *Fish House* opened to the public. From the beginning, it provided an attraction of the first order.

In essence, the *Fish House* consisted of rectangular containers extending along the longitudinal walls; at the middle of the hall stood a series of relatively small, open basins on tables. Glass was used only where the containers faced the public; the other three sides were slate. Within the displays, sea flora and fauna had been arranged into an idyllic landscape of coral reefs and seaweed groves, a 'museum of living nature'.[19]

From London started an international tide of large, public aquariums. P. T. Barnum (1810–91) opened an aquarium as part of his established Barnums's

American Museum in New York in 1856 (it burned to the ground in 1865). Frankfurt followed suit in 1858, Boston and Paris in 1859, and Vienna and Hamburg in 1860 and 1864, respectively; then came Hannover (1866); Brussels, Le Havre and Cologne (1868); Berlin (1869); Brighton and Naples (1872); Washington (1873); Manchester and Southport (1874); Yarmouth and Westminster (1876); Edinburgh (1878); Amsterdam (1880); San Francisco (1894); New York (1896); and Sebastopol 1897 – to mention only the largest.

Essentially, the London Aquarium was just a big hall with a number of glass pools standing on tables. In contrast, the Paris Aquarium of the *Jardin d'Acclimatation* offered an array of enchantments to play upon visitors' fancy. As we observed above, a look into underwater spaces in the 1850s seemed to many as foreign as catching sight of the dark side of the moon. Quite unlike the sobriety on display at the London *Fish House,* the Paris Aquarium presented the hitherto unknown dimension as an infinite playing field for the imagination – a hybrid work featuring the metamorphoses of paradise, phantasmagoria and opium dreams all at once. It was a gallery in neoclassical style, forty metres long and ten metres wide; along one wall ran fourteen tanks, almost cubic in proportion, which presented a succession of paintings, as it were. The gallery lay in semi-darkness: the only light came through the top of the fishtanks. Ten of them contained sea creatures, and four held animals from rivers. Here, for the first time, the water – some 900 litres per tank – no longer needed to be replaced by hand; a pumping system invented by Lloyd saw to it that it remained fresh and circulated as if by ocean current.

The light trickling in through the glass tanks suffused the hall with an unreal, greenish glow. This effect is undoubtedly the reason why aquariums were perceived as theatre – and light shows from the inception – after all, they enacted the principle of nineteenth-century light design that illuminated the stage and plunged the audience into darkness. Wholly charmed, Théophile Gautier (1811–72) wrote that he felt he was seeing an intoxicating 'ichthyological drama in fourteen acts moving past'.[20] Mangin spoke of *tableaux vivants*, living polyramas:

> This system of lighting has a very impressive effect, and produces a singular illusion. The gaze not being distracted by surrounding objects, our attention concentrates itself entirely on the living polyorama before us; and, as the idea of magnitude is only relative, the tableaux soon assume in the eyes of the spectator greater and yet greater dimensions, or rather, their real dimensions disappear to give place, in everybody's perception, to those which the imagination is willing to lend them. The decoration of these novel theatres, where the drama of submarine life is seriously enacted, is, moreover, exceedingly well-designed Grottoes of pebbles, domes of shells, rocks of diverse character, with the most fantastic and mildly varied outlines, are clothed with marine plants and anthozoaires.[21]

The Grotto Labyrinth and Natural Science, Art Nouveau Style: The Aquariums of Berlin

The architecture that proved decisive for nineteenth-century aquariums was the grotto. The first examples were the aforementioned freshwater aquariums at the Paris World Fair in 1867 – a modern marvel in the eyes of many visitors. Hans Christian Andersen (1805–75), the Danish author of fairy tales, fell in love with the 'midsummer night's dream' set on the ocean floor; without further ado, he erected a poetic monument to it in *The Dryad* (1868).[22]

That same year, the *Musée-Aquarium d'Arcachon* adopted the Parisian grotto aesthetic; Redon was likely one of its most enthusiastic visitors.[23] In turn, the aquariums in Le Havre, Sables D'Olonne and, finally, Berlin (Unter den Linden) did the same. A few years earlier, its director, Alfred Brehm, had taken part in building the Hamburg Aquarium – which *Die Gartenlaube* euphorically described as a 'Marine-Aquariums-Tempel' (marine aquaristic temple) and a 'Meeres-Feenschloß' (fairytale castle under the sea).[24] But while the Hamburg Aquarium had been designed as a vault and only the pools were shaped like natural grottoes, the Berlin Aquarium presented the main exhibition space as a single, multi-storied cave underwater.[25] On the lower level, the panes of the tanks looked like chance openings in the walls of a cliff, excavations made by a perpetual ebb and tide that had never really happened. Visitors were meant to feel as if they were underwater themselves and enjoy the illusion of observing ocean inhabitants in their 'natural environment'. This mise-en-scène, staging a simulated underwater realm and a theatre of natural history, aimed for the effect of magical immersion; it was meant to disclose an alien world and, at the same time, 'hollow out' this space.[26] It was further heightened by the gothic flair of construction – and design elements introduced by the architect, Johann Heinrich Wilhelm Lüers (1834–70). In addition to the visual impression of colours and forms hitherto unknown, then, a form of spatial experience came increasingly to the fore: the intimation that one was really diving into the oceanic dimension and wandering at the floor of the sea. Accordingly, for example, the path leading from the freshwater gallery passed over a staircase made of rock, which presented the way down to the lower levels as a descent into the depths, suffused in the blue glow of twilight; when visitors reached the bottom, they now stood below sea level, 'in' the ocean aquarium.

The Berlin Aquarium offered two further attractions: reconstructions of two of Europe's most beautiful and imposing works of nature, the *Grotta Azzurra* on Capri and Fingal's Cave, a basalt formation in the cliffs of Staffa in the Southern Hebrides. Key elements of the speluncular décor lent the space its overall fairytale quality. In the first three months alone, more than a hundred thousand curious visitors had journeyed to the Berlin Aquarium. As a 'riesiger Felsentempels lebendiger Naturwissenschaft' (a huge stone temple of spirited natural science)[27] it proved exemplary, setting the style for the display-aquariums in Leipzig (1879) and Munich (1881), as well as El-Gezireh Island in Egypt (1902), which were all built in the grotto-style.

At the turn of the century, the design of aquariums changed again. If the initial concern had been to draw the public's notice to the abundance of ocean life, the focus now fell on presenting, as fully and as systematically as possible, the immense array of species and shapes to be found in the sea. New construction technology incorporating steel and concrete made it possible to build bigger structures with more levels. The grottoes vanished as more clearly defined and functional buildings took the stage. The display pools no longer represented *féeries*; instead, they were meant as authentic recreations of the spaces that animals inhabited in nature. Thus, for an exhibit of life on the Adriatic coast or Heligoland, rocks were quarried on location, shipped to Berlin and reassembled there – a process that required infinite care and detail work.[28] At the same time, the space through which visitors wandered remained darkened, as if to recall the theatre.

The aquarium that opened in 1913 in the Budapester Strasse in Berlin exemplified new ambitions and expectations. Under the direction of its future curator, Dr Oskar Heinroth (1871–1945) – the founder of the research area of ethology – and in collaboration with the architects Zaar & Vahl, a three-storey building was erected, fifty-three metres in length and thirty-five metres wide. It held generously proportioned tanks of fresh – and salt water on the ground floor – respectively, eleven and fourteen display pools three-metres long, in addition to twenty-five small, artificially illuminated basins. The next level housed an expansive terrarium. An insectarium was located on the top floor. For all that, the real sensation was the crocodile exhibit; situated at the centre of the building, it encompassed all three levels. Measuring 27 x 10 metres, it offered the first indoor space that visitors could walk through – a revolutionary idea for the time. One entered on the second floor and went along a bamboo bridge. From here, visitors beheld a jungle river and a landscape of exotic plants; just a few metres below, on sandy banks that were heated artificially, dosed the great armoured lizards. The warm, wet air recreated the atmosphere of the tropics and the sluggish, ever-yawning creatures induced a sense of foreign languor. But at ground level, visitors had the uncanny experience of standing at eye level with the antediluvian creatures. Through a thick, round pane of glass that seemed to open onto a black void, they could see the crocodiles and alligators swimming by, grey demons propelled by vigorous and elegant strokes of their tails.

In particular, the building's design stressed the multiplicity of its occupants, which came from all corners of the world, and their evolutionary origins. The aquarium was meant of provide a site of synthesis and exchange for technical, artistic and biological thinking, as well as a model for the community life of its alien inhabitants. Accordingly, planners passed on a stylized architectural design for the building itself in order to focus on varied scientific and artistic ornamentation.

The façade already fascinated visitors and passers-by.[29] Dinosaur-reliefs made in accordance with the latest scientific findings adorned the *parterre*; at the level of the second floor stood glazed porcelain depictions of prehistoric creatures. By the garden entryway loomed an iguanodon in its original dimensions, five metres tall; from 1927 until 1943, the aquarium was home to this dinosaur's most imposing and exquisite descendant, a Komodo dragon named 'Moritz'.

The stairway featured large and colourful windows designed on the model of the art forms Haeckel had discerned in nature. The oval windows above the pools in the aquarium-hall displayed similar motifs. Two great murals in the stairwell showed how ocean creatures had evolved from prehistoric times until the present day. The corresponding pole was constituted by a fossilized ichthyosaurus from the environs of Holzmaden in Württemberg.[30] All in all, no expense had been spared in creating a spectacular, scientifically advanced and intellectually sophisticated monument to the wondrous origin of life in the water and to the heroic races that had vanished in the course of evolutionary history.

The Berlin Aquarium counted as one of the most beautiful and stylistically varied buildings in the world. It was a *Kulturinstitution*. 'One' went to the aquarium as one attended the theatre – in order to relax or for stimulation. That, at any rate, is how famous 'regulars' such as Josephine Baker (1906–75), Joachim Ringelnatz (1883–1934) and Max Pechstein (1881–1955) described it.[31] In the Second World War, it provided the stage for the flipside of modern miracles and scientific-technological progress: war and catastrophe. In the night of 23–4 November 1943, a bomb hit the crocodile hall dead centre and the Berlin Aquarium was completely destroyed. An eyewitness described the dying of the marine creatures, especially the giant snakes, crocodiles and alligators which writhed down the grand hall in pain, like a vision of Dante's inferno.[32]

The dying giant lizards are one of many images that symbolize how, in the twentieth century, the aquarium lost any pretension to natural 'innocence'. Henceforth, aquariums could no longer be viewed as artificial paradises or miniaturized oceans; power politics and economics – inscribed in them from the beginning – shone forth, plain as day. Now they appeared as precarious spaces: the result of a contingent and decontextualized process of selection and exemplary self-regulating systems (which, for all that, relied on sophisticated technological mechanisms of control). The crocodile hall – the first walkable 'biotope' – going up in flames presents a scene epitomizing the uncertainty of existence. Accordingly, it expresses the fantasies of insatiable hunger for power and imperialist designs, on the one hand, and, on the other, an appeal, along biblical and utopian lines, to assure the survival of our blue planet. Public aquariums no longer provide a stage for the dances of mermaids or a looking-glass for the opium-filled rêveries of narcissistic flâneurs. Now they serve the purpose of democratic, public instruction and stand as so many Noah's Arks made of glass. It proves impossible to detach aesthetic experience from awareness that humankind, now as then, knows little about the vast ocean, even if, for some hundred years now, it has been ravaged by instrumentalizing discourse and industry.

Gardens of the Ocean: The Aesthetics and Epistemology of the Home Aquarium

More or less in parallel to public enthusiasm for large-scale aquariums, the popularity of aquariums in laboratories and homes emerged.[33] Historians still disagree about whom to credit for their invention, but the following stands firm.

The idea that an equiblibrium exists between animal and plant life, based on chemical principles, has a long, interdisciplinary prehistory. Presumably, it was Joseph Priestley (1733–1805) who first determined that plants emit oxygen under certain conditions. Not long afterward, Jan Ingenhousz (1730–99) discovered that the leaves of water plants generate oxygen when exposed to sunlight; thus began research on photosynthesis.[34] Studies of water animals and plants by Martin Frobenius Ledermüller (1719–69) warrant mention in this context, too; in the 1760s, he succeeded in keeping them at a relatively stable equilibrium in vessels.[35] In France, approximately seventy years later, Charles DesMoulins (1798–1876) made a study of turbellaria he kept in a glass container and, in so doing, determined that plants are able to keep water clean. That said, the report in the *Actes de la Société Linnéenne de Bordeaux* 1830 failed to attract much attention.[36] His countryman Felix Dujardin (1801–60) applied DesMoulins's findings to his own research on marine animals; in 1838, he managed to preserve moderate stability in a saltwater aquarium. For all that, his media-technical 'invention' also passed largely unnoticed by the scientific community.

At around the same time, the French zoologist Jeannette Power de Villepreux (1794–1871) was researching argonauts – emperor nautiluses – in Messina (Sicily). She kept the delicate specimens in wooden boxes built expressly for this purpose; initially, they lay anchored at sea, but she later set them up in her laboratory near the coast and supplied them with seawater by means of simple pumping system. For the rest of her days, Power claimed to have invented the aquarium; indeed, the wooden boxes – *cages à la Power* – were an integral component of her systematic study of living organisms.[37] That said, because she had no awareness of the balance between oxygen and nitrogen that is necessary in aquariums, her boxes do not really qualify as prototypes.

It was in England that aquariums experienced a breakthrough and became the object of epistemic and aesthetic delight for the educated classes and others interested in natural history. This occurred through a felicitous concurrence of marine research, chemistry and horticulture, as well as other favourable circumstances such as the abolition of the glass tax in 1845 and easy railway access to coastal regions.

Early works known to us include the scrupulous investigations of Sir John Graham Dalyell (1775–1851), who kept various marine animals in individual glass tanks from about 1790 onwars.[38] He saw to it that the water was changed daily; to this end, sizeable quantities of seawater were delivered a good three times a week. In consequence, some specimens enjoyed a lifespan of eight to ten years; indeed, one anemone, affectionately named 'Granny' by members of the household, stood in the family's possession from 1828 to 1887 – outliving Sir John himself by thirty-six years.

Marine invertebrates in glass tanks also enlivened the garret of Edward Forbes, who, in the 1830s, conducted research on marine biology at the University of Edinburgh with his students John (1814–67) and Harry (1816–47) Goodsir and George E. Day (1815–72). Many enthusiastic and entertained visitors spread word of the new research method, even over the Atlantic. In 1842, in *History of British Sponges and Lithophytes,* Forbes's colleague George Johnston (1797–1855) reported an experiment that had successfully preserved little corals, a seastar and a few mussels in sealed glasses for eight weeks without undue damage.[39]

In 1846, Anna Thynne (1806–66), an avid geologist and the wife of the dean of Westminster Abbey, collected some hard corals (madrepore) from the tidepools at Torquay on the southern coast of England.[40] She wanted to take them to London for research purposes and had them sent in glasses packed with seawater via coach. Then, she transferred her wards into large glass tanks and fed them with tiny pieces of shrimp. Obtaining fresh seawater proved a problem, however. Since it was too expensive to send for new water every week, a maid was given the task of aerating the tanks every day – that is, the servant spent hours on end at an open window pouring water from one vessel into another an inconceivably time-consuming procedure.

In spring 1847, Thynne decided to create a more stable home for the marine creatures, and she had sea plants brought in from Torquay. In a few weeks' time, she determined that kelp was particularly well suited to maintaining the balance of oxygen and nitrogen in the tanks. Some two years later, Thynne was the proud owner of a fully equilibrated sea aquarium whose inhabitants had enjoyed robust health for three years. The matter of copyright becomes somewhat complicated at this juncture. Historically, there is no doubt that Anna Thynne was the first person to have made the aquarium into a world apart in miniature – and, moreover, to have used it for a program of experimentation. That said, she never called any of her seawater tanks an 'aquarium', nor did she publish anything about them. Additionally, she was basically interested in one organism alone, madrepore, and did not care about the epistemological implications of her arrangement. Rebecca Stott, Anna Thynne's biographer, has claimed that Anna has been denied due recognition for inventing the aquarium and makes the further charge that the individuals credited with this discovery were scheming plagiarists who stole the idea without so much as mentioning the poor woman in passing.[41]

To be sure, even educated women such as Anna Thynne had no access to the epistemic culture of the sciences of their day. All the same, however, the allegation of 'intellectual theft' must be proven.

It emerges from Thynne's notebooks that, for two or three months in spring 1849, she showed her invention to various different people – 'professed naturalists and other persons interested in natural history', as she writes.[42] Stott speculates that visitors included the London chemist Robert Warington (1807–67) and a figure we have already encountered: Philip Henry Gosse, the extremely success-ful author of works of popular science. Both of these men are viewed by many historians as the inventors of the aquarium, even though Gosse never claimed as much for himself. Their reports of experiments undertaken with aquariums begin in the early 1850s. From a purely chronological perspective, then, they might qualify as plagiarists. But for all that, Stott fails to offer any proof for her charge. She claims that natural scientists and amateurs flocked from all over the world to see Thynne's madrepores at Westminster Abbey, yet she provides no names. Thynne's menagerie is supposed to have been the 'talk of zoologists';[43] unfortunately, however, the reader never learns who said what, where, or when. In this context, it is significant that Gosse always documented his sources and the development of his ideas in painstaking detail. *The Aquarium: An Unveiling of the Wonders of the Deep Sea*, which appeared in 1854, devotes many pages to noting who performed experiments with water plants or animals first, what suc-cesses they charted and, finally, how the idea gained acceptance that the proper combination of flora and fauna would produce a balance of oxygen and nitro-gen. The book's second edition, which appeared in 1856, mentions Thynne, who, the same year, had given Gosse her notes on the reproductive life of the madrepores she studied so carefully. Gosse did not employ her manuscript for his own research, as Thynne had intended, but he saw to it that the prestigious *Annals and Magazine of Natural History* published her findings under the title, 'On the Increase of Madrepores'.[44] Finally, in light of Anne Thwaite's biogra-phy of Gosse, which is based on extensive research, it seems altogether unlikely that he would have suppressed, for years on end, the intoxicating sight of thirty marine creatures living in a saltwater aquarium at Westminster Abbey.[45]

At any rate, the key matter does not concern a criminal action and pirated inventions. Instead, what commands notice is the strange fact that the successes charted by Johnston and Thynne did *not* spread like wildfire. This is all the more remarkable given the fact that marine biology had become a hot topic among English scientists during the 1830s and 1840s. Between 1839 and 1850, the *Brit-ish Association for the Advancement of Science* sponsored numerous expeditions to dredge up marine organisms, and collections of specimens grew enormously from year to year.[46] For all that, there is no sign that efforts were made to keep the creatures alive in order to study them.

The decisive impulse to view the situation with fresh eyes seems to have come from the field of botanical research. In 1830, Nathaniel Bagshaw Ward (1791–1868), a botanist and surgeon, determined that plants known for their delicate nature, certain ferns and mosses, could flourish even when sealed in airtight, glass containers. Here, a stable microclimate developed which protected plants from polluted air and allowed them to grow without suffering from variations in outside temperature. Curiously, Ward did not share his findings until 1837, in a report for the *British Association for the Advancement of Science* that proposed that the 'economy' of air circulating between animals and plants be studied.[47] In 1842, he published *On the Growth of Plants in Closely Glazed Cases* and expressly declared 'that the animal and vegetable respirations might counterbalance each other'.[48] Independent of theoretical considerations, Ward's little greenhouses soon went into production. *Wardian Cases* became fashionable as the notorious *fern craze* raged; in no time at all, no household in Britain was complete without 'plumy emerald green pets glistening with health and beadings of warm dew'.[49]

Ultimately, it was not Ward who connected marine biology and the emergent field of physiological chemistry, but Robert Warington (1807–67), a member of the London *Society of Apothecaries*. Until now, 'aquarium owners' such as Dalyell, DesMoulins, Johnston and Thynne had been concerned primarily with creating conditions for the study of marine animals. Warington, on the other hand, sought to investigate the principles underlying these same conditions. In 1849, he inaugurated a process of systematic experimentation to understand how animals and plants interact underwater. He filled a fifty-litre tank about half way and put two goldfish inside; sand covered the bottom,

> and some loose fragments of limestone and sandstone, so arranged as to afford shelter and shade. A small specimen of *Valisneria spiralis* [common tape or eel grass] was at the same time planted in the mud, and kept in place by a stone. The whole was then left undisturbed.[50]

When Warington remarked that some of the older leaves on the eel grass had begun to rot, he added a few pond snails, which ate the decaying vegetation; additionally, their eggs provided sustenance to the fish.

> Thus the success of the experiment was established, and an Aquarium was formed in fresh water; which has continued to prosper to the present time; the animals and plants maintaining each other in healthy life, and the water preserving its purity unchanged.[51]

In 1852 – at about the same time that Gosse was doing the same – Warington began to experiment with tanks containing seawater. He reported on his findings in the *Annals of Natural History* in November 1853; Gosse had already done so in 1852.[52] Whereas Warington published his research only in scientific journals,[53] Gosse caused a sensation with his popular-scientific bestseller, *A Naturalist's Rambles on the Devonshire Coast* (1853). Both men have the merit of having coined the term 'aquarium'.

Prior to 1852, aquariums were mostly cylindrical glass containers with a few fish or invertebrates inside. After this date, they came to be viewed as the living space for a marine community and an idealized miniature landscape. In 1854,

Gosse published a book that proved especially rich in consequences, *The Aquarium: An Unveiling of the Wonders of the Deep Sea*, which triggered a veritable craze. A newspaper article from 1856 entitled 'The Aquarium Mania' declared:

> In London itself, the mania is raging just now at fever point ... In West End squares, in trim suburban villas, in crowded city thoroughfares, in the demure houses of little, unfrequented back streets, and inside the flat, silles windows of wretched Spitalfields and Bethnal Green, everywhere you see the aquarium in one form or another.[54]

Now, aquariums were fixed appointments in British homes.

> Since the British mind was all alive and trembling with that zoological fervour excited by the appearance of the hippopotamus in Regent's Park, no animal has touched it to such fine issues and such exuberant enthusiasm as the lovely Sea-Anemone, now the ornament of countless drawing room, studies, and back parlours ... at once pet, ornament and 'subject of dissection', the Sea Anemone has a well established popularity in the British family circle; having the advantage over the hippopotamus of being somewhat less expensive, and less troublesome, to keep.[55]

In sum, it is impossible to provide a definitive answer to the question who, exactly, came up with the idea of a self-sustaining aquarium. For decades, natural scientists had kept marine animals in water tanks at home. It seems the invention was 'in the air', as it were. In 1856, Emil Adolf Roßmäßler, one of the most well-known popularizers of natural history – and of the aquarium, in particular – wrote in the *Gartenlaube* that he hardly understood why the idea of an aquarium had not occured to him.[56]

At the same time, however, it is never satisfying to hear that the times were simply 'ripe' for a given 'discovery'. Standard historical accounts also foster the impression that, for the longest time, it had hardly interested anyone to keep fish. For people to lose their heads and start speaking of worms and spider crabs as 'mexikanische Prinzen' (Mexican princes),[57] it is necessary for imaginations to be fueled, and for a viable media-aesthetic design to develop and achieve widespread attention. Undoubtedly, the aquarium 'boom' occurred because of other factors, as well, such as the expansion of railways and the abolition of the glass tax in England in 1845. But for aquariums to be viewed as epistemic objects of the first order, it was necessary, above all, for their occupants to become attached to the intellectual currents and predilections of the day.

The Ocean-Lover and the Menagerie

In the early years of its history, the aquarium was commonly associated with metaphors of voyages charting unknown regions of the world.

> We can scarcely poke and pry for an hour among the rocks at low-water mark, or walk with an observant downcast eye along the beach after a gale, without finding some oddly-fashioned, suspicious-looking being, unlike any form of life that we have seen

before. The dark, concealed interior of the sea becomes thus invested with a fresh mystery; its vast recesses appear to be stored with all imaginable forms, and we are tempted to think there must be multitudes of living creatures whose very figure and structure have never yet been suspected.[58]

The treasures and curiosities drawn forth from these unexplored regions called for suitable presentation. The aquarium provided a ready medium for bringing the immeasurable and sublime sights of submarine worlds to the surface, and it transformed them into kaleidoscopic and intoxicating visions in miniature. The aquarium enriched the repertoire of scientific representation[59] and, moroever, connected in many ways to the cultures of display in an age characterized by the fetishism of objects. As a 'world behind glass', the ocean menagerie was tied to the realm of the rarefied and exquisite; it was surrounded by an aura of the sacred, in keeping with sentiments of natural religion that were widespread at the time.

What proves remarkable – almost paradoxical – is that aquariums did not house rare or unique objects as a rule. Unlike early modern cabinets of wonder and curiosity, they mostly contained native plants and animals. Nor were they the exclusive possession of the ruling class. Instead, they represented one of the first 'mass media' and stood in the service of newly popular practices of collecting, observing and exhibiting. As such, aquariums are to be viewed as part of the larger historical process whereby the conception of uniqueness achieved a new symbolic dimension; now, commonplace organisms could provide a sensation in practically any home. Uniqueness ceased to refer to a material quality, but was rather a matter of perception, or feeling, that resulted from modes of interaction and mise-en-scène:

> Uniqueness happens when objects are personalized in the privacy of someone's specific universe, whether it be an album, a room, or any individually articulated space ... Collectors establish a particular relation with their objects: no matter how common, an object can always be rescued from its apparent banality by the investment in it of personal meaning.[60]

It is not surprising, in this context, that one did not acquire an aquarium 'just like that'. It had to be earned – by way of patience, modesty and dedication, with refinement, *esprit* and artistic ability. The aquarium had come into being as the result of connections forged between images of the sea of long standing and observations made by collectors schooled in natural history. Their 'finds' were blessed by the spirit of discovery – *objets ambigus*, as Paul Valéry wrote half a century later of his shell collection.[61] Sea trumpets, horned helmets, abalones and tiger cowries were 'surreal hybrids' that dissolved the opposition between nature and art.[62] Accordingly, it required considerable subtlety and art, when putting together the community of animals and plants in this wonderland of erudition, to strike a harmonious balance between the paradoxical elements. As such, the process of readying and arranging an aquarium was caught up in a kind of pre-liminal dramaturgy; the 'explorer' went from novice to initiate in the course of discovering hidden mysteries. From the first steps they made on the shore, enthusiasts were surrounded by a polyphonous sea of discourse and entangled in a web of associations that modulated their attention, suggested impressions and guided the decisions they made.[63]

If one commanded the new practices of seeing, more than mere delight at the wonders of nature lay in store. One could count on a source – which would never run dry – of invigorating spectacles and boundless treasures. The world lay at the feet of an aristocratic conqueror:

> And what a shore! Precipitous walls and battlements of rock rise on each side, making a bay; before us, sharply-cut fragments of dark rock start out of the water for some distance. Every yard of ground here is a picture. The whole coast-line is twisted and waved about into a series of bays and creeks, each having a character of its own; and whether we stand on the Tors, and look along the coastor on the shore, and look up at the rocks, it is always some new aspect, something charming for the eye to rest upon.[64]

> These same rocks, objects of my former aversion, are now to me what the stairs of Holyrood are to the antiquarian, the Calton Hill to seekers after fine views, or, to expand the metaphor, each barren crag offers to me what quiet woods full of game, and fields abounding in rare flowers, offer to the sportsman and to the botanist. After a time it became a pleasure, almost bordering upon a passion, to be examining some crystal pool ... Even the commonest sea-weeds, once that I began to know something of their physiological wonders, assumed an importance they could never otherwise attain. Nor was there wanting, I hope, one of the most salutary results of initiation into any branch of knowledge upon the mind, opening it to a higher conception of the wonders of the universe.[65]

> A pleasant and healthy recreation is that of wading among the low rocks and seaweeds of the beach, regardless of shoes sopped in salt water, and occasional slips into ugly holes. Then is the time to decide whether we have 'eyes or no eyes', or whether we can perceive, as well as see, the interesting, the beautiful, the grand: the grand, in the towering cliffs on either hand and the expanse before us; the interesting and beautiful, in the several forms of active and passive life at our feet. Here, in the grotto-like hollows, are the little branching corallines and many-coloured seaweeds, spreading out their floating threads and ribbons to the light, and forming many a mimic landscape by their fanciful grouping; there, the larger algae hang in clusters over the rough blocks and tablets, sheltering perhaps some small mollusca, or some valorous crab who hides his body under it, all but the one claw with which he tries to terrify the intruder.
> If we turn up some of the stones, or look under the projecting ledges, we shall sometimes find, among other things, certain oddly shaped bodies having a leathery or fungus-like look about them; and, looking a little more carefully, we may meet with others transparent and star-like, all growing, as it were, on the rocks or surfaces of large weeds. These belong to an order of shell-less mollusca (if indeed they are rightly so classed) called 'Tunicata'.[66]

> So now we stand upon the froth-fringed margin of the sea, and at our feet the rippling waves 'Just kiss the shore, then sleep'.

A lady, the companion of our excursion, evidently inspired by the scene, is growing poetical, and exclaims, appropriately enough

"How various the shades of marine vegetation
Thrown here the rough flints and the pebbles among!
The feather'd Conferva, of deepest carnation, The dark purple Slake, and the
olive Sea-thong'.

Yes ! here they are scattered in rich variety; and among the multifarious assem-
blage we may likewise observe innumerable specimens of what the ladies call 'White
Sea-weed' and which are likewise known as 'Hornwrack' and 'Sea-mats', the FLUS-
TRA FOLIACEA of scientific authors.[67]

The day after a full or new moon, according to Gosse, offered the best opportunity for collecting water plants. In the pools exposed by the falling tide, the
extravagant majesty of different kinds of algae and kelp, as well as dahlia and
frilled anemones, could be gathered; specimens that were slender and winged,
feathery and finely wrought: *lovely, brilliant, luxurious, glowing, elegant, delicate, pretty, attractive, graceful, charming* and *tender*. Equipped with hammer and
chisel, glass containers and a basket, one ventured onto the rugged rocks and,
even hanging upside-down as circumstances dictated, extracted the sensitive
growths along with the stones they grew on. Just as quickly, the carefully packed
spoils were rushed home; then, one began setting up the aquarium.

Gosse recommended covering the bottom of the aquarium with an insulat-
ing layer of clay, then adding pebbles and sand and finally including small pieces
of rock to provide tiny bridges and hiding-places for the occupants to come.
Once the plants are installed, the miniature landscape is ready for fresh seawater.
Then, one waits for darkness to fall:

At night examine the sides of the bottle carefully with a pocket-lens, placing a candle
on the opposite side. The multitude of curious little creatures that will have crawled
out, and will be found mounting the walls of their prison, is quite surprising. Minute Mollusca, both bivalve and univalve, uncouthformed Crustacea, tiny Starfishes,
and expecially Annelida, will pretty certainly reward the investigator. The last-named
Class occurs in remarkable abundance and variety; while if, after you have gone round
the glass, noticing particularly the very edge of the surface-line, you pass your eye,
assisted by the lens, carefully over the surfaces of the bits of stone, you will probably
find many more creatures, such as tube-dwelling Annelides, the smaller Zoophytes,
and several species of the delicate Bryozoa.[68]

The following day proves decisive. One must be prepared for the first encounter
with the aquarium's inhabitants:

[It] is one of the great charms of natural history collecting, that you never know
what you may obtain at any moment. The expectation is always kept on the stretch;
something new, or at least unthought of, frequently strikes the eye, and keeps the
attention on the *qui vive*.[69]

[C]urious, varied, and abundant are the creatures he [the natural scientist] discovers'.[70]

Because the sea is a site of 'unknown and unimagined treasures',[71] Gosse reports having encountered some hundred animals in the tank on the evening of the first day. The initial cast presented a colourful array:

1 Fifteen-spined Stickleback (*Gasterosteus spinachia*)
1 Scrobicularia
7 Grey Mullet (young) (*Mugil capito*)
1 Corkwing (*Crenilabrus Cornubieus*)
1 5-beard Rockling (*Motella 5-cirrata*)
1 Great Pipefish (young) (*Syngnathus acus*)
1 Deep-nosed Pipe (*Syngnathus typhle*)
2 Worm Pipe (*Syngnathus lumbriciformis*)
2 Ashy Top (*Trochus cinerarius*)
1 Navel Do. (*Trochus umbilicatus*)
3 Common Periwinkle (*Littorina littorea*)
2 Common Cockle (*Cardium edule*)
2 Hermit Crab (*Pagurus bernhardus*)
3 White-lined Worm (*Nereis bilineata*)
2 Thick-horned Anemone (*Actinia crassicornis*)

1 Black Goby (*Gobius niger*)
1 Purple (*Purpura lapillus*)
3 Yellow Do. (*Littorina littoralis*)
1 Anomia
1 Do. (*Pagurus Prideauxii*)
2 Ascidia
5 Daisy (*Actinia bellis*)
1 Prawn (*Palæmon serratus*)
3 Crown worm (*Serpula triquetra*)
3 Weymouth Do. (*Actinia clavata*)
2 Parasitic Do. (*Actinia parasitica*)
4 Sand Shrimp (*Crangon vulgaris*)
6 Plumose Do. (*Actinia dianthus*)[72]

The group acclimated quickly and was joined, a week later, by further personnel – starfish and urchins, above all.[73] With that, the roles were filled in the main. From this point on, the task was to watch over and care for the delicate menagerie. Finally, after a few weeks, a wondrous event lay in store:

> [We] have accumulated a marvellous store, – almost enough, indeed, to set up a little provinical museum. Brittlestars and urchins; cucumbers great and small; bivalve and univalve mollusks; swollen ascidians, smooth and warty; active, shuffling, sucking fishes; heaps of mossy Bryozoa; long bristling tufts of Hydroid zoophytes, naked worms, twining and writhing amidst the mass, gleaming in purple and pearl; seamice, armed in gold', like Virgil's Orion; tangled masses of Serpulapipes, every one with its scarlet-crowned tenant; these, and multitudes of creatures besides, come up from the teeming sea-floor, and all at once claim our bewildered attention.[74]

And so, even though contemporary accounts tend to represent setting up an aquarium as a *rite de passage* that the nature lover experienced at the seashore, a flourishing trade soon developed. In short order, simple glass tanks assumed an array of forms, becoming finely worked appointments for the domestic salon with complicated designs; now, they might evoke the submarine universe with fountains, incorporate a birdcage or include luxuriant jungle flora. Some aquariums were coupled with aviaries, others presented a stately rectilinear or octagonal structure and still others, for veritable enthusiasts, occupied an entire wall.

Great temples of aquatic commerce opened in the European metropolises. They included *Lloyd's Aquarium Warehouse* in London. This establishment was surely the most impressive of its kind; some 15,000 animals are said to have been available for purchase every day, and fresh seawater could be bought on tap. Merely by existing, such stores drew in the public and fascinated passers-by:

> A crowd has gathered on the pavement outside the double glass shopfront of the Aquarium Warehouse, recently opened by William Alford Lloyd ... Faces, mouths open in astonishment and wonder, are reflected in the panels of window glass, through which the crowd stares at a range of aquariums, some domed like the glass jars used by taxidermists, others hexagonal or intricately laced with ironwork, some with wrought-iron stands, others with fountains, all fully stocked with rocks, gravel, seaweed and strangely shaped marine creatures, all seeming to perform for the crowd.[75]

For those who wanted the aesthetic pleasure of a home aquarium but did not wish to undertake such elaborate preparations, there existed the possibility of ordering, by catalogue, exquisitely wrought denizens of the sea made out of glass. The famous figurines manufactured by the Dresden firm of Leopold (1822–95) and Rudolf Blaschka (1857–1939) were sent all over the world; the artistry employed evoked the greatest achievements of Baroque cabinets of wonder.

Accounts differ as to how long the aquarium-craze lasted. There was surely marked variation from region to region, but research indicates that a quick decline occurred. It might seem that aquariums counted as *passé* after 1880. Such a view is too superficial, however, inasmuch as it only takes stock of manic enthusiasm and overlooks the fact that when it subsided, aquariums had been established as epistemic and aesthetic objects. As the nineteenth century drew to a close, aquariums stood on firm scientific footing and also represented a solid component of visual entertainment culture. The First World War interrupted the inaugural wave of public aquariums; trade in aquarium supplies, on the other hand, has flourished to this day. 1882 witnessed the founding of the first association for aquarium aficionados in Germany (Gotha), aptly christened *Aquarium*. In 1893, the *Humboldt* followed in Hamburg. In 1895, the *Nymphaea alba* was founded in Berlin; the next year, it was joined, in the same city, by *Triton*, which remained the largest organization of its kind for decades. The first German magazine devoted to the subject had begun publication in 1890, *Blätter für Aquarien- und Terrarienfreunde*. Some seven thousand visitors attended the exhibition organized by the *Verein der Aquarien und Terrarienliebhaber* at the *Grand Hotel Alexanderplatz*, which included exhibitors from Austria, Bavaria, Saxony and Thuringia – as well as the steering committee of the Ichthyological Division of the Russian Society for Acclimation in Moscow.

Who, Paul Valéry asked, has not sought to penetrate the ocean depths in spirit? In this sense, the aquarium is a dream of the ocean given material form. Aquariums stand among the most successful medializations of the sea; the number of visitors who show up whenever an enormous oceanarium is opened today – as occurs time and again – reflect as much. According to figures provided by the World Watch

Institute in Washington, some five to six hundred million fish are fished out of the sea yearly to staff these theatres made of glass. As early as 1907, Gosse's son, Edmund, already noted the disastrous effect of this self-service mentality:

> These rockbasins, fringed by corallines, filled with still water almost as pellucid as the upper air itself, thronged with beautiful sensitive forms of life, they exist no longer, they are all profaned, and emptied, and vulgarized. An army of 'collectors' has passed over them, and ravaged every corner of them. The fairy paradise has been violated, the exquisite product of centuries of natural selection has been crushed under the rough paw of well-meaning idle-minded curiosity. That my Father, himself so conservative, had by the popularity of his books acquired the direct responsibility for a calamity that he had never anticipated became clear enough to himself before many years passed, and cost him great chagrin.[76]

Evenings at the Microscope

> To open the path to the myriad wonders of creation, which, altogether unseen by the unassisted eye, are made cognisable to sight by the aid of the Microscope, is the aim and scope of this volume. Great and gorgeous as is the display of Divine power and wisdom in the things that are seen of all, it may safely be affirmed that a far more extensive prospect of these glories lay unheeded and unknown till the optician's art revealed it. Like work of some mighty genie of Oriental fable, the brazen tube is the key that unlocks a world of wonder and beauty before invisible, which one who has once gazed upon it can never forget, and never cease to admire ... The author has swept rapidly across the vast field of marvels.[77]

Aquariums provided aesthetic objects and, at the same time, allowed scientists and laymen alike to come to understand ocean-dwellers fully, at their 'stillen, geheimnisvollen Arbeit' (silent, secretive work).[78] To this voyeuristic end, the key instrument was the microscope. Under its lens, the most banal objects transformed into precious gems and breathtaking works of art: 'I am absolutely filled with wonder and in an ecstasy of deligth at the structure and contrivance of some of the extremely minute species ... !' exclaimed John Eddowes Bowman, an avid 'gentleman scientist', whose collection of ferns and fossiles still constitutes an important component of the holdings at Oxford, Manchester, Kew and Merseyside today.[79]

But for all that, the microscope did not occasion only curious and delighted amazement; it afforded unsettling discoveries, too. On the one hand, it revealed delightful curiosities and the sublime textures of diatoms and radiolarians; on the other hand, the microscopic environments contained bizarre, menacing monstrosities. Particular interest attached to seemingly hybrid entities, so-called intermediate forms, which subverted the order of nature and appeared to occupy a position between animal and plant, for instance. Such beings gave rise to questions that proved central to the nineteenth century: What is life? Where does it come from? How did the multiplicity of species arise? How and where does spirit pass to matter?

Underwater, the confusion proved especially great. Aquariums offered a the-atre of exotic forms that were perpetually changing, a continuum of the strangest metamorphoses and paradoxical symbioses – a 'freak show', so to speak, of sensa-tional modes of reproduction and interaction. No one seemed able to understand these enigmatic relations between sexes and states of being, much less explain them. The 'hard facts' often heightened confusion more than they suggested a fixed pattern. A famous example is the acorn barnacle (*Balanus balanoides*), whose life – and reproductive cycle was first described by Darwin himself.

Acorn barnacles are hermaphroditic, although they cannot fertilize them-selves. They live in dense colonies, so that the penis, which is approximately twice as long as the body, can inseminate neighbours. The fertilized eggs remain within the calcite shell until the young hatch and swim free as nauplius larvae (the first larval stage). A few weeks later, they become cyprid larvae and the animals seek out a spot at which to spend the rest of their lives. To this end, the acorn barna-cle attaches itself, upside-down, to a mussel or crab shell, the hull of a ship or a snail shell and cements itself fast. After doing so, the creature sheds several times within its carapace in order to become adult; in the process, it generates six calcite plates that form a cone-shaped 'house' with a lid. Needless to say, such compli-cated transformation, displayed by a single animal, provoked general excitement:

> Marvellous indeed are these facts. If such changes as these, or anything approaching to them, took place in the history of some familiar domestic animal; – if the horse, for instance, was invariably born under the form of a fish, passed through several modi-fications of this form, imitating the shape of the perch, then the pike, then the eel, by successive castings of its skin; then by another shift appeared as a bird, and then, glue-ing itself by its forehead to some stone, with its feet in the air, threw off its covering once more, and became a foal, which then gradually grew into a horse; – or if some veracious traveller, some Livingstone or Barth, were to tell us that such processes were the invariable conditions under which some beast of burden largely used in the centre of Africa passed, – should we not think them very wonderful? Yet they would not be a whit more wonderful in this supposed case than in the case of the Barnacle, in whose history they are constantly exhibited in millions of individuals, and have been for ages, – even in creatures so common that we cannot take a walk beneath our sea-cliffs, without treading on them by hundreds![80]

As the example of the acorn barnacle demonstrates, the taxonomy of individual body parts posed a delicate problem for the understanding of nature. Thus, Darwin's description states that the adult animal 'glue[s] itself by its forehead to some stone' while its 'feet' remain free in the water. During its larval stage, in turn, its 'mouth' is situated between its legs; a second 'eye' is located above the stomach inside the body.[81] Needless to say, when one seeks to apply concepts of human anatomy such as feet, arms, eyes, mouths to marine invertebrates, incongruous images result.[82] Instead of discovering homologous principles or an 'original' anatomical form, the visual culture of marine research teemed with bizarre palimpsests of mouths and strange bodily apertures – a radically mocking, obscene and anarchic welter of 'living grotesques'.[83] In consequence, the acorn barnacle's anatomy and reproductive process yielded a burlesque comedy with disquieting sexual implications. One way of dealing with the problem – which Darwin employed, too – was to play down the obscenity with an anthropomorphizing wink. For example, in a letter to Charles Lyell dated 14 September 1849, he wrote of a polygamous specimen:

> The other day I got a curious example of a unisexual instead of a hermaphrodite cir-ripede, in which the female had the common cirripedal character, and in the two halves of her shell had two little pockets, in each of which she kept a little husband; I do not know of any other case where a female invariably has two husbands.[84]

Two 'little husbands' in two 'little pockets'. This scene of cozy family life sounds much less threatening than an image of immobile, parasitic masculinity – a widespread phenomenon underwater, which gave rise to considerable unease among nineteenth-century land dwellers. On the one hand, invertebrate marine animals, embodying the fragility of biological and anatomical borders, were precisely what proved central for the development of evolutionary theory; on the other hand, their manifold forms and metamorphoses is what made them, after the ape, into the chief icon of prospective degeneration.

A second mode of handling the grotesquely physical nature of many denizens of the deep was to relate them to past epochs and thereby transfer them to new registers of discourse. Cnidarians, entoprocta and tardigrades mutated into adventuresome creatures of wonder, such as had once inhabited the outer reaches of the known world in the travel accounts of John Mandeville, Gervase of Tilbury and Marco Polo.

> [T]he Periwinkle (*L. littorea*) [is] marching soberly along beneath his massive mansion, stopping to munch the tender shoot of some Algae, or leisurely circumambulating the pretty tide-pool which he has chosen for his present residence. You may tell that all his movements are marked by gravity and deliberation, for if he does not let the grass grow under his feet (I beg his pardon, he has but one foot; though, as that is somewhat of the amplest, he is not deficient in understanding) he lets it grow over his head. It is with goodly Ulva or other sea-weed that has taken root on the summit of his shell, so that he habitually sits under the shadow of his own roof-tree.[85]

The similarity between the common periwinkle with the one-legged fellow from Mandeville's universe, who uses his leg as an umbrella, is unmistakeable. The practicing of referring to marine organisms as 'foreign peoples' was also widespread. In many cases, ocean regions were compared to the wild, unmapped areas of Africa. Even a tidepool on the southern coast of England could turn into an African jungle:

> Its upper side is a whole forest of see-weeds. Large and small; and that forest, if you examined it closely, as full of inhabitants as those of the Amazon or the Gambia ... Countless ages before we appeared on earth the depths of the old chalk-ocean teemed with forms as beautiful and perfect as those, their lineal descendants, which the dredge now brings up from the Atlantic sea-floor.[86]

The rhetorical move at work – as in many other accounts of the day – rests on the equation of 'primitive' and 'prehistoric'. The 'primitive peoples' inhabiting the sea stand in for the natives of Africa and East Asia, whom anthropological discourse viewed as the prototypical ancestors of Westerners. The study of 'primitive' peoples and their cultures was supposed to open understanding of a past that the Europeans had long left behind. Analogously, peering into the aquarium was discussed as the unveiling of a prehistoric landscape, on which the firstborn of creation wandered, and of an unknown continent whose primitive cultures were to be studied and colonized. Time and again, oceanographic works stressed the vast range of strange physical attributes; in particular, mouths and genitals received attention. The monstrous appetite of sea creatures was remarked above all – corresponding to colonialistic narratives that featured motifs of cannibalism and vampirism to describe what was other, foreign and exotic.[87] Kingsley mentions the orifices of marine organisms over forty times. For example, his description of a madrepore's mouth and digestive process:

> [A]t last the gentle creature, after swallowing and disgorging various large pieces of shell-fish, found viands to its taste in 'lean of cooked meat and portions of earth-worms', filling up the intervals by a perpetual dessert of microscopic animalcules, whirled into that lovely avernus, its mouth, by the currents of the delicate ciliae which clothe every tentacle. The fact is, that the Madrepore, like those glorious sea-anemones whose living flowers stud every pool, is by profession a scavenger and a feeder on carrion; and being as useful as he is beautiful, really comes under the rule which he seems at first to break, that handsome is who handsome does.[88]

From the 1860s on, the imperialistic note became more pronounced in marine literature, albeit without predominating entirely. Just as often, one might read a theatricalized account of ocean life as an anthropomorphic comedy played by exalted performance artists. Compare the following descriptions of an anemone:

> It is under the veil of night that the Anemones in general expand most readily and fully ... While the glare of day is upon them, they are often chary of displaying their blossom beauties, but an hour of darkness will often suffice to overcome the reluctance of the coyest.[89]

> [The anemone] has no pretty ways to captivate our hearts – a mere drawing-room beauty, large, lazy, lymphatic and unintellectual.[90]

The design and understanding of the world, as presented in nineteenth-century aquariums, bore carnivalesque traits à la Bakhtin. In contrast to both classical conceptions of the body and the naturalistic, individualized forms of early modernity, grotesque physicality blurs the borders between the world and its inhabitants by making them mobile and hyperbolic to the point of virtual madness. Phenomena are shown in the course of transformation and metamorphosis, coming into being and vanishing again. Two bodies appear as one; alternately, each body brings forth the other. From this perspective, the aquarium, which had promised a clear view through its glass walls, reflected a world exactly matching the artistic logic of the grotesque. As it arranged the field of vision, the aquarium presented the ocean realm as a sphere where differences disappeared and transitory phenomena achieved paradoxical stability; everything seemed assembled for an illustrious feast and yielded an opulent *tableau vivant*. Here, creatures evidently thrived on being indeterminate and incapable of remaining still – on flowing transitions that actually made the extreme their goal. The underwater obscenities, contradictory couplings and seemingly impossible forms revealed, to a rapt public, the relativity of their own circumstances and the bourgeois world. Scientific observation and the accounts offered in popular journals depicted a carnivalesque inventory of grotesque anatomies; here, the fundamental distinctions between life and death, male and female, yielded to a submarine poetics of metamorphosis. Blurring standing boundaries and dissolving discrete forms, the protagonists of aquariums took the stage as 'grotesque bodies' that cycled through all the variants of corporeal integrity – and the limits which define it. At the mid-nineteenth century, literature and visual culture were altogether fascinated by the interplay of natural and artificial elements. The love life of candy-coloured bristle worms, say, or the meals of 'green-haired' urchins brought out the aesthetic component of scientific observation in alluring, capricious and provocative manner.

Aquariums played the part of highly curious objects in the epistemological discourse of the late-nineteenth and early-twentieth centuries. Any knowledge they afforded also contained an aesthetic dimension; they were *the* key medium of the age, promising to fulfill the dream of lifting the green veil of the waves in order to reveal the depths of the sea. As such, aquariums were employed both for experiments and as mobile laboratories; indeed, it seemed they represented the paragon of the scientific method: a little world apart, where researchers could vary events at will. On the other hand – and at the same time – they were not just epistemic objects and sites for generating evidence, but rather subversive media-aesthetic designs of fantasy: an intoxicating kaleidoscope of the imaginary and a room for projecting jet-black thoughts. As a libidinous technology of observation and (re)production, aquariums transformed their occupants from items of study into poetic objects, *objets ambigus* or *objets à réaction poetique*,[91] serving invention and inspiration alike by offering a source of enigma. The impression that one was witnessing scenes of a world underwater caused scientific perception and aesthetic experience to merge. Often without even noticing as much, famous scientists and educated amateurs, time and again, would whisper in hushed tones of awe as they sat before their little tanks.

Tableaux Vivants: On the Mediality of the Aquarium

Aquariums were new, image-generating media in an age marked as a whole by the multiplication of new image technologies. As part of a new mode of seeing, they contributed to the transformation of the visual unconscious in Western culture and shaped the framework of perception. Specifically, they effected a shift from static images to dynamic arrangements, presenting a world of water, glass, pumps and living bodies in contrast to the tradition of optical projections and arts of illusion. In so doing, they were of a piece with the structural changes that occurred as a fixed and feudal world changed into a universe of motion and speed. Telegraphs and railways had radically changed the organization of space; what formerly appeared solid seemed to become immaterial and flowing. The arts – especially Impressionism – responded by breaking down defined contours and making particles of colour float through the atmosphere of pictorial space. Just as frequently, one encounters metaphors of currents and liquid motion in the epoch's efforts at self-understanding. 'All is streaming, flowing, in transition'; the aquarium epitomized the shimmering changes in perception like no other medium.

The aquarium presented the underwater world as a symbol for the flux of matter and sensation. The submarine perspective involved a complex cultural construct: observational space that mobilized the gaze on the basis of a mise-en-scène which was both theatrical and cinematic. As set designs of modern media and technology, aquariums offered the ocean world as the actualization of one of many options, as the continuous production of possibilities. Multidimensional processes of imitation, illusion, fiction and simulation gave rise to artificial worlds that made a central aspect of scientific research visible in a particular way, to wit, the 'Ästhetik des Vorgehens' (aesthetics of procedure)[92] as well as the medial construction of spaces and objects of insight that it entails. In this context, the submarine world posed questions about relations between knowledge, science and aesthetics, in addition to the role that emotional, sensory and somatic experience play in the production of knowledge. As holds for media in general, aquariums involve a heightened form of experience. They condense and multiply what, supposedly, is 'merely' observed into lived and stimulating reality; here, natural and medial space combine – or dissolve – in such a way that standing differences between reality and fiction are levelled. *One* consequence is that habits of watching detach from the panorama-perspective fixed by modern orders of observation and epistemology; they start to 'float' – or, alternately, 'dive'. Jean-Luc Nancy has called this experience the 'Fluidität des Tauchens' (fluidity of diving) – 'ein Gefühl der Immersion' (the experience of immersion) at eye-level with the deep.[93]

Cinéma avant la lettre: The Aquarium as a Moving Image

Aquariums offered a new means of image-production. By presenting condensed images of motion, they created a hybrid, electrifying form of reality in which the spaces of observation and representation intersected. This experience, which could prove inspiring or disquieting, essentially derived from the verticalization that both scientific and artistic modes of attention underwent in the course of the nineteenth century. As a simulation of the ocean floor, the aquarium seemed to afford the observer a standpoint 'under the water', at eye-level with the unknown, strangely stripped of its actual, empirical location.[94] The significance and cultural reach of such virtual immersion are evident in the way that genre painting and scientific illustration adopted authentically maritime perspectives. Following the work of Martin Rudwick, Stephen Jay Gould has remarked that appreciation for aquariums increased in tandem with the popularization of underwater views.[95] That said, the innovative dimension of this mode of visualization appears only in historical retrospect. Today, it seems obvious that the underwater world can only be depicted as being underwater. However, prior to the advent of the aquarium, the standard practice of scientific representation was to arrange marine animals outside of, or above, the water – that is, to show them from the perspective of an observer standing on the shore. Needless to say, it was quite difficult to adopt a standpoint that did not match conditions of actual human existence; interestingly, however, this same, 'non-human' perspective has come to represent the only natural one since it was first introduced. For the purposes at hand, the point is that aquariums generated a space for investigation that possessed phenomenological autonomy from the inception. Nor did this occur simply because the observer was now 'deprived' of a fixed position and the visual axis plunged vertically, into the deep. Just as importantly, perception underwent temporalization – a process that connects the aesthetics of the aquarium with the way that images were experienced in the arts preceding the cinema. The play of colour and motion in aquariums involved the experience of time, above all; as such, they fit seamlessly with nineteenth-century efforts in the arts and sciences to sound the depths of motion, in its many dimensions, as the universal principle of life. Aquariums were also called *tableaux vivants*, living polyramas. On the one hand, contemplating the image entailed the loss of self-governed time; on the other, what was seen yielded dynamic continuity. Together, they unfolded as cinematic magic *avant la lettre*. To this very day, the elementary sense of amazement that occurs when an image moves represents the decisive connection between the visual culture of film and that of the aquarium.

The cinematic quality of the aquarium suited the great *mysterium* of early fascination with the sea perfectly: its power of metamorphosis. It was heightened by the feature, inherent in every moving image, of being ungraspable. Accordingly, the visions offered by aquariums could induce states bordering on hypnosis. It is reasonable to assume that the 'ocean on the table' could only be partially described as a mobile research-station or a cross section of nature under glass. More than all else, it amounted to the aesthetic mise-en-scène of cinematic dynamism: an infinitely 'blank' space – not a place at all – which, as soon as it became perceptible to the human senses, transformed into a field of play for images and the imagination.

Etienne-Jules Marey

In this light, it is hardly surprising that the history of the 'subaquatic cosmos of pictures' was closely tied to the development of film.[96] The connection between the underwater image and the moving image was established in the work of the French doctor and physiologist Etienne-Jules Marey (1830–1904). Following his attentive study of how human beings, horses, birds and projectiles move, Marey, at the end of the 1880s, turned to marine organisms, ocean waves and currents in the water and air. He founded an experimental institute for hydrodynamic research in Naples in 1890; inspired by Anton Dohrn[97] and the culture of experimentation at the *Stazione Zoologica*, he had an aquarium built into the outside wall of his villa in the neighbourhood of Posillipo. Here, against a luminous background, he could contemplate the complicated sequences of movement described by the grey-black silhouettes of jellyfish, sea spiders, feather stars and seahorses, or study the infinitely delicate, swimming 'flight' of rays.[98] In fact, it was practically a matter of cinematic 'exposures'; unlike the chronophotographic method – serial snapshots for recording animals, plants and human beings – Marey used strips of photosensitive paper and transparent film for his work with the aquarium. The sequence of images that resulted could be passed in front of a projector to offer the impression, if only for a few seconds, of a fish in motion.

Between 1890 and 1892, Marey recorded various invertebrates and fish; seven of the sequences are preserved at the *Cinémathèque française*. They number among the earliest examples of film, and they are strictly scientific; they were not conceived for the broader public, but meant to enrich marine research by providing studies of currents and movement. They exemplify an art of observation supported by technological means, which brought forth unprecedented views laden with epistemic and aesthetic effects. Marey employed the camera exclusively as a scientific instrument for disclosing the facts of reality inaccessible to the human eye. For him, experimentation involved visualization first and foremost; accordingly, what his experiments generated was, above all, 'ein visuelles Wissen, ein *savoir voir*'.[99]

It was widespread practice in Marey's day to feel one's way forward visually, to produce *expériences pour voir*.[100] In consequence, points of interference arose where art and science met up – even if Marey never acknowledged as much. He sought to break down movement into a mass of equal and discrete units, thereby transforming the flow of time into a sequence in space. But the language of phenomena is not always as clear as Marey wished.[101] Inasmuch as he encountered movements of marine organisms as moving images, movement itself increasingly proved a 'wonderful fact'.[102] Many of Marey's contemporaries experienced his 'decomposition' of time, space and perception in a manner wholly different than the one he intended. They did not view the matter in positivistic terms; instead, it gave them the opportunity to approach new and foreign realities – that of images and imagination, for example, or the subjective nature of temporal perception.[103]

Thus, even though he thought he was performing another disciplinary function entirely, Marey exercised an immense effect on the development of film and the visual arts. In the second half of the century, his work played a decisive role in the advance of the underwater world – which sublimated so many different aspects of life and experience – to the preferred medium for imagination, dreaming and unconscious visual phenomena. Simultaneously – and possibly without being remarked – the aquarium entered the history of cinematic perception as the model for a 'fluid universe': a reality that proves much more mobile than what the senses readily perceive.

The Aquarium as a Media-technological Apparatus of Naturalness

And so, from the first, the aquarium proved an influential medium for giving shape to the moving image. A pane of glass always separated the world it presented and the observer, yet it also connected the two spheres. One may consider this situation symptomatic for the emergence of new positions of power bearing on visuality. Marey's contemporaries marvelled at the continuously changing views and mysterious arrangement of light; as a result, the aquarium quickly secured a central place both in the tradition of pictorial entertainment and in the technologically mediated regime of seeing. Observers spoke of a new kind of theatre, a *laterna magica,* or the stage of an opera; aquariums were liked to panoramas and dioramas, and the public delighted in its magical, fairytale effects. In brief: aquariums were viewed in analogy to all the technological and artistic achievements distinguished by the generation of bright illusions and spectacular *amusement.*

It is all the more surprising, then, that aquariums were also viewed as natural habitats. As we have seen, they symbolized – and more than any other media-technological dispositive – the emergence of a new paradigm for investigating, and experimenting with, processes of life. One reads in contemporary accounts over and over that it was now possible to bring to the surface, unveil and decipher the mysteries that had hitherto lain inaccessible in the ocean deep; indeed, the visual records provided by researchers such as Marey seemed to double the 'revelation'. However, as Wolfgang Schivelbusch has demonstrated, glass restores the aura that has gone missing. By nature, it asserts its present between inside and outside spheres and beckons observers to extend their vision beyond rational limits.

On this score, the visual culture of aquariums connects with the aesthetics of the vast glass palaces at world exhibitions, and with the striking window displays of vast department stores and arcades.[104] Moreover, from 1900 on, and by way of the metaphor furnished by metropolitan shopping centres, aquariums met up with the cinema, which evolved in keeping with impulses that often proved contradictory: on the one hand, it was held to open a perspective on reality that existed independently; on the other hand, it clearly offered sights that have no existence apart from what we perceive. Given the clamour of such varied and paradoxically modulated lines of discourse, people failed to notice that their exquisite glass houses afforded neither a primal scene of life's origins nor a representative sample of the ocean world. There were no currents, seasons, migrations, schools of fish, pressure differences or complex food chains; nor were there elaborate processes of growth and reproduction. Nor does nature have glass walls, pumps, fixed mealtimes or gawking visitors. Topics of lively debate such as natural selection and the survival of the fittest went missing entirely – the very matters that were supposed to come to light by viewing the origin and development of life in the aquarium and the behaviour of its occupants. The *tableaux vivants* – living pictures – were not supposed to turn into a battlefield; instead of natural selection, the artificial ecology of an idyll ruled the scene. But who or what was one actually observing, then?

To sum matters up, we may affirm that aquariums represented a site of aesthetic confusion for quite some time. Otherwise, it is impossible to explain the paradoxical statements that occur in scientific studies, for example, references to 'fairy castles' and 'Mexican princes'. What is more, research praxis was attended by contradictory assessments: on the one hand, it was a matter of shaping the pictorial space that the aquarium provided – a fun and creative mode of interaction readily understood in today's culture of digital images; on the other hand, the aquarium was supposed to yield a self-regulating system representing natural conditions of marine life. To be sure, isolated voices raised objections – for example, Emil Otto Zacharias (1846–1916), who founded the *Biologische Forschungsstation für Süßwasserforschung am Plöner See* (Research Station for the Study of Freshwater Biology at Plön Lake)[105] and pioneered limnology, pointed out the clear difference between conditions in aquariums and bodies of water in nature. But all the same, the prevailing view until well into the twentieth century held that animals in captivity behaved just as they would in the wild. Only later, by subsequent generations of researchers – including Oscar Heinroth (1871–1945), the director of the second Berlin Aquarium and the founder of ethology, as well as behavioural scientists such as Konrad Lorenz (1903–89), Niko Tinbergen (1907–88) and Irenäus Eibl-Eibesfeld (b.1928) – was the constructedness of the aquarium duly noted: the fact that it occupies a space somewhere between nature and culture and the influence that the observer and his position have on phenomena under examination. Lorenz formulated the matter in concrete terms:

Two aquaria, separated from each other by only a few inches, have individual characters just as sharply defined as two lakes many miles apart. That is the attractive
part about a new aquarium. When one is setting it up, one never knows how it will
develop and what it will look like by the time it has reached its own particular stage of
equilibrium. Suppose that one establishes, at the same time, and with the same inor-
ganic material, three containers which one places close together on the same stand
... : in the first, a dense jungle ... may soon be flourishing ... in the second the opposite may take place and, in the third, the plants may harmonize, and apparently from
nothing there may spring a perfect life community.[106]

'The aquarium is a world',[107] Lorenz fitly remarked; after receiving the Nobel
Prize (with Tinbergen and Karl von Frisch (1886–1982)) for research on the
organization and causes of individual and social patterns of behaviour, he used
the money to install a tropical saltwater aquarium at his home in Lower Austria.
He would often go swimming in it himself and took great pleasure in examining
the fish up close, without any immediate epistemological concerns.

Evolutionary Dreams, Spirit Animals, and Nightmares: The Aquarium and the Arts

Aquariums opened an echo chamber for social, scientific and artistic issues. Their
ubiquity assured that they became a vital source of fantastic motifs connected
to the deeps of the sea and images of floating and weightlessness. In particular,
they gave rise to ideas and theories that addressed states and dynamics of transition – for example, points of contact and exchange between the physical and
psychic realms or the material and the spiritual worlds. That is, aquariums not
only promised an aesthetic perspective on what was 'out there' in nature, but also
functioned as windows that opened onto the 'inside'. In consequence, they generated a visual culture of inexplicable, miniature worlds steeped in dream-like,
sometimes nearly psychedelic colours, which, in their encapsulated self-suffi-
ciency, seemed to adhere to a kind of narcissistic aesthetic. This cosmos of images
yielded one of the central models of *fin de siècle* art;[108] in this setting, aquariums
were used to create allegories of invisible inner life, dreams, exhilarating mental
states and the workings of the unconscious mind. During the Enlightenment,
such states had been viewed in opposition to the mind and reason; now, they
came to constitute fields of scientific research in their own right. In the aquarium
they fused with Darwin's theory of evolution, new microbiological experiments,
and the latest reports from great deep-sea expeditions. In this context, the diverse
and heterogeneous discourses that came to bear on aquariums are less remarkable than the unique aesthetic symbiosis that occurred, joining the mysteries of
life in artificial worlds with mysteries in the depths of the human soul.

The Deep, the Unconscious and the Dream

> Thus, European culture is inventing for itself a depth in which what matters is no longer identities, distinctive characters, permanent tables with all their possible paths and routes, but great hidden forces developed on the basis of their primitive and inaccessible nucleus, origin, causality, and history.[109]

Delving into the depths of the ocean, one does not simply abandon familiar sights; one also encounters a world with a different logic. In this respect, experiencing the ocean is similar to life lived in the realm of dreams – to states of exhilaration, transport and even mania; it resembles times when we are 'not ourselves', erotic ecstasy, and the unconscious realm as described by poetry and psychoanalysis.

The Abyss of Self-Contemplation

In general, the profundity of the unconscious mind is associated with the name of Sigmund Freud.[110] The psychoanalytic project of bringing a repressed truth to light seems to connect naturally to sounding hidden recesses. The phrase 'depth psychology', which Eugen Bleuler (1857–1939) coined to describe Freudian theory in 1910, makes as much plain. This terminology – which Freud himself adopted subsequently – refers to two things: first, the matter of relating the patient's present problems to what happened in childhood, and second, the analyst's task of penetrating unconscious psychic dimensions. 'Depth' represents a key feature of unconscious processes, referring both to remote time and to decisive, effective intervention 'at the root' of disturbances. In a general sense, it involves cognitive mechanisms of soul-searching and self-exploration that lead to the authentic experience of what stands at the core of one's personal being.

Freud did not invent the idea of inner processes located in deep seclusion, however. The notion goes back to antiquity. The earliest cultural records available to us already feature 'depth' as a synonym for what is inestimable, difficult to explore and concealed. Freud's key essay, *The Uncanny* (*Das Unheimliche*), which was published in 1919, might seem to suggest that many psychoanalytic concepts derive from Romanticism, but if so, the depth of the soul is not one of them. Among others, Georg Christoph Lichtenberg (1742–99) had viewed depth as a metaphor for the soul; likewise, Johann Gottlieb Herder (1744–1803) had compared the 'spiritual' element to a 'sea of depth, where wave upon wave stirs'.[111] It forms the 'whole dark foundation of our soul, in whose unfathomable depths unknown forces sleep like unborn kings',[112] he wrote, in what may be the first evocation of the unconscious. Still, Romanticism did introduce a scientific foundation for musings on, and 'investigations' of, depth. For example, Gotthilf Heinrich von Schubert (1780–1860), in *Ansichten von der Nachtseite der Naturwissenschaft* (Views of the Dark Side of Natural Science) (1808), described psychological anomalies – for example, somnambulism, clairvoyance and ecstatic and magnetic states – as the persistence of prehistoric conditions, which he related to contemporary discussions of 'deep time'. It is no accident that mining and subterranean journeys provide a central motif of Romantic literature and thought; such descents, which reach deep into nature, afford new perspectives on the earth's geological eras.[113] If fossils provide evidence of archaic life forms, emotions perceived deep in the inner world of human beings corresponded to a state of mystic harmony with nature outside of time. The Romantic effort to uncover a new dimension in the ego and extract real knowledge was closely tied to contemporary geognosy, which studied deep time as recorded by geological layers of the earth. In either case, the point was to 'rediscover' inner and prehistoric worlds – the 'cosmos' that stretched far below and far within. In what is probably the most famous articulation of the Romantic project, Novalis wrote:

> We dream about travelling through the universe – but is not the universe within ourselves? The depths of our spirit are unknown to us – the mysterious way leads inwards. Eternity with its worlds ... is in ourselves or nowhere.[114]

Needless to say, the path to the unconscious had not yet been charted by psychoanalysis. The inner depths were not held to have resulted from repression enacted by the ego; they existed in analogy to geological formations shaped by catastrophes and dramatic events in the earth's history. Whether metaphorical or material, floods, maelstroms, chasms, earthquakes and ice ages had created rises and falls; what had sunken into oblivion needed to be brought to light. Until the late-1880s, phrases such as 'the depth of infinite space', 'the invisible', 'the unknown' and 'the unknown force' were used to describe aspects of the unconscious, which finally received a theoretical articulation at the end of the century.

Nocturnes – The Poetry of the Aquarium

In Romanticism, the unconscious – in the broad sense of a hidden psychic dimension of depth – entered an aesthetic alliance with contemplating the ocean. Heinrich Heine, a master of such sentiments and visions, wrote in 1825 on the island of Nordeney:

> I love the sea, as my own soul.
>   I often feel as if the sea were really my own soul itself, and as there are in it hidden plants, which only rise at the instance in which they bloom above the water, and sink again at the instant in which they fade; so from tme to time there rise wondrous flower forms from the depths of my soul, and breathe forth perfume, and gleam, and vanish – Evelina![115]

Over the course of the nineteenth century, the ocean became the spiritual landscape of the ego: the soul derives its form from the abyssal depths of the sea. A quarter century later, in 1857, Charles Baudelaire (1821–67) would exclaim, 'The sea is your mirror; you contemplate your soul', in 'Man and the Sea' – a poem that connects unfathomable nature with the abyss of the heart.[116] By way of metaphors of this deep, dark and fluid element, discourse about the sea gradually merged with representations of the 'oceans of the unconscious' as Jules Laforgue put it.[117] Hereby, the ocean came to stand as a 'Symbolraum' occupied by all that is infinite, abyssal and unsoundable; finally, the nocturnal aspects of dreams had found a fitting aesthetic and poetic vehicle. For Victor Hugo, who described himself as a 'somnambulist of the sea',[118] the ocean symbolizes the dark world of abyss, sleep and night:

> To gaze into the depths of the sea is, in the imagination, like beholding the vast unknown, and from its most terrible point of view. The submarine gulf is analogous to the realm of night and dreams. There also is sleep, unconsciousness, or at least apparent unconsciousness, of creation.[119]

The aquarium emerged as the paradigmatic model for the convergence of all three currents: the sea, dream and the unconscious. Because it condensed and abbreviated the ocean realm, as it were, it became the symbol of contemporary aestheticism for writers such as Joris-Karl Huysman (1848–1907), Jules Laforgue (1860–87), Maurice Maeterlinck (1862–1949), Émile Verhaeren (1855–1916), George Rodenbach (1855–98) and André Gide (1869–1951).[120] 'The dream-world is the Aquarium of Night',[121] Hugo affirmed.

During this epoch, the dream stood for a particular facet of the imaginative and fanciful.[122] From the 1840s on, it occupied the centre of a discourse where scientific and artistic perspectives met. Whereas scientific encounters focused on the 'laws' governing dreams – questions of perception, memory and potential meaning – artists were more concerned with capturing their ephemerality, what eluded or defied rules.

> People attempted by many different means – by recording dreams and in more or less scientific treatises, but also through poetry and through the quill, the etching needle, and the paintbrush – to grasp the logic of this alien, nocturnal realm of experience, in which the apparent memories and familiar images in the mind of the dreamer provide material for constantly evolving new topographies, imaginary constructs, processes, and fragmented stories.[123]

The manifold, oneiric shapes the ocean presented – colours shimmering with cinematic dynamism – emerged as an inexhaustible reservoir of a-logical and non-decipherable aesthetic inspiration. Their whimsical and poetic quality seemed to give form to 'other' states of consciousness; these phenomena appeared to inhabit the same, liquid element as dreams. In either case, one encountered pictorial impressions that, as Alfred Maury (1817–92) – a French dream researcher – put it, merged like the canvases of a moveable panorama.[124]

'There, life must glide away like a glad dream', wrote Michelet in *La Mer* (*The Sea*),[125] intoning a hymn to the flowing sea in oneiric language that abounds in metaphors of metamorphosis, phantasmagoria, monstrosity, hybridity and ambiguity. In *The Carnival of the Animals* (1886), Camille Saint-Saëns (1835–1921) presented a 'musical aquarium' conjuring up a meditative atmosphere of floating fish and gentle undulations. That same year, Jules Laforgue described the aquarium as a refuge of calm and, more still, as the aesthetic model for portraying the optical unconscious.[126] *L'Aquarium*, represents the visitor's descent into the subterranean labyrinth, along underwater worlds behind glass, as a dive into innermost psychic space, a return to a harmonious state of unthinking contentment that all but amounts to a primitive condition preceding human evolution.[127] Attended by the languorous drifting of the fish all around, Laforgue's sleepwalker is led on his journey by the 'All-Einigen'.[128] Ultimately, it brings him down to the native habitation of the first human being; when he reaches the ocean floor, he lapses into Nirvanic bliss. Laforgue's term for this happy experience of becoming one with the world is *se madréporiser* – 'changing back into a madrepore'.

Then, in the early twentieth century, the aquarium acquired a new dimension. Texts such as Marcel Proust's (1871–1922) *A la recherche du temps perdu* (1913–27), Louis Aragon's (1897–1982) *Le Paysan de Paris* (1926) and Alain Robbes-Grillet's (1922–2008) *Les Gommes* (1953) expanded the poetic discourse of the aquarium from a look into the soul's inner depths to an interpretation of the universe as a whole; the introspective *aquarium mentale*[129] was joined by an exhibitionistic theatre of the world.[130] Hans Christian Andersen had already fashioned a model for such a change of perspective when he described the nocturnal journey of his little chestnut-dryad through the 'Höhlen der Meerestiefen' of the Parisian aquarium at the world's fair in 1867:

> In the fresh water grew water-lilies, nymphaea, and reeds ... Fat carps stared at the glass wall with stupid eyes ... They had come to see the Exhibition, and now contemplated it from their fresh or salt-water position. They looked attentively at the crowds of people who passed by them early and late. All the nations in the world, they thought, had made an exhibition of their inhabitants, for the edification of the soles and haddocks, pike and carp, that they might give their opinions upon the different kinds.[131]

Although the switch between observer and observed – toying with the question of which side of the glass pane is actually occupied by the public – proves an exception in Andersen's fairy tale, when society takes the underwater stage in Proust fifty years later, it attests to the diffusion of the aquarium as a model and metaphor for human relations on a general scale. A hotel dining room is described as 'an immense and wonderful aquarium' looked in upon by the working classes of Balbec, the fishermen, and even middle-class families pressed against the windows, in an attempt to see the luxurious life of these denizens, glowing amid the golden sway of the eddies, all of it as weird and fascinating for the poor as the existence of strange fish and mollusks (but whether the glass barrier will go on protecting forever the feeding of marvelous creatures, or whether the obscure onlookers gloating toward them from the outer dark will break into their aquarium and hook them for the pot, therein lies a great social question).[132]

Here, the aquarium offers a refuge for *haute société*, a zone of exclusion and election; the glass wall guards against entry and instrusion on the part of outsiders, whose interest is all the more roused by the separation. In *Recherche,* the Faubourg Saint-Germain turns into a fish tank; a soirée at the Paris opera presents the scenery of a submarine grotto with 'radiant daughters of the sea' and 'bearded tritons'.[133]

References to contemporary biological discourse opens a new repertoire of images. Proust offers an *ichtyologie humaine*: women glistening in the light, who seem to be halfway in the water and half outside, like fish caught in a net. In the narrator's recollection of past Christmas celebrations, brightly lit windows turn into little aquariums; here, in golden radiance, cheerful figures are seen playing. In contrast, Robbe-Grillet's *Les Gommes* (*The Erasers*), published in 1938, makes a fat, greenish fish farmer the centrepiece of the aquarium-café which serves as the sluice between past and present in the novel. Here, in the dark depths behind a curtain of algae, lurk the spirits and ghosts of prehistory and traces of primordial ooze. Instead of a plurality of images, associations and settings, there is only one aquarium, which synthesizes all that occurs in the world, on the stage, or within the soul; it represents conscious and unconscious psychic phenomena in and out of time, offering a metaphor for the world and reality in equal measure.

The connection that echoes in Laforgue's association of dreams and the unconscious on the one hand, and evolutionary biology on the other, proved extraordinarily fascinating for the modern arts. Here, the aquarium functioned both as a medium of introspection and self-knowledge and as a mirror of, symbol for and window onto social relations. Both as a figure, informed by evolutionary theory and marine biology, for the dreaming, unconscious mind in troubled times and as an aesthetic projection of the immeasurable secrets of the ocean, it symbolized far more than what actually fit inside its four walls.

Redon, Kubin and Klee: Three Painters Visit the Aquarium

Aquariums made a world otherwise inaccessible to human beings present and perceptible and, in the process, aestheticized it. As sites where the latest scientific insights, utopian ideas, fears, desires, gender-fantasies and aesthetic issues crossed and took form, they disclosed the potential lying within the foreign ocean world by enabling the most varied kinds of connection. What united the often highly disparate cultures of knowledge and representation was the search for structures of origin, elements of a first beginning, and signs of Creation. Visual artists, in particular, viewed the aquarium as an examplary space of observation; its aesthetic phenomenology seemed to link the internal dynamism of artistic activity and ongoing processes of generation and decay in nature. Thus, painting brought forth an opalescent universe from medializations of the sea, where interweaving, wandering movement and productive encounters between visual motifs achieved visibility.

The French painter Odilon Redon is known for his symbioses of dream-formations and enigmatic beings that teem on the ocean floor.[134] From childhood on, he took a keen interest in natural history and enthusiastically pursued botanical, geological, and zoological studies. He was close friends with Armand Clavaud (1828–190), an expert on water plants who directed the botanical garden of Bordeaux and was a resolute proponent of Darwin's evolutionary theory. Even though the concept of evolution did not enjoy wide popularity in France until the final years of the century, Redon already took due notice of Darwinian thinking in 1862, when *The Origin of the Species* appeared in translation.[135] In short order – and inspired by Clavaud's experiments with starry and bearded stoneworts[136] – he began to fill his sketch- and notebooks with evolutionary-historical and anatomical drawings, especially after he moved to Paris in 1871 and began spending his afternoons at the *Muséum d'Histoire Naturelle*. Redon also drew inspiration from Michelet's depiction of the sea as a protean paradise; likewise, he paid attention to the pictures, which were popular at the time, of microscopic 'worlds in a drop of water' featured in illustrated magazines, journals, light shows and theatres of science. Redon is especially famous for his *Noirs*, a series of black lithographs – particularly the forty-two pages of illustration that interpret Gustave Flaubert's *Tentation de Saint Antoine* (1888, 1889, 1896). These pictures show bizarre floating beings, will-o'-the-wisp phantoms, jellyfish-like whirls with delicate tentacles and swarm-clouds of tiny spherical animals like protozoan hot-air balloons. Seahorses, chivalrous crustaceans and long-lashed sea cucumbers phosphoresce in the underwater night; they cast a subtle shimmer on a world of first beginnings, where the viewer beholds a state of uncanny incompleteness. Especially the images in the final series – *The Beasts of the Sea, Round like Leather Bottles* and *Different Peoples Inhabit the Countries of the Ocean* (1896) – are representative of Redon's late work and display his artistic conceptualization of the aquarium. The Trocadero aquarium at the Paris world exhibition of 1878 had made a deep impression on Redon and the *Musée-Aquarium d'Arcachon* counted him among its 'regulars'.[137] From 1904 on, his most fascinating creations emerged from this constellation, the so-called aquarium paintings, including *Le Rêve* (1904), *L'Allegorie* (c.1907), *Oannes* (c.1910) and *Vision sous-marine* (1910). Redon pictures the hour of Creation in an iridescent, magical sea, as the movement of light and glowing colour: ultramarine, lemon yellow, coral red, sea green, shell white and deep purple. It is is a realm of non-objectivity that gives free rein to fantasy, a representation of the unrepresentable.

Redon always rejected claims that the creatures he made were the outgrowths of delirium bubbling up from the depths of the unconscious – demonic miscarriages or spiritist visions. To be sure, it was a matter of the creative power of imagination, yet he insisted that his chimerical inventions were fundamentally based on the empirical study and scientific observation of nature.[138] In his primal scenes and visions, which present the 'firstborn of creation' in a prehistoric, cosmogonic light, one beholds water-beings that are biologically plausible yet undefinably fantastic and seem to have sprung from the 'other side' of evolutionary history.

The sea was a neverending fount of inspiration for Redon, who always stressed that he faced a world that was indeterminate when practicing his art. In this realm of the flowing, ambivalent and undefinable, where borders between animals, minerals and plants were unfixed and porous, a universe of coming-into-being and passing-away, Redon found the means to represent the unretrievable moment of origin – to borrow the words of Hermann Hesse, 'the magic that dwells within each beginning'. The ocean world, so ambivalent in its aesthetic dimensions, seems to mirror, paradigmatically, one of Redon's central experiences: that the artistic process defies all certainty and cannot be reconciled to standing categories.

'Toute genèse garde un peu d'ombre et de mystère',[139] Redon wrote in *À Soi-Même*: something dark and mysterious remains in every act of creation. This *sens du mystère* is not to be confused with the currents of occultist and esoteric thought that fascinated so many of his contemporaries. For Redon, it stood for the riddle of artistic creation and the mysterious immortality of aesthetic power in nature.

As Redon wandered through the halls of the aquarium in Arcachon, hundreds of kilometers to the east, in Upper Austria, a young artist was also looking for new sources of inspiration.[140] Like Redon, he had a great interest in matters of natural history, and readings in Darwinian theory exercised an equally great effect on his work. However, Alfred Kubin was not guided in his pursuits by a natural scientist like Clavaud. Instead, he was influenced by Haeckel's monistic religion of nature – especially the artful world of form and colour presented in the latter's *Kunstformen der Natur*. Michelet's declaration of love to the sea, *La Mer*, did not stand on his desk; instead, his library was filled by the works of Immanuel Kant, Arthur Schopenhauer (1788–1860) and Friedrich Nietzsche (1844–1900), as well as theosophical texts, occult lore, esoteric volumes, studies in psychology and Wilhelm Bölsche's sex-obsessed accounts of an ocean teeming with erotic activity.[141] What is more, Kubin's readings in natural history occurred in a mythological framework from the get-go and were shot through with wild creatures of biological fancy. Likewise, the spectrum of form and colour in his paintings emerged from the visual tradition of works of popular nonfiction, weeklies and illustrated magazines. He visited galleries and museums to study the canvases, of course, but he also immersed himself in the pictorial language presented in *Die Gartenlaube, Die illustrierte Welt*, and Grandville's *Un autre Monde* (1844). In his 1926 memoirs, he published an account of a key episode in the 'half-forgotten land' of his childhood:

> Gerne lag ich an einer abschüssigen Stelle des Seeufers oder am Kanal im Moor und schaute lange in die durchsichtige Tiefe. Durch einen Sonnenstrahl wurde die Unterwasserlandschaft magisch beleuchtet, und als mein aufmerksamer Blick Fische, Molche, Schwimmkäfer, Ruderwanzen zwischen den Steinen und Pflanzen da unten erblickte, zog mich das rätselhafte Element ganz unbegreiflich an. Die zahlreichen Unterwasserbilder, die später entstanden, dürften wohl ihren Ursprung diesen Eindrücken verdanken.
>
> (I liked to rest on the shore of the lake and look into the transparent deep. The submarine landscape was magically lit by a ray of sunlight, and as I caught sight of fish, newts, diving beetles, water boatmen between plants and stones down there, I strangley felt attracted to that mysterious element. The numerous underwater paintings, which I did later, might have originated in those early impressions).[142]

Spring 1905 – at least as Kubin represented it – marked the beginning of a new orientation. His library was filled with bestsellers such as Alfred Brehm's *Illustrirtes Thierleben* (1876), Gustav Jäger's *Das Leben im Wasser und das Aquarium* (1873), Wilhelm Haacke's *Die Schöpfung der Tierwelt* (1893) and bound volumes of the illustrated journal *Über Land und Meer*; these works of popular science were all known for their illustrations of the underwater world. But Kubin names another example as the one he followed: the grand public aquaria and the microscope with which he uncovered fairy-like structures and enthusing organisms .

Marsha Morton has speculated that Kubin was probably inspired by the aquaria of Berlin and Paris as he travelled to both cities (in 1902–3 and 1906, respectively). It would certainly have made sense to journey to Paris to visit the atelier of Redon – who was quite old by now and whom Kubin greatly admired – and study his underwater views 'on location.'[143] At any rate, once back at home in Zwickledt, Kubin did not content himself with memories. He installed 'die schönsten Aquarien' (the most beautiful aquaria), as he put it in a letter to his friend Wilhelm Hausenstein (1882–1957) years later; lying on his belly he began to hunt even the most minute species and carefully observed their metamorphoses under a microscope.[144] The 'release' he experienced proved productive and profound in equal measure. Even though all manner of aquatic beings had always inhabited Kubin's works, now, instead of striking poses as symbolist riddles, they floated freely in the shadowy depths – forms in a state of transition between emergence and vanishing; by the same token, the thick, pasty consistency of the colours on the paper (or, alternately, cardboard) produced 'surfaces' that 'evoked the primeval, rocky floors of the sea and fossils.'[145] *Tiefsee, Meeresgrund, Helm im Meer, Ochsenfisch, Verpuppte Welt, Feuerfisch* were all created in 1906; the paintings reflect a creative frenzy in which Kubin combined the world of zoological illustration – which was strange enough on its own – with imaginary, fairytale beings and hybrid monstrosities to bring forth 'biologische Malerphantasien (a painter's biological phantasies)'.[146] It seems, in part, that one is peering into an aquarium of evolutionary enigmas; equally, one has the impression of contemplating a cosmic sphere of monuments to natural religion. In either case, the works represent the sustained effort to get behind the 'fließend-veränderlichen, rätselhafte[n] Urwesen des Lebens' (fluid-changing, enigmatic primordial beings of life).[147]

Kubin engaged with evolutionary theory and sought to reach the 'Dämmerungswelten' of 'vorhistorischen Menschheit' (gloaming worlds of prehistoric mankind)[148] by way of jellyfish, urchins, sea spiders and deep-sea fish; in contrast to Redon's works, his were marked by monistic and occultist thought and incorporated elements of popular superstition. Whereas Redon, by combining scientific and artistic perspectives, explored possibilities of aesthetic and epistemological insight, Kubin incorporated marine-biological and microscopic forms to engineer harmony between the world of science and that of myths and fables. It was a matter of combining the most varied levels of reality: scientific observation, imaginary creatures and spheres of life occupying radically opposed dimensions, from the microscopic and tiny to the oceanic and vast.[149] Throughout, the constant was the realm of the supernatural and mysterious, and the central motif of Kubin's creative process was exploring its reaches and depths.

The 'Seifenblasenwelt' (soap-bubble world) he describes in numerous personal accounts – inhabited by ghosts, wizards, genies, fairies, witches, kobolds and the like – follows the lines drawn by contemporary psychological and psychoanalytic research on myths, especially *The Interpretation of Dreams*; it represents an 'empirical' effort[150] to lend form to the land of 'Seelenzwielicht' (spiritual twilight).[151] In this respect, Kubin was an altogether modern artist, and he thought it important to situate his grotesque figures in terms of the science of the day, as archaic spiritual forces and emanations from a world of dreams.[152]

For a few years, Kubin seems to have viewed the ocean depths as the site where the past could be preserved and held secure for the future, in both a scientific and a spiritual manner. After 1908, submarine phenomena are hardly featured in his works. All the same, he appears to have considered the ocean a realm of fantasy until the end. At any rate, a quarter of a century later, he looked back on his life as a nautical voyage, drifting along screened off lights with his freight of paintings.[153]

In the course of the long nineteenth century, the aquarium emerged as a self-evident component of general visual habits. Redon and Kubin saw in it, above all, a hybrid combining cultural artefact and living nature, reflecting the ambivalence of the sea itself. A third mode of encounter viewed the aquarium as a communicative relay-station that sparked creativity.

In March 1902, Paul Klee (1879–1940) and his friend, the sculptor Hermann Haller (1880–1950), visited the famous aquarium at the *Stazione Zoologica* in Naples. Klee was already an avid fisher and avowed fish-lover; a year earlier, in spring

1901, he had produced a series of aquarelles devoted to these themes.[154] At the time, Naples – or, more precisely, the *Stazione Zoologica* – was in the process of turning from an 'insiders' tip' into a 'place to be' for artists. For Klee, it was a revelation.[155]

Indeed, the institute of marine biology at Naples was a noteworthy site, in large part because of its incomparable aquarium. Here, Carl Vogt wrote excitedly in the *Gartenlaube*, even the palette of a Makart would fail[156] to depict the wondrous beings. In the words of the art critic Julius Meier-Graefe (1867–1935), it housed a crowning achievement of German painting – 'a wonder' to be found in the institute's library on the second floor.[157] In the space of a few months, as if in a single moment of transport, Hans von Marée (1837–87) had created a monumental cycle of frescos, five murals of the sea and marine landscapes.

Connections between art and scientific culture were a solid feature of life at the institute.[158] Anton Dohrn, its celebrated founder and director, kept the doors of his own home, *Casa Dohrn* in Rione Amedeo, open for writers, painters, actors, musicians and philosophers. Indeed, from the inception he had engaged artists and friends in shaping the activities of the *Stazione Zoologica*. Besides his two assistants, Nikolaus Kleinenberg and Hugo Eisig, the inner circle included the sculptor Adolf von Hildebrand (1847–1921), the Scottish poet – and translator of Kleist – Charles Grant (1841–89) and Hans von Marée himself. Hildebrand assisted Dohrn in architectural matters and in designing the façade of the building, above all; later, he sculpted detailed busts of the well-known embryologist Carl Ernst von Baer and Charles Darwin for the fresco room. Initially, Dohrn viewed his visionary undertaking as a risk, and he often likened his colourful crew to the Bremen city musicians.[159]

Toward the end of the 1870s, the illustrious company was joined by Arnold Böcklin (1827–1901), the painter mentioned by Klee.[160] For many years already, his unconventional, hybrid beings had been romping through ocean scenes; to the joy (or, alternately, dismay) of contemporary audiences, their anatomical accuracy proved uncanny. The *mare Böcklinianum*[161] lay just off the coast of contemporary zoology and marine biology.

Because of the attention to minutiae and zoological classifiability – it is easy to discern scales, iridescent patterns, detailed fins and even hairs – Böcklin's mermen and mermaids, along with the accompanying retinue of sea serpents, seals, seagulls and deep-sea creatures, call to mind the tradition of early modern *Fischbücher* by Gesner, Belon, and Rondelet, as well as the antiquarian and mythologizing illustrations that accompanied the travel accounts of the day.

Böcklin was captivated by the multiplicity of colour and form displayed by marine fauna and flora. Therefore, when he first came to Naples in 1879, he did not do so only because he wanted to see the frescos of his friend Marée; he also wanted to visit the *Stazione Zoologica*. Böcklin, who was interested in natural history and had pursued intensive studies of his own by reading the latest scientific accounts, found exactly what he wanted right away. A year later, he gladly accepted an invitation from Dohrn to take part in a four-day boat excursion to the island of Ponza. From this point on, Böcklin counted as a frequent and welcome visitor. He often took the *vaporetto* – a small steamboat – from the *Stazione Zoologica* out onto the sea and impressed many a researcher with his prodigious powers of observation and memory for colours and forms.[162] And so, over the years, he and Dohrn participated in a lively exchange about matters of evolutionary theory; the two men also drew close because they were both passionate swimmers.

Although Böcklin visited Naples often and painted his sea creatures with great attention to zoological detail, his oeuvre offers only one example of an 'aquarium motif': the lower part of his altarpiece, *St Anthony Preaches to the Fish* (1892). Here, with a rare degree of irony, Böcklin thematizes the competition between Darwinism and Christian doctrine. The main image above features a grey-blue shark folding its fins above a wreath of seaweed in an act of humble piety; in contrast, the underwater view below represents the Darwinian reality of natural selection. The greenish-grey light makes the possibility of sudden movement manifest and lends a menacing quality to the cat shark, swordfish and tub gurnard. In this meeting of hagiography and shockingly realistic ravenousness, Böcklin achieves a singular allegory connecting Darwinian evolutionary theory and the dispositive of the aquarium: a new world of images and space for innovative experimentation.

Not everyone returned from the Naples aquarium a changed man, although Meier-Graefe reports that many artists were fascinated by the colours and movements of the animals – this 'Fundgrube künstlerischer Formen' (treasure trove of artistic form). In his memoirs, Igor Stravinsky, for example, describes how he and Pablo Picasso spent long hours before the fascinating flickering of ocean fauna in spring 1917. But if only the fewest visitors experienced this as a magical moment of initiation, Klee was one who did; for him, the pilgrimage to the aquarium represented an artistic awakening. Marine life – its fantastic forms and fairytale hues – seemed purer than paradisiacal beings; as he wrote to his fiancée Lily Stumpf, it was wholly saturated by light, a veritable outpouring of divine humour and fantasy.[163]

Klee's debt to Italian art and architecture is well known. For all that, his encounter with the 'transparent-spiritual' element he saw for the first time behind the panes of the Neapolitan aquarium, which opened onto the mysterious world of primordial, mythical forms, proved equally productive.[164] Some of his most enigmatic paintings – for example, *Goldfisch* and *Fisch Zauber* (1925), or *Um den Fisch* (1926) – are magnificent reflections of this experience; symbolically, they offer insight into a mythical and mysterious night, the heart of all things, now revealed. In this incomprehensible realm 'in-between' it seemed Nature had abandoned the borders of the merely natural and become become visible as Creation itself.[165]

Klee was not enamoured with birds alone; also – and above all – he loved fish. They swim through his works on every level and occupy an exceptionally prominent position in the created world:

> Simple of shape and bold of hue, seemingly unhampered in their movements, and passing their secretive Lives out of sight of much of the rest of nature, they appear to have been seen by the artist as among the most fortunate and liberated of beings.[166]

Klee viewed fish as magical beings in an endless history of nature; in them, the secrets of nature were most fully unveiled. It appears that in the aquarium of Naples Klee first struck upon what would shape his art as a whole: the quest to lay hold of the life-processes at the heart of nature, the formative impulse that gives rise to invisible archetypes and the multiplicity of species.[167] The study of submarine life provided a key inspiration for this endeavour for as long as he lived.

In the course of half a century, the aquarium transformed from a small glass box in which local coastal water dwellers were kept into an exemplary heterotopic space. Aquariums were ocean museums, aquatic exhibitions, sites of experience and representative locations where the epistemic category of life met the aesthetic category of creative nature. As such, the aquarium acquired a theatrical air as spatial art, epistemological *mise-en-scène* and a cabinet of wonders for mass culture. As a habitat for fantastic, immoral, enchanting and terrifying beings – a microcosm and simulation of the ocean in one – it stood as the medium of deep space, staging an unreachable world in spectacular manner. On the one hand, aquariums were thought to afford insight into a 'piece of nature'; on the other hand, they were matters of design – moving worlds of images that could be manipulated to achieve experimental ends. Aquariums were not only constitutive for the genesis and spread of new research facilities and knowledge in the fields of oceanography, evolutionary, behavioural biology and ecology; they also played a constitutive role in the development of artistic perspectives. Ultimately, aquariums were both prototypical models within a discursive order of nature and objects of a differentiated system of media practices. By turns they reflected and shaped the early culture of modernity – sometimes as philosophical and poetic metaphors for understanding nature and the self, sometimes as a complex amalgamation of art and entertainment culture, and sometimes as experimental arrangements and time machines, *la vie merveilleuse en miniature.*

5 MISE-EN-ACTION: ASLEEP IN THE DEEP

Like the aquarium, diving poses questions about the relation between media-aesthetic and scientific-technological practices. Here, too, a glass interface for vision is involved – or, as Rozwadowski und Keuren put it, 'an almost constant intercession of technology between the observer and the observed'.[1] Of course, human beings can hold their breath and look around underwater without protection for a certain span; however, just as surely as the body longs for air, submarine worlds prove resistant to human curiosity. We can remain underwater only to the extent that we replace natural surroundings with technological and artificial conditions; here, the 'extraordinary sensuousness' of the sea is given in mediated form, at best. In this sense, Cindy van Dover, a deep-sea biologist, describes all dives as 'unnatural act[s]',[2] and the ethnologist Stefan Helmreich speaks of a 'ball of culture submerged in the domain of nature'.[3] Penetrating regions of nature hostile to human life represents a double challenge, both in terms of aesthetics and *aisthesis*: a view of the new as well as a novel way of seeing. By definition, visibility involves a specific mysteriousness that is best described phenomenologically – in the words of Maurice Merleau-Ponty: 'Nothing is more difficult than to know precisely *what we see*'.[4]

Submarine Visions

Eugen Ransonnet-Villez and the Discovery of Indeterminacy

On a sunny spring morning in 1865, the few coastal inhabitants of the shallow waters off the island of Ceylon (Sri Lanka) witnessed a strange sight. A big, black metal box was suspended beneath the water's surface, gently swaying between them. Two human feet peeked out underneath; thick sacks were hanging from the box's four corners and a long tube extended from the interior up to a boat. The bearded face of a man peered through two glass windows – an unusual spectacle, indeed, which seemed to interest the fish.

The man in the box – an accomplished natural scientist – was '[e]ntzückt von diesem Schauspiel und konnte [sich] nicht enthalten, wenn auch im Wasser sitzend, einen Theil dieser Herrlichkeiten sorgsam abzuzeichnen' (ravished by the spectacle and could not abstain, even though sitting in the water, from diligently copying some of the splendour).[5]

The curious diver was Eugen Freiherr von Ransonnet-Villez (1838–1926), a great lover of natural history and painting.[6] He came from a family of Austrian civil servants; his father, Karl von Ransonnet, was vice-president at the highest Court of Auditors in imperial Austria; accordingly, he had received a comprehensive education at the Academy of Arts, as well as training in law at the University of Vienna. In 1858, he entered the Foreign Ministry as a ministerial official; from this point on, his diplomatic career made it possible for him to go on extended voyages. In 1860, he travelled through Greece and Turkey, and two years later over Palestine, Upper Egypt and Arabia, where he began to investigate the submarine world of the Red Sea. On the coast at Tor (also spelled Tur on older maps – today, Al-Tür or El-Tür), a small settlement of eleven houses and one hundred inhabitants at the foot of Mount Sinai, he made his camp. From the very first, the underwater submarine flora and fauna seemed to present an Elysian paradise, a fallen part of the heavens, so delicate as if woven of foam.[7]

Soon enough, Ransonnet could not resist the 'wunderbaren Gefüge' (miraculous arrangement) stretching out in the crystal-clear water under his boat. On his second excursion, he yielded to the temptations of the deep and jumped into the waves:

> Wie eigenthümlich sieht sich' da unten im Meere. Allerdings kann man die Formen in der Tiefe nicht mehr genau unterscheiden, allein wie schimmert da Alles in schöner und fremdartiger Beleuchtung! Braun, violett, orange, in gelbem und blauem Lichte, leuchtet's dem Taucher entgegen.

> How peculiar things appear under water! Though one cannot exactly distinguish the contours in the deep, yet everything gleams in beautiful and strange illumination! Brown, violet, orange, in yellow and blue light, everything glows towards the diver.[8]

What might seem like a trifle today was a sensation in 1860. For more than two thousand years, human beings had practiced diving in order to work or fight, but hardly anyone had had the idea to go swimming in order to investigate the marine world. Even though legend held that Alexander the Great (356–23 BC) descended to the ocean floor in a glass bell, and Halley's diving experiments with a diving bell were a matter of record, the list ended there. (What is more, Halley's dives did not occur to study flora and fauna; the window of the vessel was not designed for looking out; instead, it was meant to capture light: the metamorphoses of colour *inside* the bell were what interested Halley.)[9] And so, even though almost every book on the history of diving opens with the bold declaration that human beings have always dreamed of descending to the depths of the sea to investigate its mysteries, historical documents suggest a wholly different picture. Until about 1700, prevailing notions held that an underworld of dangers lay in store; only gradually did such views yield to an aesthetics of the sublime, which culminated, some two hundred years later, in a paradise of magical beauty. Thus, until late in the nineteenth century, the prospect of a submarine stroll would have been more readily associated with fear than curiosity. An account made by an unknown diving 'tourist' in 1859 reads:

> I had not the slightest intention of walking about; I was much too dismayed at the impressive silence and gloomy solitude of the water, in which I seemed to myself lost ... Wishing to carry back with me a tangible proof of where I had been, and a souvenir of my excursion, I stooped down and picked up a pebble from the bottom of the sea ... This being done, I gave the signal to hoist me up to the surface.[10]

Under such circumstances, the exceptional nature of Ransonnet's learned plunge emerges even more clearly – especially if one considers that the modern world had already been taking 'pleasure in the sea' for more than a hundred years. After all, ocean fever was widespread at mid-century; people were enchanted by aquariums, collecting curiosities on the beach and the secrets hidden in the depths. Why, then, had neither scientists nor amateurs not sought to wander on the ocean floor? Since the 1830s, the *Standard Diving Apparatus* designed by Charles Anthony Deane (1796–1848) and his brother John (1800–84) had stood at their disposal; later, the German-born engineer August Siebe (1788–1872) made improvements and marketed it successfully.[11]

Still, Ransonnet had *one* famous predecessor. Already in 1844, Milne-Edwards had dived with his colleague Armand de Quatrefages off the Sicilian coast to explore the underwater landscape of the Mediterranean.[12] With the help of an open diving helmet with glass windows, a lined leather jacket, lead soles on his feet, a lifeline and an air tube,[13] Milne-Edwards strolled around for half an hour at a good four metres' depth in the harbour of Milazzo; here, Quatrefages reports, he sought out the animals of the sea in their most secretive hiding places.[14] In the bay of Taormina, he even reached a depth of approximately eight metres, where he discovered a kind of shellfish that, until now, had been known only through its abandoned bivalve housing.[15] Milne-Edwards was aware of the value of his field research from the outset: 'My opinion as to the inadequacy of research made on animals preserved in alcohol was further confirmed. The observation of living animals allows you to recognize the cause of errors ... and to correct them.'[16]

Some thirty-five years later, Anton Dohrn likewise observed that the clear and warm waters of the Mediterranean afforded optimal conditions for submarine scouting missions.[17] After initial efforts in the harbour of Kiel with his friend Werner von Siemens (1816–92) in 1878, Dohrn started to explore the Gulf of Naples on his own in summer 1879. Equipped with a scaphander suit supplied by the Italian government, his initial trials already reached depths up to thirteen metres. Here, he discovered many new kinds of algae and sponge, as well as three new types of planaria (turbellaria). He enthusiastically noted that, with the new diving technology, 'unsere Macht gegenüber den Geheimnisses des Meeres' (our power over the secrets of the ocean) could be raised by 60 to 70 per cent.[18] Indeed, just two years later, the *Stazione Zoologica* was in the position to present initial findings on the new scientific field of ecology and the relationship between algae growth and the incidence of light.[19]

Over the follow years, research with scaphanders became an important component of activity at the station in Naples. The botanists Paul Falkenberg (1848–1925) and Gottfried Berthold (1854–1937) established the principles of marine ecology – in paricular, psychology – while zoologists such as August Weismann (1834–1914) and experimental physiologists like Émile Yung (1854–1918) investigated the evolutionary connections between mutations and environmental factors. That said, a serious accident just at the turn of the century put an initial stop to this productive phase. Reinhard Dohrn (1880–1962), the founder's son, fell unconscious during a dive due to a technical error; the crewmembers noticed too late and started the rescue without making necessary pressure adjustments. His life was saved, but he suffered the consequences of decompression sickness for years and was never able to dive again.

Underwater research was dangerous to be sure. What is more, equipment was expensive and difficult to use. It is all the more surprising, then, that Quatrefages and Milne-Edwards cheerfully tell how they enjoyed complete freedom of movement in their suits – especially since the professor tangled up the lifeline over and over and nearly drowned on several occasions. This might be one reason that it did not occur to more than a handful of marine researchers to go diving. Alternately, it might be that the aquarium, as a successful model for experimentation, ruled the contemporary landscape of research. But still: dives *were* performed, and some of them yielded groundbreaking results. No one questioned the incomparable value of diving or the wholly new opportunities and fields of research it opened. At midcentury, diving no longer counted as a mad endeavour. Hundreds of helmet divers were at work in Europe, laying the foundations for lighthouses and bridges, destroying enemy ships, salvaging sunken wrecks or harvesting sponges. Nor were these heroic deeds performed in secret; they occurred in the same harbours where marine researchers shipped in for their various expeditions. Opening a newspaper, one might read the most recent exploits of Alexander Lambert (*c.* 1837/43–1892), the bravest diver of the day; he was said to have captured a tiger shark in the Indian Ocean and to have crawled hundreds of metres through the flooded shaft of the Severn Tunnel in south Wales to close a sluicegate that was stuck.[20] Popular journals regaled their astonished readers with terrifying stories about divers retrieving corpses from under the ice or salvaging stolen treasures; theatregoers in London attending *The White Heather* could witness a dramatic underwater duel pitting Good against Evil. In short, over the course of the nineteenth century, diving came to constitute an important undertaking in both military and economic terms; it had a place in visual and popular cultures and represented a 'viable' idea when leisure activities were discussed. For all that, one was unlikely to encounter interest in underwater landscapes or lyrical effusions about what they offered.

Ransonnet-Villezs's dive appears all the more remarkable in the context of the seeming 'hydrophobia' of marine biologists, zoologists, physiologists and ecologists. However, its exceptional significance is not just the fact *that* he dove – especially since, as we have seen, he was neither the first nor the only one to do so. Rather, it is *how* he dove. Unlike Milne-Edwards, Dohrn or guest researchers at the Zoological Station in Naples, Ransonnet reflected on the conditions of possibility of his excursions under the sea. 'How peculiar things appear under water!' he had written of free-diving in the Red Sea, whereby he thematized how acoustics and relations of light changed. Two years later, he returned to the ocean off the shores of Ceylon – now equipped with diving equipment he had designed on the model of Franz Kessler's (1580–1650) 'water armour' (1616). His gear consisted of a 'diving bell' – approximately three feet high, weighing about eighty pounds – two glass windows – one on the side and one on the ceiling – provided a splendid view and four weights hanging from the corners kept the bell steady. At the open bottom end an iron bar provided a kind of seating.

Ransonnet made his first dive on 25 November 1864.[21] Seated in his bell, he managed to observe the underwater world without disruption for an extended period, even making sketches on location. The drawings subsequently provided the basis for four lithographs, which were published with his account of the voyage; they were also incorporated into a large oil painting exhibited at the Museum of Natural Science in Vienna together with animal specimens he had obtained for quite some time afterward. Not only do they stand out for their exceptional colour – a mysterious effect of the submarine realm that Ransonnet had studied intensely – but also for the singular atmosphere they radiate. Ransonnet wrote in the foreword to his account of the dive off Ceylon that he wanted to give his readers 'dem Taucher gleich, einen Blick auf wirklichen Meeresgrund' (a true sight of the ocean ground as if seen by a diver).[22] In the judgment of Hans Hass (1919–2013), the Austrian diving pioneer, Ransonnet captured the defining quality of being underwater in a way that modern photography did not afford.[23] Indeed, the purpose of Ransonnet's pictures is not to depict, but to convey. Those who contemplate them are meant to behold the ocean floor 'like the diver'. This concerns not just *what*, but, above all, *how* things appear underwater – 'another kind of seeing', Ransonnet wrote, which one must learn. His manner of translating this principle was both simple and ingenious: he conceived an aesthetics of submarine indeterminacy that represented the water's material resistance to penetration by the observing scientific gaze.

To be sure, Ransonnet did not 'invent' blurred lines or backgrounds that run together. According to Wolfgang Ullrich, the aesthetic potential of indeterminacy [*Unschärfe*] was already broadly employed in landscape painting at the turn of the century – most effectively, perhaps, by William Turner (1775–1851), who sought out extreme situations that defied standard viewing practices.[24] The same holds for Ransonnet's underwater views, even if the visual form they assume is distinct from artistic landscapes inasmuch as they follow the aim of scientific precision (or, alternately, insight). In his metal cocoon, Ransonnet had studied his surroundings as exactly as possible, with the keen eye of a taxonomist; he made the conscious decision for indeterminacy as a way of making the conditions of visibility as such aesthetically perceptible. The exceptional, 'artistic' reflection of sensory dimensions makes Ransonnet's views of the underwater world stand out among scientific images.

Remarkably, Ransonnet's pictures did not command much attention at the time. Despite a small print run, the accounts of his expeditions were noted, but it would be decades before others proved 'mad' enough to go underwater with paper and pens themselves. Ransonnet's effort to gain insight into the medial conditions of submarine research and to engage in aesthetic reflection on the results themselves is evident only in hindsight. As in the case of the aquarium, more than half a century passed before marine researchers saw fit to consider the media and technology that had made their investigations possible in the first place. The few underwater images produced by those who followed in his wake reveal that Ransonnet, in pursuing this mode of representation, was not yielding to a merely subjective impression; rather he had identified the epistemological core of conditions of insight. Painters who had never visited the ocean floor, such as Edward Moran (1829–1901), imagined their submarine landscapes as if the water were something else.[25] Their fish circle strangely in the 'air' or seem to float on invisible wires in a theatre. In contrast, artists who were familiar with visual conditions underwater incorporated blurry elements – for example, the French painter of submarine scenes, M. Durrand-Brager.[26] His works include two remarkable pictures dating from 1872 – when efforts were made to recover the treasures of the Spanish silver fleet, which had sunk in Vigo Bay in 1702 – with the help of the underwater observatory designed by Pierre-Antoine-Ernest Bazin (1807–78). This was the first time the underwater light Bazin had invented was employed; for long afterward, the pictures that resulted were considered photographs because of their realistic lack of focus.

Louis Boutan, Francis Ward and Underwater Photography

In spring 1893, the first pictures were taken underwater at Banyuls-sur-Mer.[27] The photographer, Louis Marie Auguste Boutan (1859–1934), was a passionate zoologist and assistant to Henri de Lacaz-Dutons (1821–1901), a famous zoologist, anatomist and biologist of the day, as well as the founder of *Laboratoire Argo*, the research centre there.[28] Lacaz-Dutons had encouraged his young associate to use one of the institute's scaphanders to study the invertebrates that interested him so. Boutan learned to use the equipment and dove in the bay of Banyuls during August and September 1886; he enthused:

> The strangeness of these submarine landscapes made a very deep impression on me, and it seemed a lamentable fact that they could not be reproduced in any other way than in a description which, however exact, was necessarily imperfect. I was filled with the desire, therefore, to bring back from these submarine explorations a more tangible souvenir. But, however good a diver one may be, it is scarcely possible to make a drawing, or even a sketch, under water. I then resolved to try to see if I could not obtain a photograph of this hidden region. As it is not difficult to take a landscape in the open air, why, I asked myself, could I not succeed in making a photograph at the bottom of the sea? Though it is quite certain that water is a much denser medium than air, still, as the eye can distinguish objects in the midst of water, there should be, I argued, no insuperable obstacle in the way of a photographic plate receiving an impression under the same conditions.[29]

Not much is known about events over the next four years. In 1891, however, Boutan led an expedition to the Red Sea to study the flora and fauna; it also seems that he, thirty years after Ransonnet-Villez, spent a while in Tur because the underwater world there proved especially fascinating.

Boutan continued his photography project in France. With the help of his brother Auguste, an engineer, he designed waterproof housing out of brass and rubber to protect a 9 x 12 cm plate camera. It included two levers for operating the shutter and changing plates; an elastic balloon served to balance pressure between the inside and the surrounding water. *Alvergniat*, a firm in Paris, was commissioned to build the apparatus. To Boutan's dismay, there were delivery problems, and his experiments had to wait until spring 1893 to start. But then, everything went smoothly. In May, Boutan was already able to send initial prints to Lacaz-Dutons; by the summer, he had produced a series of exquisite pictures. In the process, the skills of the institute's mechanic, Joseph David (1869–1922) proved invaluable; time and again, he made improvements, and he also constructed the first underwater flash device, which used magnesium powder. Until this point, photographs taken at depths up to eleven metres had required an exposure time between ten and thirty minutes; now, Boutan obtained the same results much more rapidly. Finally, on 31 July, Lacaz-Dutons presented Boutan's report, *Sur la photographie sous-marine*, to the *Académie des Sciences*. Countless periodicals and journals made it a worldwide success.[30] The picture of the marine researcher in his diving suit capturing the wonders of the underwater world on the camera plates in the glow of the magnesium lantern became a symbol of scientific and technological triumph at the *fin-de-siècle*.

But Boutan was not content to stop there. The artificial light caused a lightning storm underwater, which frightened the animals and did little to improve the quality of images. What is more, the photographs lacked depth of focus. Even though this produced a 'painterly' effect, Boutan thought it unacceptable: he wanted the photographs to show the underwater landscape as it appeared to a diver. But what, exactly, did the divers of the time – encumbered by bulky outfits – see? Did Boutan, from inside his Denayrouze helmet, have a clear view of submarine activity?

There can be no doubt that Boutan knew perfectly well that the medium of water has properties entirely different than air (for which cameras were originally built): water filters light, changes colours, makes objects appear larger and, moreover, possesses viscous materiality in the form of thousands of salt crystals, microorganisms, grains of sand and particles of dirt swirling about. Moreover, everything in the sea is constantly moving: divers, animals and plants float and sway incessantly. None of this blocked Boutan's ambition:

> L'expérience de ces trois dernières années m'a prouvé que le milieu n'était nullement le coupable et qu'il suffisait de perfectionner l'appareil photographique pour perfectionner les images obtenues.[31]

It was clear to Boutan that underwater photography would become an important tool for the marine sciences. If one could manage to observe and photograph ocean creatures living 'freely', it would yield important biological insights and increased understanding of events on the ocean floor that still proved mysterious.

And so, in 1895, Boutan and David started a new series of experiments. After three years and a few misfires, they ultimately managed to produce images worthy of notice; David proved the more skillful photographer of the two. With a 18 x 24 cm plate camera, they even obtained pictures taken at depths of up to three metres. Boutan's report to the *Académie des Sciences* – *L'instantané dans la photographie sous-marine* – not only attracted international attention, but also landed him a significant commission: to prepare some submarine photographs for the world exposition scheduled in Paris. With generous financial support from a Parisian firm for optical instruments, Boutan acquired two large arc lamps (twenty amperes each) that were powered by sixty batteries and protected by cast-iron housings. Boutan intended to use these vast resources of energy and light to penetrate fabulous depths: hundreds of metres, where the inhabitants of the eternal darkness would be revealed for the first time. Late in the summer of 1899, all was ready. Unfortunately, his initial efforts, with David, yielded discouraging results. When they attempted their record dive, it proved more difficult than anticipated to make the equipment, which weighed almost a ton, actually sink. To avoid putting divers' lives at risk, Boutan stopped the operation at just fifty metres and brought the equipment back to the surface. All the same, the image they obtained – only one had worked out – became a sensation. On 25 July 1900 Boutan presented his most recent underwater views at the world fair in Paris. Afterward, he gave up underwater photography.

In the years that followed, Boutan returned to his zoological studies. There is no record whether he continued to dive, but his students are said to have called him the 'sea-wolf' because he never missed a chance to make an excursion with them and bring new forms of life out from the depths. He likened his work to research conducted by a visitor from the moon, encountering an atmosphere as impenetrable as water proves for marine biologists:

> What must he do if he wishes to know something of what exists below the layers of clouds which hide our globe from his view? He would do as our naturalists have done – construct dredges and nets, and, having weighted them, would let them down like the anchor of a ballon, and try and pull them along the surface of the earth. Do you think that with such primitive instruments he would obtain very precise ideas of the terrestrial globe?[32]

Boutan's submarine photographs exude a special magic. They present a fascinating, alien world that the fewest earthlings had ever experienced first hand at the turn of the century; what is more, they reveal themselves, so to speak, as media attesting to another order of seeing. This self-reflective component, which sets them apart from the gesture underlying strictly scientific images, is condensed in the figure of the researcher-as-diver, vividly enacting his unsuitedness to the conditions under the water. We – like Boutan's voyager from the moon – see everything only in mediated fashion, through 'layers of clouds' that, no matter how we may seek to arm ourselves technologically, will always leave their imprint on whatever we record. Boutan was thoroughly aware of this 'problematic' aspect – after all, water is 'denser' than air; he characterized the indeterminacy it entailed as a 'cloudiness' inherent in the medium.

> There still lingers in the background of the proofs a slight mistiness, due to the medium being denser than air ... This peculiar cloudiness of the background, this sort of mist which settles over distant objects, seems to me to be the characteristic feature of submarine photography ... and the submarine horizon will always be limited to about one hundred meters.[33]

Boutan's legacy includes the first collection of underwater photgraphy, *La Photographie sous-marine*, a volume that summarizes the technological possibilities of the time. In the conclusion, the author declares that he has paved the way; now it is time for others to follow and find paths of their own. Boutan calls on amateurs, in particular, to take up underwater photography and conduct their own studies during vacations on the seashore.

It is difficult to tell, today, how many enthusiasts heeded Boutan's advice. It could not have been more than half a dozen, however; little record exists of divers who also took pictures until the 1920s. The work of the English amateur ichthyologist, avid fisherman and keen Boutan-reader Francis Ward stands out among the handful of 'shore photographers' who would stand knee-deep in the water and hold their cameras underwater. Ward also took an interest in the material properties of water, but he sought to go beyond what had been attempted until now by means of a significant change of perspective: Ward wanted to know what life underwater looked like to a fish.

Like many an amateur fish-researcher of the day, Ward was enthusiastic about aquariums. Over the course of time, however, he became convinced 'that fish should be watched and photographed while swimming free in natural environments, and illuminated as in nature'.[34] To this end, he set up a large pond on his property in Ipswich. At the bank, he built a subterranean observation room; through a window, he could watch and photograph, unobserved, the life of its inhabitants. In 1911, he published the lavishly illustrated *Marvels of Fish Life: As Revealed by the Camera*, which appeared in many editions. The book addresses the perspective of the deep – *fish as seen by fish* – as a matter of the incidence of light and disguise. As a point of contrast to photographs taken under natural conditions, he offered pictures that were made with artificial side lighting in front of an aquarium. Whereas the latter provide exemplary clarity, the former display the fascinating indeterminacy of the underwater perspective; more still, the milky atmosphere makes the watery medium available to perception. The secrets of camouflage are revealed as effects of light, first and foremost. Ward describes in detail how the colouring of scales reflects and absorbs light; in this way, fish seem to melt into the background and become practically invisible. Ward was one of the first to employ the autochrome plates developed by the Lumière brothers; his impressive pictures provided the theoretical basis for observing – and understanding – the habits of fish, their behaviour when hunting or fleeing.

*Figure 5.1: **Observation Pond Empty.** F. Ward, Marvels of Fish Life: As Revealed by the Camera, 2nd edn (London: Cassell, 1912).*

Ward hypothesized that specific visual relations determine the behaviour or fish. He focused on motion and bearing, as well as changes in colour and pattern on the animals' skin. Until this point, pictures of underwater landscapes had been conceived along the lines of a still life; Ward's photographs, however, resemble film stills – that is, snapshots from life that condense the whole drama of 'intentions and emotions'[35] in an archetypal situation. Instead of an idyll outside of time, they offer suspense; the characters are *in action:* annoyed, aroused, vigilant, disappointed or doubtful according to Ward's interpretations of their bodily expressions. Even if these were anthropomorphic projections more than actual emotional states of the animals, Ward discovered a sensational new science, which conveyed to readers the surprising impressions made by personal encounters. His photographs suggest participation on the part of the observer and, by means of diffuse relations of light, generate the impression of visibility that is immediately given. The fluid contours within the frame convey a feeling of presence that cannot arise when representation is clearly modulated and defined by objects. Ward's famous oval 'portrait' of a pike, emerging from nebulous depths and seeming to stare the observer straight in the eye, epitomizes the sensation (see fig. 5.2). This frozen moment oscillates between astonishment at the unusual sight and the novelty of being looked at; at the same time, it alternates between the ephemeral unreality of the fish and the surreal clarity of its gaze. As an epistemological corollary, suspicion arises when the eye of the observer penetrates nature without a human element; the operative distinction between subject and object – and therefore the basis of knowledge – is rendered uncertain. The situation's lack of clarity – who is studying whom? – gives rise to 'methodische Unschärfe' (methodological indeterminacy);[36] however unintentional it may be, this corresponds, in aesthetic terms, to the pike's lack of photographic definition, the way its form dissolves into the depths of pictorial space. The singular quality of the fictive encounter in the photograph proves just as palpable as the factual and irreducible otherness of human and ichthyological *aisthesis*.

*Figure 5.2: **Pike Following up a Roach. Photographed in Pond.** F. Ward, Marvels of Fish Life: As Revealed by the Camera, 2nd edn (London: Cassell, 1912).*

Ward also investigates visual relations – the impression of a 'watery appearance'[37] – in his late work, *Animal Life Under Water*, which was published in 1919: 'Underwater appearances differ so greatly from the usual conception that it will be necessary first to explain the general principles of an under-water scene'.[38] If Ward focused largely on a horizontal perspective in his first book, now the vertical axis predominates. Above all, he examines the optical laws of refraction as they relate to views from the sea floor to the surface, as well as the question of the image presented by intruders from the airy world above. In so doing, Ward takes leave of familiar perspectives and explores a world that seems to obey an entirely different logic. In consequence, the reader sees the stomachs of ducks, the legs of herons, cormarant bills, swimming beavers and diving otters – as well as blurry fishermen – often strangely doubled and oddly distorted. In particular, Ward concentrates on the surface of the water as a border between two completely different lifeworlds – an exceptionally ambivalent membrane, a kind of 'skin'. Sometimes it mirrors events underwater, sometimes it leads one's gaze up into the clouds or presents the creatures there as hazy shapes. By employing perspectives that hold so many aesthetic surprises in store, Ward makes the significance of the observer's standpoint and its optical conditions plain; more still, he illustrates the categorical otherness of the underwater view in striking fashion. His unusual pictures are an eloquent, early document of insight into the relativity of human vision, the media that condition it and its culturally constructed nature.

Showtime: *With Williamson Beneath the Sea.* On the History of Early Underwater Film

On a lovely summer day in 1913, John Ernest Williamson (1881–1966), a cartoonist, reporter and photographer for the *Virginia Pilot*, was leafing through a copy of the *London Illustrated News* at the barber's.[39] He happened on an article about the *merman* Zarh Pritchard (1866–1956), one of the most celebrated underwater painters of the day, whose pictures were already hanging in many collections – the Boston Museum of Fine Arts, for example, and the American Museum of Natural History in New York.[40] Paintings of underwater landscapes had been quite popular since the aquarium craze, but Pritchard's works were unique in that they were actually made under the water. The artist employed a simple diving helmet and special colours; his paintings did not convey marine-biological details so much as the atmosphere of the ocean by means of a metaphysical use of light and dreamlike washes of colour. Another article in the same journal told of an English gentleman who had had the whimsical idea of building a subterranean observation chamber at his pond in Ipswich; in this way, he could observe and photograph the fish unnoticed. 'So, Ernie cautiously pinched the magazine', Williamson's biographer Thomas Burgess reports, 'and then, not so cautiously, sent a wire off to Denver. There was an idea here and it was getting considerably warmer.'[41] Burgess describes the slapdash success story of two guys who wanted to become rich and famous – soon after receiving the telegram, Ernie's brother George is supposed to have arrived at their parents' house where the whole family worked out a plan. Ernest, however, in the biography he published in 1936, recalls a different – and much more monumental – setting in which he received inspiration and beheld a vision:

[O]ne day, [I stepped] into a magic world that later became a reality. It happened in the old seaport of Norfolk, Virginia ... Long, mysterious shadows filled the space between the ancient buildings looming ghostly and unreal against the glow of the setting sun. Silence reigned. The place seemed utterly deserted and forgotten... Above the crooked roofs and sagging chimney was a fathomless green sky, and a strange sensation of standing on the bottom of the sea among the ruins of some sunken city came to me. I knew it was visionary ... Standing there ... I visualized these [sunken] cities once peopled by humans and now the haunts of creature of the sea. What wondrous stories they held! What astaunding pictures they would present if photographes! Perhaps there would be wrecked ships, loaded treasure galleons ... I was seized with a sudden inspiration to make photographs of the world beneath the sea.[42]

John Ernest and George Williamson helped the wonders of the sea make a grand entry into show business. Their father, Captain Charles Williamson, was the actual inventor of they diving apparatus they used: an elastic tube made out of iron rings suspended from the hull of a ship, which opened onto an observation chamber just large enough for one man to sit and operate an array of gripping mechanisms. Captain Williamson had had rescue operations, harvesting sponges and salvaging wrecks in mind. His sons, however, thought that the observation chamber would be an ideal studio for filming and photography. They retooled the construction in a few days and took it to the Chesapeake Bay; here, Ernest obtained underwater images made with artificial light at a depth of thirty metres. The photgraphs were first published in the *Virginia Pilot*. A little later – blown up to 1.8 metres – they were featured at the first International Motion Picture Exposition at the Grand Central Palace in New York, where they quickly became the main attraction. The brothers quickly found a host of sponsors to finance their 'submarine moving pictures' project. They set off for Nassau, in the clear coastal waters of the Bahamas, with Keville Glennan, Ernest's former editor-in-chief, and Carl Gregory, a cameraman, in tow. In the meanwhile, the old observation chamber had been replaced by a roomier cabin with a round window 1.5 metres in diameter; the total weight of the contraption was four tons – including ten waterproof mercury arc lamps. Ernest dubbed it the *Williamson Photosphere*; in the history of diving devices, it represented a further gain in 'enclosure and verticality'[43] – a model for 'armchair travellers'. Williamson held fast to the fundamental separation of man and the sea for all his life; it never occurred to him for a second to jump into the water with a camera himself.

The fantastic world of coral reefs in the Caribbean made a lasting impression on the team. Reports on and about the filming did not stint on vocabulary of the wondrous, dreamlike and exquisite. Ernest seized the opportunity to style himself as a discoverer whose groundbreaking invention now, for the first time in human history, would reveal the marvels of the sea to a broad audience. The publicity worked: prominent scientific and political figures made a stop in Nassau, had a look at the *Photosphere* and marvelled at what was happening underwater.

'They were amazed, and marvelled at the remarkable sights that they saw from the observation chamber', one reads in an account Victor Allemandy provided of the expedition, 'and declared that the Williamsons were without doubt the pioneers of a project which would startle the entire world with its wondrous possibilities, and which would one day be worked out to the benefit of the universe.'[44]

The exultant tone of grandiloquent self-promotion was also mirrored by the dramaturgy of the film. It showed an idyll of 'waving sea-gardens' and 'gorgeous coloured fishes', 'a spirit land, where all things were at peace',[45] divers collecting sponges and children diving for coins. To Ernest, however – and clearly for many of his contemporaries – it seemed that a decisive element was missing: 'The large blue shark that infests the waters surrounding the West Indies ... the most dreaded of all the monsters of the deep'.[46]

In order to stage the mythical battle of 'man against monster', the crew suspended a dead horse in the water as bait. A throng of native divers had been hired to fight with sharks – *ocean tigers*, as they are called in the film. Indeed, one of them managed to kill an animal with a knife. Unfortunately, this happened out of focus. Ernest decided to take the matter in hand personally:

> He ... leaped overboard into the sea He made for a big shark which had seen him and was making for him with mouth wide open, showing his six sets of vicious-looking teeth. At that moment he was right within the range of the camera, but he missed his stroke, and came back the barge. He dived again, and finally obtained what he wanted. He urged the big fellow right up to within a few feet of the glass window of the observation chamber, and drove his knife into the belly of the shark, as Gregory, thrilling as he looked, ground out his film.
>
> Those who watch this exciting battle in the submarine picture may well know that here was an example of daring which has never been approached in the history of film production.
>
> This thrilling incident brings to a close the wonderful story of how the sea has given up its secrets, and has at last bowed to the restless genius of man.[47]

The Williamsons's first films offer a history of the appropriation of the marine world, which they present as a playground for adventure standing at the ready. As is so often the case with wonders, the matter involves massive destruction and taking-possession without the slightest concern for, or curiosity about, the other. An alien world provides the backdrop for 'conquerors' to stage themselves in grippingly heroic mises-en-scène, as veritable athletes providing incomparable spectacles of entertainment. Accordingly, the wounded shark's desperate battle with death is presented as a 'monster's rage' (which proves unsuccessful, needless to say); likewise, an Atlantic ridley sea turtle twitching helplessly in the grip of its tormentor is supposed to signify the power and superiority of a human hunter. A characteristic feature of this modern way of staging the wondrous is the rhetorical combination of record-making and sensation; talk abounds of the 'greatest', 'most beautiful', 'unique', 'world-famous', 'never-before-seen', etc.

By way of the fight against the 'monster', the Williamsons brought the heroic incarnation of the diver, which had already proven successful in the nineteenth century, into cinematic history. Ernest had performed his dive 'naked', but the subsequent films he and his brother made – as well as those of their successors – showed a preference for an armoured hero of the deep. Hardly does the latter arrive at the ocean floor but a fearsome octopus threatens him with gigantic tentacles, two or three of which he duly chops off. Just as soon, a giant clam closes on one of his boots, as a ravenous sharks circles in the background. Invariably, the diver has descended either to perform a rescue or for reasons of greed: the 'treasures of the deep' always means a sunken chest of booty. The imagery remains popular to this day; in hundreds of variants, one beholds scenes of a foreign world filled with magic and terror for the courageous to conquer.

In 1914, *Williamson Expedition Submarine Motion Pictures* made its theatrical debut in the United States and unleashed waves of enthusiasm.[48] According to the *Boston Independent,* the fascination derived chiefly from the new medium of the moving image: 'These ocean meadows and forests look very different from the pictures we have seen in books and the specimens we have seen in museums, for we realize that it is life we are looking at',[49] the journal noted. Shortly thereafter, the film went on tour and garnered rave reviews the world over. The American film industry was just starting to take off in earnest, and the aesthetics of the underwater world became quite fashionable in the United States. In addition to Zarh Pritchard, mentioned at the outset, Harry Hoffman, Charles Demuth and Glyn Philpot opened exhibitions with underwater pictures in New York, San Francisco and Paris. Nor was the *London Illustrated News* the only publication to feature articles on pond landscapes and underwater gardens; reports also appeared in *Scientific American, Technical World* and *National Geographic.* Business in so-called glass-bottom boats boomed on the coasts, and the world underwater provided a setting for fiction more and more. The brothers did not want to let the opportunity pass them by, and so they augmented the intrepid treasure-diver's adventures by adding the seductions of beautiful mermaids. Having achieved the reputation as the best submarine cinematographers, they experienced a further breakthough when Universal Studios commissioned them, just a short while later, to film the underwater scenes for the adaptation of Jules Verne's *20,000 Leagues under the Sea.*[50]

Ernest's filming of Captain Nemo's underwater adventures set a new standard. The scenes surpassed all that had been presented onscreen until this point, and the press lauded the movie as 'das aufregendste Fotodrama, das bis zum heutigen Tag produziert wurde' (the most exiting photographic drama, which had been produced until this day).[51] Viewers were particularly impressed by the sight of Nemo and crew strolling on the ocean floor in diving helmets and autonomous breathing devices. Williamson had hired practiced divers who wore oxygen tanks underneath their costumes and could therefore move without a tangle of tubes connecting them to the surface. For all that, the *pièce de résistance* was the dramatic sequence when a pearl diver takes on the giant *kraken*. '[It is] one of the rarities of the camera', actor John Barrymore is reported to have said after a screening, '[t]here is no question of fake or deception'.[52]

Williamson had his 'sea monster' patented in 1921. Concealed in the creature's rubber head, a diver operated a device that made the inflated tentacles (which were equipped with tennis balls cut in two for suckers) thrash about and grab victims in wild movements. In addition, the beast was designed to spray black ink and glide, seemingly without effort, over the sea floor – a hair-raising spectacle.

Over the following years, Williamson made a series of other films, but none of them even came close to the overwhelming success of *20,000 Leagues*.[53] In 1917, he and his brother George experienced a disagreement, and the two men parted ways. Ernest stayed on in the movie business, and he also participated in a few research expeditions. They included a voyage to Andros Island (1924–34), where he and Roy Waldo Miner (1875–1955) blasted dozens of tons of coral reef from the sea in order to build a diorama for the *American Museum of Natural History* and the *Field Museum of Natural History*; ultimately, the greatest part of the Andros Reef found its way to the United States.[54]

Even though Williamson spent his entire life collecting flora and fauna from the ocean, and making photographs for purchasers, scientific curiosity remained foreign to him. Now as then, he gladly went on shark hunts and – when he was not busy blowing them apart with dynamite – thought octopodes were related to anemones (after all, they both had tentacles). He embraced the role of an emissary from Neptune's Court. In 1932, *With Williamson Beneath the Sea* hit the screen: a collage about an hour in length presenting his adventures; here, his favourite exploits – sharks, sunken treasures, foundered vessels and doughty divers – made a final, dramatic appearance. Even the old rubber octopus had a 'terrifying' cameo; it appeared between chests scattered in the deep and draped with skeletons and duly frightened grim-faced treasure hunters.

Despite these seemingly bizarre staging elements, *With Williamson Beneath the Sea* was viewed as a documentary film and shown around the world until the late-1950s. Audiences loved seeing Ernest with his wife, Lilah, and their little daughter Sylvia, who crawled around the floor of the *Photosphere* as a baby. Papers the world over spread the tale of the first little girl to visit the fish in their own home. The happy family under the waves proved a popular element of contemporary visual culture; little Sylvia, the child of 'submarine sweethearts', would be rocked to sleep in her cradle, 250 feet beneath the surface. The film was promoted as an 'adventure among the mysteries and monsters of the deep', in 'a lost world fathoms below recovered in savage splendor'.[55]

To be sure, Williamson's significance does not derive from the fact that he combined monsters and mysteries in the cinema. He did not advance the technology of underwater cameras or diving equipment. Nor did his activities lead to new scientific insights. Instead, he deserves credit for having stimulated the medium of underwater film in decisive fashion. First and foremost, he contributed to the visual culture of the submarine realm through the media-technical arrangement of the *Photosphere*, which afforded a spaceship-like view of alien worlds and, at the same time, a domestic idyll. This little family travelling through the ocean depths cocooned in a waterproof living room was the exact opposite of the elite crew aboard the technologically sophisticated *Nautilus* – which, with all its dining rooms, libraries and natural-historical collections, had represented the Europe of the late-nineteenth century. In addition, Williamson's films provide striking examples of a change in the medialization of the sea, which became the stage of mass-media entertainment more and more – a 'happy hunting-ground', as Williamson summed up his experience with the *Photosphere* in his autobiography.[56] In this form, voyage into the depths no longer occurred for the sake of truth and knowledge; instead, it displayed playful and athletic qualities. In the following years, a ludic culture of submarine amusement emerged; now, the craziest things were attempted, as if one could simply make oneself at home on the ocean floor.

In Paradise and the Netherworld with William Beebe

Submarine Promenades – The Pleasures of Helmet-Diving

Needless to say, not every one is keen on hacking the arms off an octopus or losing a boot to a murderous clam. A somewhat different version of the diver came into fashion during the 1920s. Instead of taking the stage as hero of the perilous depths, the helmeted figure now strolled through an underwater paradise as *homo ludens*. There is scarcely a better representative than William Beebe (1877–1962), a natural historian, avid diver and the director of the Department of Tropical Research of the Zoological Society in New York.[57] Beebe made his first dive in 1926 off the Galapagos Islands; he continued in Bermuda, above all. He wrote the bestsellers *Beneath Tropic Seas* (1929) and *Nonsuch: Land of Water* (1932), as well as exceptionally popular articles for *National Geographic* such as 'A Wonderer under Sea' (1932). His entertaining accounts of encounters with the curious sea-dwellers whisked readers saddled by economic crisis and mass unemployment off into an Arcadian utopia, where everyone could join his 'Society of Wonderers Under-Sea'.[58]

Beebe had a singular knack for public relations, and he was a genius when it came to raising money for expeditions. 'I wanted to make one brief sea venture, then one millionaire gave me a yacht, another millionaire gave me a yacht, and the Governor of Bermuda gave me an island. I spent ten years under water'.[59] Beebe was a man of the *Roaring Twenties* – 'a dashing figure,'[60] as his biographer Brad Matsen writes – and he managed, with consummate ease, to make marine research a field suited for dinner parties. But independent of his efforts, submarine motifs were already all the rage in the United States at the time. Underwater themes had already conquered the Art Decoratif-movement in France; American designers took them up and mass-produced fabrics, carpets and porcelain bathroom furnishings. Simultaneously, a new wave of aquarium fever was washing from coast to coast; this time, the stars were tropical fish housed in extravagantly soaring glass containers, which lent the new style an extra something. Beebe's articles fit the contemporary aesthetic perfectly, not just in terms of the metaphors of wonderlands they employed – 'Here [beneath the sea] miracles become marvels, and marvels recurring wonders'[61] – but also because of his ingenious choice of illustrations and pictorial material.

Beebe was convinced that the fine arts played a key role for understanding alien, marine worlds. On his expeditions, he regularly brought along painters and scientific illustrators – for example, Harry Hoffman (1871–1964) and the aforementioned Zarh Pritchard, whose paintings he collected. In all likelihood, however, Beebe's most felicitous acquaintance was the German-born nature painter Else Bostelmann.

[W]ithin days of her arrival in Bermuda, [she] dazzled him and everyone else with her renditions of the strange creatures of the abyss. She had a gift for visualizing the dead or living specimens that were her subjects as though they were swimming in the deep ocean. Their anatomies signales motion and vitality to Bostelmann, and her interpretations of them leaped from her easel.[62]

Bostelmann illustrated most of Beebe's essays, placing a pictorial aesthetic as congenial as it was unmistakable alongside his prose. Her wonderfully colourful style is why Beebe's 'underwater people' of the Galapagos Islands look so striking (*c.* 'A Wonderer under Sea', plate IV), the ocean world off the coast of Alaska resembles a realm of kobolds and fairies (*c.* 'A Wonderer under Sea', plate I) and deep-sea fish fly across the page as if they were sparkling rainbows (*c. The Depth of the Sea*, plate III). On the one hand, Beebe viewed marine life as an altogether mad world, full of paradoxes and as fantastic as the whimsical characters in *Alice in Wonderland*, his favourite book.[63] On the other hand, he considered himself a natural historian, not as a teller of tall tales, and Bostelmann worked for him as a respected scientific illustrator. As a matter of course, then, the two collaborators' ideas about an appropriate mode of representation for newly discovered flora and fauna took scientific precision as the point of departure. However, even though the scientific quality of Bostelmann's pictures is still widely recognized, they display an aestheticizing-fictionalizing component that sets them apart. On the one hand, this stems from her use of colour; on the other hand, it is matter of the pictures' dramatic construction. As indicated above, word and image interact and flow together in such a way that a cinematic impression arises. Beebe describes every ocean dweller as a dramatic figure – for example, the conservative periwin-kle, 'deep within his shell with his door shut tight',[64] the tenant in Lobster Alley, 'her antennae are for ever protruding from the window of her apartment'[65] and the amiable prawn, 'the most human of all sea creatures'.[66] In turn, Bostelmann's pictures constitute a storyboard, a way into this other world and a travelogue for all who have stayed at home or not yet made it to the sea. The combination proved ideally suited to a busy age shaped by media, which sought new and dynamic ways to express a sense of life that was modern, sporting and elegant.

Beebe represented the sea as a playground bathed in light just as helmeted diving was becoming quite fashionable. The practice first emerged as a diver-sion for members of the more elevated classes, attracting wealthy eccentrics and gentleman-ichthyologists; then, at the beginning of the 1930s, the technology and adventure associated with the activity commanded the notice of popular culture. Nothing seemed easier, more charming or more *à la mode* than to don a helmet and plunge into a mysterious world:

> The only requirements are a bathing-suit and a pair of rubber-soled shoes, a copper helmet with glass set in front, an ordinary rubber hose, and a small hand pump. A folding metal ladder is excellent, but a rope is quite sufficient. Down you go into two, four, six, eight fathoms, swallowing as you descend to offset the increase of pressure.[67]

Beebe's writings leave one with the impression that affordable diving devices might be purchased on any streetcorner. Indeed, if one takes a look at contempo-rary periodicals, headings such as 'Build Your Own Diving Helmet' occur with great frequency. Practical examples include: 'Build A Diving Helmet from a Water Heater' (1932) and 'Glass Cookie Jar Becomes Diving Bell'.[68] Beebe him-self dove with a 'real' diving helmet. However, if one looks at the photographs, it is plain enough that it is basically a simple (and rather battered) bucket with an observation window and a hose attached for air. And so, what proved far more important than costly equipment were a pioneering spirit and a sense of vision to make the most of one's visit to the depths.

> After you have made a dozen descents you will wish to do something more than stand amazed, or vainly try to catch the fish which swim close to the glass and look in at you. As we have done, you can begin to devise all sorts of new apparatus. You wish to make notes, so get sheet zinc of pads of waterproof paper, find a comfortable block of coral and write as easily as if you were sitting in the boat. Motion pictures can be taken, down to twenty or twenty-five feet, by placing the camera in a thight brass box with a bit of glass in front. If you wish to paint, weight your easel with lead, water-proof canvas or skin, and sit down with your palette of oils. You will have to brush away small fish from time to time, for some of the paints give forth an alluring odour and your palette will sometimes be covered with a hungry school of inchlings.[69]

Beebe describes the lordly life of a country squire or prosperous city-dweller on hol-iday, and he fêtes diving as an amusement fit for *soirées* with breezy playfulness. As a matter of course, one must go on the hunt in order to set up a garden of one's own:

> [C]hoose some beautiful slope or reef grotto and with a hatchet chop and pry off coral boulders with waving purple sea-plumes and golden sea-fans and great parti-caloured anemones. Wedge these into crevices, and in a few days you will have a sunken garden in a new and miraculous sense ... hosts of fish will follow your labours great crabs and starfish will creep thither, and now and then fairy jelly-fish will throb past, superior in beauty to anything in the upper world, more delicate and graceful than any butterfly.[70]

Eventually, Beebe was convinced, every house would have an underwater gar-den for hosting cocktail parties. His writings reflect the enthusiasm of an age that was mad about diving pioneers playing instruments underwater, painting pictures, setting endurance records and competing in submarine races. In 1927, the American Gustav Kobbe set the world record for speed-walking underwater; he covered five miles in two hours and twenty-seven minutes – and, in the pro-cess, beat the reigning champion, the Englishman William Smith.[71] In 1934, the Department of Marine Zoology at the University of Miami relocated some of its lectures to the ocean floor.[72] Judging by contemporary gazettes, any number of engineering hobbyists were abroad in the American oceans and coastal waters in one-man-submarines or underwater cars and trucks they had built. Submarine cruising, hunting, strolling and treasure-seeking had become a solid component of middle-class amusement culture.[73]

Aesthetics and Aesthesis of the Deep: Beebe and Trips in the Bathysphere

For all that, Beebe owed the greatest success of his career not to underwater strolls, but to a diving bell. On a Wednesday, 15 August 1934, he and Otis Bar-ton (1899–1992) managed to descend to 923 metres – that is, half a mile (an extraordinary depth at the time) – in the *Bathysphere*. On this record-break-ing plunge, Beebe and Barton saw animals that no human being had observed before; the two researchers were the first to have reached the ocean deep and live to tell of it. Beebe's report in *National Geographic*, 'A Half Mile Down: Strange Creatures, Beautiful and Grotesque as Figments of Fancy, Reveal Themselves at Windows of the Bathysphere', sped around the world; it was hailed as a sensa-tion comparable to landing on a distant planet.[74]

The story of the expedition has been recounted at length many times – start-ing with the account made by Beebe and Barton themselves.[75] Barton, like his companion, was an avid diver, but he was also an engineer. The *Bathysphere* was essentially built according to his design – and with his financial resources. At 1.5 metres in diameter, it offered cramped room for two men. The vessel was suspended from a steel rope on a winch aboard the carrier ship, to which it was also connected by a power cable for light and a telephone line for communication. Oxygen came directly from bottles inside, and the carbon dioxide exhaled by occupants was caught in chemical filters. Beebe and Barton made trial dives in the *Bathysphere* starting in 1928. In 1930, it was deployed off the Bermuda Islands, where it reached a depth of 400 metres. Beebe wrote several articles about the undertaking for the *New York Times Magazine, Illustrated London News, Science, Harpers, McCall's Magazine, Royal Gazette, Colonist Daily* and *National Geographic.* In so doing, he anchored the theme of deep-sea diving in the public mind and secured a place for the *Bathysphere* in visual culture.[76] From this point on, accounts of his dives were a regular feature of the country's newspapers. On 22 September 1932, a live report of a dive was even transmitted over the radio. The *National Broadcasting Company* had sent a team to Bermuda to transmit Beebe's voice from the deep – 1,500 feet beneath the waves – all over the world. For twenty-five minutes, he told rapt listeners in the United States, England, Wales and northern France what organisms he encountered on his way to the deep. This broadcast is said to have been one of the biggest blockbusters in early radio history.[77] Two years later, in 1934, *National Geographic* sponsored another expedition; this was the occasion for the record mentioned at the outset. Thanks to sensational press reports and the celebrated account he authored, *A Half Mile Down*, Beebe achieved fame the world over.

A net of metaphors runs through Beebe's writings on the ocean depths, rich in association and scenic elements of visions emerging and floating by. The *Bathysphere's* porthole opens onto an infinite theatre, where Beebe has creatures originating in another world take the stage. He refers to 'Davy Jones's Locker', 'mysteries', 'figments of fancy', 'Hades', 'dragons', 'magic', 'portals of eternal obscurity' and the 'mouth of Hell' – that is, to an array of components from premodern cultures; if they represent the author's effort to convey knowledge to the public, they are also, as the historian of art and science Martin Kemp defines the term, 'deep structures of intuition'.[78]

Here, too, Else Bostelmann's illustrations accompany the text. The painter herself never dove in the *Bathysphere*; accordingly, her pictures resulted from an interesting combination of medial transformations and representational practices. To begin with, Beebe took notes and made sketches during his trips. Secondly, he stood in contact, over the telephone, with the mother ship the whole time; all that he reported was duly taken to protocol. Another element Bostelmann incorporated were specimens caught in nets that Beebe had cast; deep-sea fish were caught and photographed back on the surface. In sum, the artist was given a host of heterogeneous medial sources and practices – text, spoken language, sketches, photographs and specimens – which she then had to synthesize visually.

The extreme conditions of the observational situation undoubtedly exercised a formative influence on the constitution of knowledge. In contrast to helmet-diving, the steel ball that weighted tons and sealed Beebe in a kind of cocoon involved a different media-technical arrangement; on the one hand, he was afforded uninterrupted vision that could be varied at will, but for the most part he saw things only once. Organisms floated past his porthole whose dimensions he could not determine, and they often disappeared promptly into the night of the deep never to be seen again. Cold, tight space and fear shaped the experience of diving, which was essentially a matter of unexpected occurrences in isolation, comprehensible only in terms of metaphors of depth and darkness. Even though Beebe knew the flora and fauna he encountered on sunlit underwater strolls, the animals now appeared in their own environment. What is more, 'unprecedented' creatures came and went – phenomena that neither he nor any other marine researcher had ever seen. The experience of seeing for the first (and only) time raised the epistemological question of the process of learning to see and recognizing what, until now, had been entirely unknown.

Merely speaking about, and describing, the perception of something new involves multifaceted medial transfers and transformations. By means of a progressive, aesthetic-poetic approach, the marine world becomes aligned with the demands of scientific investigative practice. In this way, encounters prefigure how ocean dwellers are viewed and understood from this point on. Beebe's accounts and Bostelmann's illustrations prove remarkable inasmuch as they enable one to examine the initial stage of research, which is only rarely visible, whereby experience that is overwhelming and indeterminate generates a field of objects for scientific examination. The concurrent operation of the most varied processes of visualization yields what comes to stand as fact; thereby, its constructed nature is made plain. That is, a combination of measures for producing evidence and achieving aesthetic mediation is required so that what is fleeting and mobile will 'congeal' into stable images; visualizations of underwater space henceforth unite epistemological and aesthetic dimensions. This is why it is so revealing to examine medial modelizations as they progress from the initial encounter to the finished picture – to ask what transformations the object undergoes in the process and the affective and cognitive energies invested in it.

All of these aspects bear closely on the constructed nature of the underwater realm as a field of insight: it can only be observed and experienced via media-technical mediation, as an artifact. To be sure, the same holds for all epistemic spaces, but the sea in general – and the deep sea in particular – brings the matter to a head. To begin with, the *medium diaphanum* (to employ a term of the classical doctrine of *aisthesis*) is not air, but water. Secondly, the ocean depths lack a fundamental condition for human observation: light. Accordingly, the observation of nature underwater occurs in a space of perception that is alien to the human sensorium; here, all questions concerning perception and epistemology involve wholly different relations between the observing subject and its object and the same holds for the epistemic status of what is perceived.

Beebe notes a striking visual disturbance that happens at the very outset of every dive:

> As I have said, the first plunge erases to the eye all the comforting warm rays of the spectrum. The red and the orange are as if they had never been, and soon the yellow is swallowed up in the green. Although the cheerful rays are only one-sixth of the visible spectrum, yet, when they are winnowed out at a hundred feet or more, all the rest belongs to chill and night and death.[79]

Metre by metre, the traveller remarks how the colour of his surroundings changes and his vision progressively adapts to the sights of a gloomy underworld. He is moving through a realm of the dead distinguished by radiance, as it were. At 600 feet, the sea appears as a dark blue glow, 'an indefinable translucent blue quite unlike anything I have ever seen in the upper world'.[80] This incredible hue provokes an extraordinary feeling of being touched and moved – a form of startled perception in which, to borrow Bernhard Waldenfels's phrasing, 'Eigenes und Fremdes, Selbstaffektion und Fremdaffektion sich miteinander verschlingen' (intimacy and strangeness, self-affection and extrinsic affection intermingle).[81]

> [T]he blueness of the blue, both outside and inside our sphere, seemed to pass materially through the eyes into our very beings. This is all very unscientific ... but there it was ... I think we both experienced a wholly new kind of mental reception of color impression. I felt I was dealing with something too different to be classified in usual terms.[82]

400 feet lower, the sea appears black-blue or dark grey. A little later, the culminating point is reached: 'The sun is defeated and colour is banished forever until a human at last penetrates and flashes a yellow electric ray into what has been jet black for two billion years'.[83]

In Beebe's account, the central points of reference are shifting colour and darkness. These features are joined by the indeterminacy and fleetingness of impressions when animals dart by at a great distance or too quickly. Over and over, he brings up the inadequacy of his visual faculties and the limitations of language. With utmost precision, he records his physical sensations and what he senses in his surroundings; in so doing, he draws attention to the process of perception itself, as well as practices of description and recording. Such self-reflectivity makes it possible to read his account as a historical case study about the influences on, and conditions of, scientific observation – that is, it raises the question of *what* defines scientific observation in a given situation (in this instance, marine research).

A mere ten centimetres beneath the water's surface, one must already ask what the real colour of a fish is. What is more, hearing, touch and smell hardly factor into perception underwater. Since the human sensorium comprises a polymodal arrangement and components do not operate independently, submarine research poses the question of the extent to which minimized hearing and touch influence visual perception. So long as there is still sunlight, the problem does not seem so urgent: seeing underwater proves different, yet light conditions in air and water approximate each other for human beings. As depth increases, however, an artificial source of light proves necessary; the submarine world is not made for human beings and their preferred sensory organ, the eye. In the deep, acoustic and olfactory sensors, as well as perception of temperature and water pressure, define the parameters of *aisthesis*. And even when vision plays a role, a particular world of light and colour prevails in the deep, the actual properties of which can only be discerned in the dark – or, more precisely, not discerned at all. Curiously enough, camouflage is important even at the deepest layers of the submarine world.[84] This is why black and silver colours predominate among deep-sea dwellers – whereas prawns display red 'clothing', for seawater takes this hue out of the colour spectrum at just ten metres' depth. That said, the velvety void that does not reflect light often occurs only on animals' backs, whereas their sides cast a silver shimmer. In this way, the flanks produce the effect of mirrors reflecting the light that remains in the 'twilight zone' above 800 metres with the same brightness and colour – a veritable feat: the fish look invisible! For a similar reason, the underbelly of many deep-sea fish is light in colour. Predators, which often swim below, look for the silhouettes of prey against the dim light falling from above; a bright stomach conceals the animal from its enemy lurking in the deep.

However, the light design of inhabitants of the deep does not just fuse the effects of light and shade; it also features phosphorescence, bioluminescence and even blinding mechanisms. Of the approximately 1,500 kinds of deep-sea fish, about one thousand possess a 'lantern', feet that glow green, or firecracker-red feelers – that is, colourfully shimmering photophores that indicate species, sex and behaviour. Some of these features are used to lure prey or radiate sex-appeal; others are arranged in rows on the fish's sides to act as a means of disguise, what is known as 'backlighting'. Thus, lanternfish have an internal dimmer to adjust their photophores to even the slightest change in light, which makes their silhouettes invisible to predators underneath. Many prawns and cephalopods, but also *searsia koefoedi*, a kind of smelt, can blind enemies by becoming 'flamethrowers'. They secrete a luminous substance that can glow for up to four seconds, and they use this time to escape into the dark. Beebe was the first person to see these effects of colour and light 'live and in person' – an experience that only heightened his impression that he was floating through a heavenly landscape under the water. Needless to say, however, stars and phosphorescent organisms are visible only in the dark. If one turns on the light – as Beebe did – then the bioluminescent sea of stars pales before the electric sunrise; looking for comparable effects of colour is like examining a fish in dry air.

When light from an electric source shines on them, the creatures of the deep transform into wan phantoms and guardians of Hades. At any rate, that is how Beebe describes some of the encounters he had by the light of the *Bathysphere*.

Under the heading, 'A Fish New to Science', he presents his discovery of something wholly unknown:

> The strange fish was at least two feet in length, wholly without ligths or luminosity,
> with a small eye and good-sized mouth ... An unusual thing was the color, which ...
> was an unpleasant pale olive drab, the hue of watersoaked flesh, an unhealthy buff. It
> was a color worthy of these black depths, like the sickly sprouts of plants in a cellar.[85]

Beebe christened the 'strange apparition' *Pallid Sailfin* (*Bathyembryx isti-
ophaxma*), a Greek phrasing for 'the fish from the depths that swims with ghostly
sails'. To this day, no specimen fitting this description has been seen again. At
1,900 feet, a little later on, another discovery lay in store:

> At a moment of suspension came a new and gorgeous creature. I began ... to absorb
> what I saw: Along the sides of the body were five unbelievably beautiful lines of light,
> every one surrounded by a semicircle of very small but intensely purple photophores.[86]

Here, too, Beebe provides a name. He stresses that he does so but rarely when
encountering a fish he never seen before – only when he is sufficiently certain, despite the short duration, of its taxonomy: 'My name for it is Bathysidus
pentagrammus, the Five-lined Constellation Fish. In my memory it will live
throughout the rest of my life as one of the loveliest things I have ever seen'.[87] The
Constellation Fish has also not been sighted again; along with the Pallid Sailfin
and a few other discoveries, it lives on in the catalogues of cryptozoologists.

Beebe describes the Pallid Sailfin as a ghostly fish with a sail-like tail. Its colour,
he writes, is an unpleasant olive – yellowish grey; its skin seems unwholesomely
leathery and water-logged, sickly like mould in a cellar. When one looks at the
accompanying image, however, an undead, watery corpse does not necessarily
spring to mind. Here, the beam of the *Bathysphere* splits the impenetrable darkness
and produces a sheen of glowing blue. One sees three fish floating past, unaffected
by the light and undaunted by the craft. The light – or, more precisely, the aquamarine background – gives their bodies a golden yellow glow, and the 'sails' look
delicately transparent. The effect of light on colour is reflected in the illustration
itself, which shows other fish outside its sphere of incidence. Here, their colouration is indeed brownish-dark, and the sails are bluish-violet; for all that, they do
not necessarily give one the impression of ghosts lurking in a basement.

The fact that, to this day, no one else has seen the Pallid Sailfin does not necessarily speak against Beebe's powers of observation. The deep sea may house
500,000 – or even ten million – different species, including the Pallid Sailfin; we
do not know. Beebe's colleagues often faulted him for his 'amateurish science';
in particular, his taxonomic suggestions met with incomprehension.[88] Once he
called a fish that he had seen from the window of his craft the 'Untouchable Bathysphere Fish' and published the discovery in a professional journal. The Latin name

he provided was *Bathysphaera intacta* – a source of no small amusement in the scientific community: *intacta* means 'virginal'.[89] But all criticism notwithstanding, Beebe counted – and counts – as an important figure who identified many new species. Bostelmann's pictures are also considered important scientific illustrations to this day.[90] And so, the question that arises involves discrepancies between image and description. Needless to say, as the *Bathysphere* descended, the longish form of the Sailfish, its great fin and dark brown tones, conjured up associations for Beebe in which modern and archetypal motifs of western imagology fused.

This combination should not surprise us. After all, the vertical dimension plays a singular role in the modern repertory of metaphors and images. Fittingly, Beebe's descent in search of new insights was attended by the whole cast of occidental voyages to the deep – mythological, geological, archeological, psychoanalytic, etc. It represented a journey into the underworld, into the land of the dead, eternal night, to the foundation of truth, the abysses of one's own soul and back to the first stirrings of life. It would be mistaken to see a lack of objectivity at work here. After all, all knowledge is formulated in a culturally specific manner. Yet such relativity does not mean that knowledge itself is arbitrary. Instead, the language that Beebe employs points to the metaphorical potential and evocative richness harboured by every scientific idiom; in particular, it attests to the imaginative intensity and amazement that accompanied and fueled nascent deep-sea research. Still today, for example, the American NOAA (National Oceanic und Atmospheric Administration) speaks of the deep as 'inner space', in analogy to the 'outer space' of the cosmos.[91] What is more, the prospect that investigating the ocean depths will bring us closer to the secret of how life on earth began is regularly renewed by further discoveries. Not long ago, Michael Russel and William Martin advanced the claim that the environment surrounding the 'Black Smoker' discovered in 1977 might present conditions like the ones that prevailed when life first started.[92]

It is worth remarking that Beebe, when presenting the Sailfin and Constellation Fish, thematizes the process of visualization that will later occur. Apropos of the Sailfin, he speaks of his effort to materialize his mute impression of the fish in spoken and written language, and ultimately to transfer it into an image 'dictated by what I had seen through the clear quartz'.[93] That is, Beebe had to lend concrete form to what had presented itelf immaterially; his gaze through the window prescribes something – in medial terms, records data – that must then undergo transformation. Of the Constellation Fish, he writes: 'Little by litte my brain fish materialized, their proportions size, colour, lights, fins interdigitated with those of my memory'.[94] Beebe describes the scientific process of visualization as a matter of embodiment and media-technical projection. In this arrangement, the ray from the *Bathysphere* does not just illuminate the the deep; it also constitutes part of a process of medial visualization. Joined with the darkness of the underwater 'screening room', the diving bell becomes a camera and projector in one – both a *laterna magica* and a *camera obscura*, part of a medial recording apparatus and a device for producing virtual phenomena. At any rate, Beebe's account suggests interpretation along these lines. On the one hand, the deep sea represents a field of scientific exploration, and its undeniable materiality could cost him his life at the slightest mistake. On the other hand, what he sees seems immaterial and unreal; he confronts 'brain fish' – abstract, ungraspable creatures, ghostly figures and dream visions that present themselves as moving images of light. Such interpenetration of practiced nature-observation and imagination at the moment of discovery – the *apparitions* that pass before Beebe's eyes – accounts, in my estimation, for the discrepancy between text and image. The *Pallid Sailfin* is the ship of the dead – this is what the Latin name means in translation; it travels on ghostly sails and appears to Beebe as he makes his way into eternal darkness. At the same time, the fish that glows gold against the aquamarine background; its mysterious comings and goings beckon onward, into the expanse of deep-sea night. As Beebe perceives it – or recalls it – the two become fused. We 'landlubbers' must decide whether we wish to believe the text or the image – at least until we see a Pallid Sailfin of our own.

At 1,900 feet Beebe was greeted by the *Constellation Fish*, brilliantly illuminated like a pulsating *Aurora borealis*, as he writes, a celestial glow of the sea. Apart from the luminous lines, the Constellation Fish displays almost the same colour as the Pallid Sailfin, except that Beebe describes it as 'decidedly brownish'[95] here. Viewed from the front, he reports, the Constellation Fish wears a mournfully pensive, contemplative expression; from the side, however, it displays its true, unbelievable beauty. For the rest of his days, the voyager writes, he will remember the vision as the loveliest he ever saw. The illustration does not make the reason for this overwhelming impression intuitively clear. To be sure, the fish are pretty and delicate – as if they came from an illustrated world of fairytales. As much is underscored by Beebe's anthropomorphizing description of their facial expression as contemplative and mournful. But at the same time, it seems that Beebe has beheld a vision of the ineffable, shimmering, pulsating and beautiful. He has had the experience of something that divides his life into a 'before' and 'after' in encountering the marine *Aurora borealis*, a wonder of nature. The fact that it displays childlike traits does not speak against the intensity of the voyager's emotion; indeed, it confirms it. In the vastness of the sea, Beebe finds exactly what he hoped to discover: a fairytale being that could have come from his favourite book, *Alice in Wonderland*.

In this context, the Constellation Fish appears as a figure of night and dream – the ocean world opens onto a land beyond the looking glass. Magical effects overlie the situation of scientific observation. In the play of submarine lights it becomes clear that looking through the quartz windows of the *Bathysphere* does not reveal a riddle so much as the contours of a gripping, new mystery. This mystery opens a new category of experience that does not belong to this world.

The book that appeared the year following the dive, *A Half Mile Down*, explicitly expresses as much by juxtaposing two photographs: the star-filled sky and marine phosphorescence. That Beebe really did feel his underwater journey to be a cosmic voyage is also made clear by the conclusion of his account in *National Geographic*:

> The only other place comparable to these marvelous nether regions must surely be naked space itself, out far beyond atmosphere, between the stars, where sunlight has no grip upon the dust and rubbish of planetary air, where the blackness of space, the shining planets, comets, suns, and stars must really be closely akin to the world of life as it appears to the eyes of an awed human being in the open ocean a half mile down.[96]

The impression that the deep sea is not of this world recurs throughout all that Beebe wrote. He speaks of 'unearthly colors'[97] and a realm 'as strange as that of Mars'[98] which is 'cosmic' and 'superhuman';[99] likewise, he refers to the *Bathysphere* as 'a lost planet in outermost space'.[100] Thus, the eternal and infinite represent points of reference as constant as night, magic and journey into the past: 'I would feel ... like a paleontologist who could suddenly annihilate time and see his fossils alive'.[101] In this discursive setting, Bostelmann's illustrations animate the Darwinian drama of life in a struggle for survival that proves merciless and inventive in equal measure. Primeval-seeming scenes of eating and being eaten predominate; joined with Beebe's commentary, they yield modern picture-stories of the wonders of nature. Under the heading, *Carnivores of a lightless world,* one beholds 'Baby Dragons ... with enormous fangs, which they use with terrible effect'[102] in an archaic universe like a cabinet of wonder. 'Armoured Knights of Neptune'[103] bearing lances strike upon 'morons of the deep',[104] 'giants'[105] and 'Big bad wolves of an abyssal chamber of horrors'.[106] However, as soon as beauty takes the stage – fish that look like rainbows, sparkling, shimmering and moving elegantly – the imagery evokes idyllic peacefullness and floating ease. In this respect, the illustrations connect to the tradition of representing the earth's evolutionary history, in which the animal world – the closer it comes to the 'arrival of man' – appears more and more graceful and noble. Even though the pictures do not provide a chronology (inasmuch as they follow Beebe's report of a dive and not the 'story of Creation'), the difference between a prehistoric period where 'survival of the fittest' prevailed and an idyllic time of cultivated forms and arrangements of colour is unmistakable. And so, Beebe and Bostelmann's connection between text and certain modes of visualization yields not only a voyage into the ocean depths, but also a journey into the abyss of time. Beebe's 'time travel' is marked by a strange paradox. On the one hand, it represents a return to 'a home which was once our own',[107] millions of years ago; here, one can say with some confidence, 'When you were a tadpole And I was a

fish'. On the other hand, this 'olden home [is] infinitely more remote and fundamental than our air-breathing life today upon the dry land'.[108] Beebe ascribes the
most contradictory qualities to the deep sea, a realm of ravenous forms of horror
and shimmering magical beings, where we encounter both absolute foreignness
and the familiar host of Western myths and monsters. On the one hand, he
describes a cosmic expanse that is as inhospitable as outer space and stands for
the wholly alien; on the other hand, the trip into the deep leads him to the birth-
place of terrestrial life, as well as zones of his own unconscious. The journey leads
back to prehistoric times, to the beginning of life, to the empire of the dead, to
a mysterious land abounding in vitality. Read in this way, Beebe's writings do, in
fact, reveal a secret: it is based on the insight that the deep sea resists the parameters of scientific construction, because it is also a mighty ocean of discourse
and a mythological echo-chamber. Here lies the foundation for modern forms
of wonder: a dynamic connection between depth and the past, outer- and inner
space, origin and return. Against this backdrop, the underwater views of another
world should also be understood as projection rooms, which both include the
pathways of knowledge and insight and involve a wondrously supple nature, as
it were. Whatever we might find, capture, measure and dissect yields a concrete
paradox as we try to comprehend it and communicate what we think we understand: the experience of, and intimate encounter with, foreignness that eludes
human grasp, familiarity with overwhelming variety and countless curiosities –
each one more otherworldly, exotic and unreal than the next.

The most remarkable thing about this strange world is that it covers 78 per
cent of our planet.

Addendum:

On 22 December 2014, the *Frankfurter Allgemeine Zeitung* reported the 'Discovery of New Fish Species in the Deep Sea'.[109] There, 8,143 metres below, the
atmosphere is 'ghostly'; the pale fish look like 'phantoms sluggishly gliding over
the ocean floor'. When these 'ghosts' were discovered, a new record was set: now,
Pseudoliparis amblystomopsis has to content itself with second place among fish
sighted at the greatest depth; a few grenadiers and 'giant' amphipodes have confirmed the unsettling enormity of inhabitants of the deep-sea world. Although
almost eighty years lie between Beebe's report and this announcement, it seems
that little has changed in many respects: the deep sea is – and remains – 'the
largest unexplored biosphere on earth' and expeditions seeking sensation and
records 'in the unknown' still go by the name 'Hades'.

EPILOGUE

'In the seas', Paul Valéry wrote at the beginning of the 1930s, 'one can really see wonders that surpass all imagination, almost touch them'.[1] Still today, many would likely agree with this assessment, even though decisive changes occurred over the course of the second half of the twentieth century. The 'Mysterious Science of the Sea' explored in the study at hand belongs to the past. On the basis of this history, however, one may reflect on age-old wisdom: There are mysteries, and then there are problems. The latter are to be solved, and they vanish in turn; the former, however, are to be experienced and incorporated into one's own life. According to Romano Guardini (1885–1968), 'a mystery solved never was one to begin with. True mystery defies explanation not because it eludes verification through some kind of trick of double truth, but because it cannot be rationally resolved in essence. Still, it belongs to the same reality as what admits explanation, and its relationship to explanation is absolutely honest'.[2] Even if many people continue to experience the seas as a world of mystery and wonder, there is no mistaking the fact that the waters covering the earth have transformed from a mystery into a problem. 'The vast ocean is, in fact, finite, fragile, and at risk', one reads at the entrance to the Lisbon Oceanarium. In vast tanks behind enormous panes of glass, visitors can study 'another world' that seems completely separate from us, which may be extraterrestrial in origin; here, other laws hold, not just the ones we know and they may contain the past of all the life on our planet. Needless to say, however, this is only half the story: as we have seen at length, contemplating the sea over time does not lead only to the depths of geological history; it also contains a utopian dimension, as it were. Just as we know that all life comes from the sea, so, too, we might recognize that there will be no future that does not depend on the oceans. And so, when we emerge again and stand on the shore, looking over the Tagus toward the Atlantic, we are gripped not just by amazement, but by profound unease, as well.

WORKS CITED

Adam, V., 'William Thompson – 100 Years of Underwater Photography?', *Focus*, 49 (September 1993), pp. 4–8.

Adamowsky, N., *Das Wunder in der Moderne. Eine andere Kulturgeschichte des Fliegens* (Paderborn: Wilhelm Fink Verlag, 2010).

—, H. Böhme and R. Felfe (eds), *Ludi naturae. Spiele der Natur in den Künsten und Wissenschaften* (München: Wilhelm Fink Verlag, 2011).

Albritton, C. C., *The Abyss of Time: Changing Conceptions of the Earth's Antiquity after the Sixteenth Century* (San Francisco, CA: Freeman, Cooper 1980).

Aldrich, F. A., 'Some Aspects of the Systematics and Biology of Squid of the Genus Architeuthis Based on a Study of Specimens from Newfoundland Waters', *Bulletin of Marine Science*, 49:1–2 (1992), pp. 457–81.

Aldrovandi, U., *De reliquis Animalibus exanguibus libri quatuor post mortem ejus editi ... de Mollibus, Crustaceis, Testaceis, et Zoophytis* (Bologna, 1606).

Allemandy, V., *Wonders of the Deep: The Story of the Williamson Submarine Expedition* (London: Jarrold & Sons, 1916).

Allen, D. E., *The Victorian Fern Craze* (London: Hutchinson, 1969).

—, *The Naturalist in Britain: A Social History* (Princeton, NJ: Princeton University Press, 1976).

—, 'Tastes and Crazes', in N. Jardine and E. C. Spary (eds), *Cultures of Natural History* (Cambridge: Cambridge University Press, 2000), pp. 394–407.

Ambrosius, *Hexameron*, ed. J. Heynlin (Opera), pp. 63–106.

Andersen, H. C., *The Dryad* (1868), trans. J. Hersholt, published online by the Hans Christian Andersen Centre: andersen.sdu.dk/vaerk/hersholt/TheDryad_e.html [accessed 12 February 2015].

'The Aquativ Vivarium at the Zoological Gardens, Regent's Park', *Illustrated London News*, 28 May 1853, p. 420.

'The Aquarium Simplified', *Home Friend*, 1 (1856), pp. 130–2.

Aristotle, *Physics, or, Natural Hearing*, trans. G. Coughlin, (South Bend: St Augustine's Press, 2005).

Assmann, P. (ed.), *Alfred Kubin (1877–1959)* (Linz: Oö. Landesgalerie, 1995).

—, 'Annäherungen an die Farbe im Werk Alfred Kubins', in P. Assmann, *Alfred Kubin (1877–1959)* (Linz: Oö. Landesgalerie, 1995), pp. 164–206.

Atz, J. W., 'The Timid Octopus: For Centuries Man has Credited This Realtive of the Oyster With Fearsome Powers which It Does Not Possess', *Bulletin of the New York Zoological Society* (1940), pp. 48–54.

—, 'The Balanced Aquarium Myth', *Natural History*, 58 (1949), pp. 72–7, 96.

Augustinus, A., *De Genesi ad Litteram Libri Duodecim*, in *St Augustine: The Literal Meaning of Genesis*, trans. J. H. Taylor, (n.p.: Newman Press, 1982).

Baader, H., 'Gischt. Zu einer Geschichte des Meeres', in W. Baader (ed.), *Das Meer, der Tausch*, pp. 15–40.

—, and G. Wolf (ed.), *Das Meer, der Tausch und die Grenzen der Repräsentation* (Zürich: Diaphanes-Verlag, 2010).

Bachelard, G., *The Poetics of Space: The Classic Look at how we Experience Intimate Places*, trans. M. Jolas (Boston, MA: Beacon Press, 1994).

Baglioni, S., 'Zur Kenntnis der Leistungen einiger Sinnesorgane (Gesichtssinn, Tastsinn und Geruchssinn) und des Zentralnervensystems der Cephalopoden und Fische', *Zeitschrift für Biologie* (1909), pp. 255–86.

Baker, N., 'William Thompson – The World's First Underwater Photographer', *Historical Diving Times*, 19 (Summer 1997), pp. 8–16.

Baltrusaitis, J., *Das phantastische Mittelalter. Antike und exotische Elemente der Kunst der Gotik* (Frankfurt am Main: Propyläen, 1985).

Barber, L., *The Heyday of Natural History: 1820–1870* (London: Cape, 1989).

Barck, K., 'Wunderbar', in K. Barck, M. Fontius, D. Schlenstedt, B. Steinwachs and F. Wolfzettel (eds), *Ästhetische Grundbegriffe. Historisches Wörterbuch in sieben Bänden* (Stuttgart and Weimar: J. B. Metzler, 2005), vol. 6, pp. 730–73.

Barnow, E., *Documentary: A History of the Non-Fiction Film* (New York and Oxford: Oxford University Press, 1993).

Barr, J., 'Why the World was Created in 4004 BC: Archbishop Usher and Biblical Chronology', *Bulletin of the John Rylands University Library*, 67 (1985), pp. 575–608.

Barton, O., *The World Beneath the Sea: The Story of the Deepest Dive Ever Made by Man* (New York: Thomas Y. Crowell, 1953).

Batchelder, E., *A Romance of the Sea Serpent or the Ichthyosaurus and a Collection of Ancient and Modern Authoritie* (Cambridge: J. Bartlett, 1849).

Bateman, G. C., *Fresh-Water Aquaria: Their Construction, Arrangement, and Management* (London: Upcott Gill, 1870).

Baudelaire, C., 'Man and the Sea', in C. Baudelaire, *Flowers of Evil*, with an introduction by J. Culler and trans. J. McGowern (1857; London: Oxford University Press, 1993).

Bauer, V., 'Einführung in die Physiologie der Cephalopoden. Mit besonderer Berücksichtigung der im Mittelmeer häufigen Formen', *Mittheilungen aus der Zoologischen Station zu Neapel*, 19:2 (1909), pp. 149–268.

Bayertz, K., 'Biology and Beauty: Science and Aesthetics in Fin de Siècle Germany', in R. Porter and M. Teich (eds), *Fin de Siècle and its Legacy* (Cambridge: Cambridge University Press, 1990), pp. 278–95.

Beebe, W., 'Down to Davy Jones' Locker', *New York Times Magazine*, 13 July 1930, p. 1.

—, 'Diving to a Depth of a Quarter of a Mile', *Illustrated London News*, 11 April 1931, pp. 594–5.

—, 'A Round Trip to Davy Jones' Locker', *National Geographic Magazine*, 59:6 (1931), pp. 644–78.

—, *Nonsuch: Land of Water* (London and New York: Puntams, 1932).

—, 'A Wonderer under Sea', *National Geographic*, 62 (December 1932), pp. 741–58.

—, 'The Depths of the Sea. Strange Forms a Mile Below the Surface', *National Geographic Magazine*, 61:1 (1932), pp. 65–88.

—, 'Descent into Perpetual Night', *New York Times Magazine*, 9 October 1932, pp. 1.

—, 'Exploration of the Deep Sea', *Science*, 76 (14 October 1932), pp. 344.

—, 'Thoughts on Diving', *Harpers*, 166 (April 1933), pp. 582–6.

—, 'Going Down', *McCall's Magazine*, 14 March 1933, pp. 51–2.

—, 'A New Deep Sea Fish Story', *Royal Gazette and Colonist Daily*, 11 October 1933, pp. 1–2.

—, 'Kingdom of the Helmet', in W. Beebe, J. Tee-Van, G. Hollister, J. Crane and O. Barton, *Half Mile Down* (New York: Hartcourt Brace and Company 1934), pp. 66–86.

—, 'A Half Mile Down. Strange Creature, Beautiful and Grotesque as Figments of Fancy, Reveal Themselves at Windows of the Bathysphere', *The National Geographic Magazine* (1934), pp. 661–704.

—, 'A First Round Trip to Davy Jones' Locker', in W. Beebe *Adventuring with Beebe. Selections from the Writings of William Beebe* (London: The Bodley Head, 1956), pp. 54–84.

Bellows, A. M. and M. McDougall (eds), *Science is Fiction: The Films of Jean Painlevé* (San Francisco, CA: Brico Press, 2000).

Belly, F., 'La Poésie de l'exposition', *Revue contemporaine*, 23 (1855), p. 162.

Belon, P., *L'Histoire naturelle des estranges poissons marins avec la vraie peincture et description du Dauphin et de plusieurs autres de son espèce observée par Pierre Belon du Mans* (Paris: Regnaud Chaudiere 1551).

—, *La Nature et Diversité des Poissons* (Paris, 1555).

Benchley, P., *Beast* (New York: Random House, 1991).

Benson, K., H. Rozwadowski and D. van Keuren (eds), *The Machine in Neptune's Garden: Historical Perspectives on Technology and the Marine Environment* (Sagamore Beach, MA: Science History Publications, 2004).

Bentheim Jutting, W. S. S. van, 'A Brief History of the Conchological Collections at the Zoological Museum of Amsterdam with Some Reflections on 18th Century Shell Cabinets and their Proprietors', *Bijdragen tot de Dierkunde* (Leiden: Brill, 1938), pp. 167–245.

Berg, B., 'Contradictory Forces: Jean Painlevé, 1902–89', in M. McDougall (ed.), *Science is Fiction: The Films of Jean Painlevé* (San Francisco, CA: Brico Press, 2000), pp. 2–47.

—, 'The Art of Science: 1902–1989', in Museum of Contemporary Art Sydney (ed.), *Liquid Sea* (Sydney: Museum of Contemporary Art, 2003), pp. 14–9.

Berman, M., *The Reenchantment of the World* (Ithaca, NY: Cornell University Press, 1981).

The Bermuda Biology Station for Research (ed.), *Bermuda 100 Years The First Century: Celebrating 100 Years of Marine Science. BBSR 1903–2003* (Bermuda, 2003).

Bernoulli, C., *Ueber das Leuchten des Meeres, mit besonderer Hinsicht auf das Leuchten thierischer Körper* (Göttingen: Heinrich Dieterich, 1803).

Berra, T. M., *William Beebe: An Annotated Bibliography* (Hamden: Archon Books, 1977).

Berrill, N. J., *The Living Tide* (New York: Dodd, Mead & Company, 1951).

—, atlantische wunderwelt (Frankfurt am Main: Bücherghilde Gutenberg, 1957).

Berthelot, M. S., *Comptes Rendus Hebdomadaires des séances de l'Academie des Sciences*, 53 (December 1861), pp. 1256–67.

Besler, B., *Fasciculus rariorum et aspectu dignorum varii generis quae collegit* (Nürnberg, 1616).

Besler, R., *Gazophylacium rerum naturalium e regno vegetabili, animali et minerali* (Leipzig, 1642).

Beta, H., 'Der Sohn des, alten Brehm', *Die Gartenlaube. Illustrirtes Familienblatt*, 17 (1869), p. 22.

Bevan, J., *The Infernal Diver: The Lives of John and Charles Deane, Their Invention of the Diving Helmet, and its First Application to Salvage, Treasure Hunting, Civil Engineering and Military Uses* (London: Submex, 1996).

Bidder, F., 'Vor Hundert Jahren im Laboratorium Johannes Müllers', *Münchner Medizinische Wochenschrift* (1934), pp. 60–4.

Biella, E., 'Das Berliner Aquarium Unter den Linden/Schadowstraße: Zur Konzeption des Rundganges und Bedeutung des Grottenstils als Ausstellungsarchitektur', *Der Bär von Berlin: Jahrbuch des Vereins für die Geschichte Berlins*, 49 (2000), p. 63.

Billardon de Sauvigny, E., *Histoire naturelle des dorades de la Chine* (Paris, 1780).

Bippus, E., 'Skizzen und Gekritzel. Relationen zwischen Denken und Handeln in Kunst und Wissenschaft', in M. Heßler (ed.), *Logik des Bildlichen* (Bielefeld: Transcript-Verlag, 2009), pp. 76–93.

Birch, T., *History of the Royal Society of London* (1756).

Bisanz, H., *Alfred Kubin: Zeichner, Schriftsteller und Philosoph* (München: Edition Spangenberg, 1977).

Bloch, M. E., *Allgemeine Naturgeschichte der Fische* (Berlin, 1782–95).

Blumenberg, H., *Der Prozeß der theoretischen Neugierde* (Frankfurt am Main: Suhrkamp, 1966).

—, *Paradigmen zu einer Metaphorologie* (Frankfurt am Main: Suhrkamp, 1998).

Böcklin, A., *Neben meiner Kunst: Flugstudien, Briefe und Persönliches von und über Arnold Böcklin*, ed. F. Runke and C. Böcklin (Berlin: Vita-Verlag, 1909).

Böhme, H., 'Umriß einer Kulturgeschichte des Wassers. Eine Einleitung', in H. Böhme (ed.), *Kulturgeschichte des Wassers* (Frankfurt am Main: Suhrkamp, 1988), pp. 7–42.

—, 'Geheime Macht im Schoß der Erde. Das Symbolfeld des Bergbaus zwischen Sozialgeschichte und Psychohistorie', in H. Böhme (ed.), *Natur und Subjekt* (Frankfurt am Main: Suhrkamp, 1988), pp. 67–144.

—, 'Das Licht als Medium der Kunst. Über Erfahrungsarmut und ästhetisches Gegenlicht in der technischen Zivilisation', in M. Schwarz (ed.), *Licht, Farbe, Raum. Künstlerisch-wissenschaftliches Symposium* (Braunschweig: Hochschule für Bildende Künste, 1997), pp. 111–37.

—, 'Das Wissen der Nacht. Warum wir ohne Geheimnis nicht leben können', *NZZ*, 20:21 (1997), pp. 65–6.

—, and G. Böhme (eds), *Feuer, Wasser, Erde, Luft: Eine Kulturgeschichte der Elemente* (München: C. H. Beck, 1996).

Bois-Reymond Du, E., 'Gedächtnisrede auf Johannes Müller' in E. Du Bois-Reymond (ed.), *Reden von Emil Du Bois-Reymond* (Leipzig: Veit & Co., 1887).

Bölsche, W., *Das Liebesleben der Natur: Eine Entwicklungsgeschichte der Liebe*, 2 vols (Jena: Diederichs, 1898).

Bousé, D., *Wildlife Films* (Philadelphia, PA: University of Philadelphia Press, 2000).

Boutan, L., 'Submarine Photography', *The Century Magazine* (May 1898), pp. 42–9.

—, *La Photographie sous-marine et les Progés de la Photographie* (1900; Paris: Jean-Michel Place, 1987).

Bouwsma, W. J., 'Work on Blumenberg', *Journal for the History of Ideas*, 48:2 (April – June 1987), pp. 347–54.

Boycott, B. B., 'Learning in *Octopus Vulgaris* and Other Cephalopods', *Publications of the Stazzione zoologica Napoli*, 25 (1954), pp. 67–93.

Boyle, R., 'Other Inquiries Concerning the Sea', *Philosophical Transactions of the Royal Society* (1666), pp. 315–16.

Brandstetter, T., K. Harrasser and G. Friesinger (eds), *Grenzflächen des Meeres* (Wien: Turia + Kants, 2010).

Braun, M., *Picturing Time: The Work of Etienne-Jules Marey (1830–1904)* (Chicago, IL and London: The University of Chicago Press, 1992).

Bredekamp, H., 'Wasserangst und Wasserfreude in Renaissance und Manierismus', in H. Böhme (ed.), *Kulturgeschichte des Wassers* (Frankfurt am Main, 1988), pp. 145–88.

—, 'Coral versus Trees: Charles Darwin's Early Sketches of Evolution', in P. H. Smith, A. R. W. Meyers and H. J. Cook, *Ways of Making and Knowing: The Material Culture of Empirical Knowledge* (Ann Arbor, MI: University of Michigan Press, 2014), pp. 357–76.

Brehm, A., 'Von der Baustätte des Berliner Aquariums', *Die Gartenlaube. Illustrirtes Familien-blatt* (1868), pp. 620–3.

—, 'Das Berliner Aquarium', *Westermann's Jahrbuch der Illustrirten Deutschen Monatshefte. Ein Familienbuch für das gesammte geistige Leben der Gegenwart* (1870), pp. 138–69.

Brinkmann-Voss, A., P. Fioroni and S. von Boletzky, *Adolf Portmanns frühe Studien mariner Lebewesen* (Basel: Schwave-Verlag, 1997).

Brockhaus, C., 'Das zeichnerische Frühwerk Alfred Kubins bis 1904', in H. A. Peters (ed.), *Alfred Kubin. Das zeichnerische Frühwerk bis 1904* (Baden-Baden: Kunsthalle Baden-Baden, 1977).

Brugman, J. and H. J. D. Lulofs (eds), *Aristotle, Generation of Animals: The Arabic Translation commonly ascribed to Yaḥyā ibn al-Bitrīq* (Leiden: Brill, 1971)

Brunner, B., *Wie das Meer nach Hause kam* (Berlin: Transit Verlag, 2003).

Bruyn, N. de, *Libellus varia genera piscium compectens* (Amsterdam: Claes J. Visscher, 1630).

Buchanan, J. Y., *Accounts Rendered of Work Done and Things Seen* (Cambridge: Cambridge University Press, 1919).

Buci-Glucksmann, C., *Der kartographische Blick der Kunst* (Berlin: Merve-Verlag, 1997).

Bullen, F. T., 'The Last Stand of the Decapods', in F. T. Bullen, *Deep-Sea Plunderings* (New York: Appleton, 1902).

Burckhardt, F. (ed.), *Charles Darwin's Letters: A Selection* (Cambridge: Cambridge University Press, 1996).

Burgess, T., *Take Me under the Sea: The Dream Merchants of the Deep* (Salem, MA: Ocean Archives, 1994).

Burnet, T., *The Sacred Theory of the Earth* (London, 1684).

Burns, R. M. *The Great Debate on Miracles: From Joseph Glanvill to David Hume* (Lewisburg, PA: Bucknell University Press, 1981).

Burstyn, H. L. and A. G. E. Jones, 'G. C. Wallich, MD: Megalomaniac or Mis-Used Oceano-graphic Genius?', *Journal of the Society for the Bibliography of Natural History*, 7:4 (1976), pp. 432–50.

Butler, H. D., *The Family Aquarium; or, Aqua Vivarium: A New Pleasure for the Domestic Circle* (New York: Dick & Fitzgerald, 1858).

Cadbury, D., *The Dinosaur Hunters: A True Story of Scientific Rivalry and the Discovery of the Prehistoric World* (London: Fourth Estate, 2000).

Caillois, R., *Der Krake. Versuch über die Logik des Imaginativen* (München: dtv, 1989).

Campbell M. B., *The Witness and the Other World: Exotic European Travel Writing 400–1600* (Ithaca, NY: Cornell University Press, 1988).

Cardinal, R., 'Victor Hugo: Somnambulist of the Sea', in P. Collier (ed.), Artistic Relations: Literature and the Visual Arts in Nineteenth-Century France (New Haven, CT and London: Yale University Press, 1994), pp. 210–20.

Carpenter, W. B., 'Preliminary Report of Dredging Operations in the Seas to the North of the British Islands, Carried on in HM. *Lightning*, by Dr. Carpenter und Dr. Wyville Thomson, Professor of Natural History in Queen's College, Belfast', *Proceedings of the Royal Society*, 17 (1868–9), pp. 168–200.

—, J. G. Jeffreys and C. W. Thomson, 'Preliminary Report of the Scientific Exploration of the Deep Sea in HMS *Porcupine*, During the Summer of 1869', *Proceedings of the Royal Society*, 1 (1869–70), pp. 397–492.

Certeau de, M., 'Writing the Sea: Jules Verne', in M. de Certeau, *Heterologies: Discourses on the Other*, trans. B. Massumi (Minneapolis, MN: University of Minnesota Press, 1986).

Chadarevian, S. de, 'Sehen und Aufzeichnen in der Botanik des 19. Jahrhunderts', in M. Wetzel and H. Wolf (eds), *Der Entzug der Bilder: visuelle Realtitäten* (München: Wilhelm Fink Verlag, 1994), pp. 121–44.

Chamisso, A. von, 'Reise um die Welt', in H. Tardel (ed.), *Chamissos Werke* 3 vols (Leipzig: Bibliographisches Institut Leipzig und Wien, 1907), vol. 3.

Chun, C., *Aus den Tiefen des Weltmeeres. Schilderungen von der Deutschen Tiefsee-Expedition* (Jena, 1900).

—, 'Ueber die Natur und die Entwicklung der Chromatophoren bei den Cephalopoden', *Verhandlungen der deutschen zoologischen Gesellschaft* (1902), pp. 162–82.

— (ed.), *Wissenschaftliche Ergebnisse der Deutschen Tiefsee Expedition auf dem Dampfer 'Valdivia' 1898–1899'*, 24 (Jena: Gustav Fischer, 1902–40)

Clarke, A. C., 'The Shining Ones', *Argosy* (December 1964), pp. 77–93.

Clavaud, A., 'Sur la Nitella stelligera', *Actes Soc. linn. de Bordeaux,* 25 (1865), pp. 348–52.

—, *Sur la prétendue Parthénogènese du Chara crinita* (Sonderdruck, 1878).

Coenen, A., *Visboock (Fish Book)* (1577–81).

—, *Walvisboock (Whale Book)* (1584–86).

Cohn, N., *Noah's Flood: The Genesis Story in Western Thought* (New Haven, CT: Yale University Press, 1996).

Collaert, A. de, *Piscium vivae icones in aes incisae et editae* (Antwerpen, 1610).

Colón, C. and D. de Colón, *Libro de la primera navegación y descubrimiento de las Indias: facsímil y transcripcion del manuscrito original de Fray Bartolomé de las Casas / introducción por Manuel Lucena Giraldo* (Madrid: Guillermo Blázquez, 2005).

Colonna, F., *Aquatilium, et Terrestrium aliquot Animalium, aliarumq. Naturalium Rerum observationes* (Rome, 1616).

Conybeare W., and H. de la Beche, 'Notice of the Discovery of a New Fossil Animal', *Transactions of the Geological Society,* 5 (1821), pp. 559–94.

—, 'On the Discovery of an Almost Perfect Skeleton of the Plesiosaurus', *Transactions of the Geological Society of London,* 1 (1824).

Corbin, A., *The Lure of the Sea: The Discovery of the Seaside in the Western World, 1750–1840,* trans. J. Phelps (Berkeley, CA: University of California Press, 1994).

Cott, H. B., *Adaptive Coloration in Animals* (London: Methuen, 1940).

Cousteau, J. and P. Diolè, *Octopus and Squid: The Soft Intelligence* (New York: Doubleday, 1973).

Cowan, Z., *Early Divers: Underwater Adventures in the 17th and 18th Centuries* (Norfolk: Treasure World, 1992).

Cowdry, E. V., *The Colour Changes of Octopus Vulgaris* (Toronto: University Library, 1911).

Cowper, W., 'Retirement' (1787); Poems by William Cowper of the Inner Temple, Esq., (London: John Johnson, 1782).

Crichton, M., *Sphere* (New York: Knopf, 1987).

Cutler, A., *The Seashell on the Mountaintop: A Story of Science, Sainthood, and the Humble Genius Who Discovered a New History of the Earth* (New York: Heinemann, 2003).

Cuvier, G., 'Sur quelques quadrupèdes ovipares fossiles conservés dans les schistes calcaires', *Annales du Muséum d'Histoire Naturelle,* 13 (1809), pp. 410–37.

Dessalines d'Orbigny, A. and A. E. d'Audebert de Ferussac, *Histoire naturelle, général et particulière, des céphalopodes acétabulifères,* 2 vols (Paris, 1835–48).

Dalyell, J. G., *Rare and Remarkable Animals of Scotland, Represented from Living Objects: With Practical Observations on their Nature* (London: Van Voorst, 1847–8).

Damon, W. E., *Ocean Wonders: A Companion for the Sea Side* (New York: D. Appleton and Company, 1879).

Dance, S. P., *Shell Collecting: An Illustrated History* (London: Faber and Faber, 1966).

Darwin, C., *A Monograph of the Sub-Class Cirripedia with Figures of all the Species* (London: The Ray Society, 1851).

—, *On the Origin of Species by Means of Natural Selection, or the Preservation of Favoured Races in the Struggle for Life, A Facsimile of the First Edition [London 1859] with an Introduction by Ernst Mayer* (Cambridge, MA: Harvard University Press, 1964).

—, *Voyage of the Beagle.* (London: Penguin Books, 1989).

Daston, L., 'The Moral Economy of Science', *Osiris*, 10 (1995), pp. 3–24.

—, and P. Gallison, *Objektivität* (Frankfurt am Main: Suhrkamp, 2007).

—, and K. Park, *Wonders and the Order of Nature* (New York: Zone Books, 1998).

—, and P. Galison, 'The Image of Objectivity', *Representations*, 37 (1992), pp. 67–106.

Daum, A. W., *Wissenschaftspopularisierung im 19. Jahrhundert. Bürgerliche Kultur, naturwissenschaftliche Bildung und die deutsche Öffentlichkeit, 1848–1914* (München: Oldenbourg Verlag, 1998).

Deacon, M., *Scientists and the Sea, 1650–1900: A Study of Marine Science* (London: Academic Press, 1971).

Dehs, V., *Jules Verne. Eine kritische Biographie* (Düsseldorf and Zürich: Patmos, Artemis & Winkler, 2005).

Delumeau, J., *La Peur en Occident (XIVe-XVIIIe sièpcles). Une cité assiégé* (Paris: Fayard, 1978).

Descartes, R., *Les Principes de la Philosophie* (1644; french Rouen, 1647).

DesMoulins, C., 'Note sur les moyens d'empecher la corruption dans les bocaux où l'on conservedes animaux aquatiques vivants', *Actes de la Société Linnéenne de Bordeaux*, 4 (1830), pp. 257–72.

Detel, W., 'Das Prinzip des Wassers bei Thales', in H. Böhme (ed.), *Kulturgeschichte des Wassers* (Frankfurt am Main, 1988), pp. 43–64.

Dezallier d'Argentville, A. J., *La conchyliologie, ou histoire naturelle des coquilles de mer, d'eau douce, terrestres et fossiles* (Paris, 1780).

Didi-Huberman, G., 'Expérimenter pour voir. Experimentieren, um zu sehen', *Kongress-Akten der Deutschen Gesellschaft für Ästhetik, Band 2 Experimentelle Ästhetik*: http://www.dgae.de/kongress-akten-band-2.html [accessed: 18 December 2014].

Dietz, B., 'Mobile Objects: The Space of Shells in Eighteenth-Century France', *British Society for the History of Science*, 39:3 (2006), pp. 363–82.

Dippel, L., *Das Mikroskop und seine Anwendung* (Braunschweig: Vieweg-Verlag, 1867).

Donald, D., *Picturing Animals in Britain, 1750–1850* (New Haven, CT and London: Yale University Press, 2007).

Dover, C. van, *Deep-Ocean Journeys: Discovering New Life at the Bottom of the Sea* (Redwood City, CA: Addison-Wesley, 1996).

Druick, D. W. (ed.), *Odilon Redon: Prince of Dreams 1840–1916, Ausstellungskatalog* (Chicago, IL: The Art Institute of Chicago, 1994).

—, and P. K. Zegers, 'In the Public Eye', in D. W. Druick (ed.), *Odilon Redon: Prince of Dreams 1840–1916* (Chicago, IL: The Art Institute of Chicago, 1994), pp. 120–74.

Ecott, T., *Neutral Buoyancy. Adventures in a Liquid World* (London: Michael Joseph, 2001).

—, *Unter Wasser. Abenteuer in einer anderen Welt* (Berlin: Argon, 2002).

Egmond, F., 'Curious Fish', in K. A. E. Enenkel and P. Smith (eds), *Early Modern Zoology: The Construction of Animals in Science, Literature and the Visual Arts* (Leiden: Brill, 2007), pp. 245–72.

Ehrenberg, C. G., *Die Infusionsthierchen als vollkommene Organismen. Ein Blick in das tiefere organische Leben der Natur. Nebst einem Atlas von vierundsechzig colorirten Kupfertafeln, gezeichnet vom Verfasser* (Leipzig: L. Voss, 1838).

—, 'Über die Natur und Bildung der Corallenbänke des roten Meeres', *Abhandlungen der Königlichen Akademie der Wissenschaften zu Berlin, Erster Teil* (Berlin: Commission F. Dummler, 1834).

—, *Das Leuchten des Meeres. Neue Beobachtungen nebst Übersicht der Hauptmomente der geschichtlichen Entwicklung dieses merkwürdigen Phänomens* (Berlin: Königliche Akademie der Wissenschaften, 1835).

Eibl-Eibesfeld, I., 'Ernst Haeckel – Der Künstler im Wissenschaftler', in E. Haeckel (ed.), *Ernst Haeckel, Kunstformen der Natur* (München: Prestel-Verlag, 1998), pp. 19–31.

Ellis, R., *The Search for the Giant Squid: The Biology and Mythology of the World's Most Elusive Sea Creature* (London: Lyons Press, 1989).

—, *Monsters of the Sea* (New York: Robert Hale, 1994).

—, *Seeungeheuer: Mythen, Fabeln, Fakten* (Basel, Boston, MA and Berlin: Birkhäuser Verlag, 1997)

Encelius, C., *De Re Metallica, hoc. est. De Origine, Varietate & Natura Corporum Metallicorum, Lapidum, Gemmarum, atq. Aliarum, quae ex Fodinis cruuntur, Rerum ad Medicinae Usum deseruientium, Libri III* (Frankfurt, 1557).

Engel, H., 'Alphabetical List of Dutch Zoological Cabinets and Menageries', in *Bijdragen tot de Dierkunde* (Leiden, 1938), pp. 247–346.

Engell, L. and J. Vogl, 'Vorwort', in L. Engell, O. Fahle, J. Vogl and B. Neitzel (eds), *Kursbuch Medienkultur: Die maßgeblichen Theorien von Brecht bis Baudrillard* (Stuttgart: Deutsche Verlags-Anstalt, 1999), p. 10.

Esswein, H., *Alfred Kubin: Der Künstler und sein Werk* (München: Georg Müller, 1911).

Federmair, L., 'Entzaubern – Verzaubern. Zu den außergewöhnlichen Reisen Jules Vernes', in B. Felderer (ed.), *Wunschmaschine Welterfindung: Eine Geschichte der Technikvisionen seit dem 18. Jahrhundert* (Wien: Springer-Verlag, 1996), pp. 236–49.

Figuier, L., *La terre avant l'Deluge* (Paris: Hachette, 1863).

Findlen, P., 'Jokes of Nature and Jokes of Knowledge: The Playfulness of Scientific Discourse in Early Modern Europe', *Renaissance Quarterly*, 43:2 (Summer 1990), pp. 292–331.

Finn, B., *Submarine Telegraphy: The Grand Victorian Technology* (London: Science Museum, 1973).

Flamen, A., *Icones diversorum piscium tum maris tum amnium* (Paris, 1664).

Fleming, I., *Dr. No* (London: Jonathan Cape, 1958)

Forbes, E., 'Report on the Mollusca and Radiata of the Ægean Sea, and on their Distribution, Considered as Bearing on Geology', *Report of the 13th Meeting of the British Association for the Advancement of Science, 1843*, 1 (1844), pp. 130–93.

—, *The Natural History of the European Seas* (1859; New York: Arno Press, 1977).

Forest, J., 'Henri Milne Edwards', *Journal of Crustacean Biology* (1996), pp. 207–13.

Forster, J. R., *Observations Made During a Voyage Around the World, on Physical Geography, Natural History and Ethic Philosophy* (London: G. Robinson, 1788).

Foucault, M., *History of Sexuality: An Introduction,* 3 vols (New York: Random House, 1990), vol. 1.

—, *The Order of Things* (New York: Random House, 1994), p. 251.

Frédol, A., *Le Monde de la Mer* (Paris: Librairie de L-Hatchette, 1865)

Frey, A., *Arnold Böcklin: nach den Erinnerungen seiner Zürcher Freunde,* 2nd edn (Stuttgart and Berlin: Cotta'sche Buchhandlung, 1912).

Freycinet, L. de (ed.), *Voyage autour du Monde, entrepris par Ordre du Roi ... exécuté sur les corvettes de ... l'Uranie et la Physicienne, pendant les années 1817, 1818, 1819 et 1820* (Paris: Chez Pillet Aîné, 1824).

Friedberg, A., *Window Shopping: Cinema and the Postmodern* (Berkeley, CA: University of California Press, 1993).

Friede, S., 'Die Welt als Aquarium – Spuren eines Schlüsselmotivs in Gides Paludes, Prousts Recherche und Robbe-Grillets Les Gommes', *Romanistische Zeitschrift für Literaturgeschichte,* 27:1–2 (2003), pp. 161–88.

Fukano, Y. and H. Strickland, *Selected References to Literature on Marine Expeditions, 1700–1960* (Boston, MA: Hall, 1972).

Furneaux, W., *Life in Ponds and Streams* (London: Longmans, Green and Co., 1896).

Galilei, G., *Dialogo di Galileo Galilei sopra i due Massimi Sistemi del Mondo Tolemaico e Copernicano* (Fiorenza, 1632).

Gautier, T., *Paris et Les Parisiens,* ed. C. Lacoste-Veysseyre (Paris: La Boîte à Documents, 1996).

Geikie, A., *The Founders of Geology* (London: Macmillan and Co., 1897)

Geimer, P. (ed.), *Ordnungen der Sichtbarkeit. Fotografie in Wissenschaft, Kunst und Technologie* (Frankfurt am Main: Suhrkamp, 2002).

Geißler, R., *Plaudereien über Paris und die Weltausstellung* (Berlin: Theobald Grieben, 1868).

Gesner, C., *Historiae Animalium. Liber 4 qui est de Piscium & Aquatilium Animantium natura* (Zürich: Froscherus, 1558).

—, *De Rerum fossilium, Lapidum et Gemmarum maxime, figuris et similitudinibus Liber* (Tiguri, 1565).

Giedion-Welcker, C., *Paul Klee mit Selbstzeugnissen und Bilddokumenten* (Reinbek bei Hamburg: Rowohlt, 1961).

Glaubrecht, M., 'Karl August Möbius: Von Lebensgemeinschaften zur Artenvielfalt', *Natur-wissenschaftliche Rundschau*, 61:5 (2008), pp. 230–6.

Gosse, E., *Father and Son* (London: Penguin Books, 1907).

Gosse, P. H., *The Ocean* (London: Society for Promoting Christian Knowledge, 1849).

—, *A Naturalist's Sojourn in Jamaica* (London: Longman, 1851).

—, On Keeping Marine Animals and Plants Alive in Unchanged Seawater, *Annuals of Natural History*, 19:2 (1852), pp. 263–8.

—, *A Naturalist's Rambles on the Devonshire coast* (London: Van Voorst, 1853).

—, *The Aquarium: An Unveiling of the Wonders of the Deep Sea* (London: John van Voorst, 1854).

—, *Handbook to the Marine Aquarium* (London: Van Voorst, 1855).

—, *Tenby: A Sea-Side Holiday* (London: John van Voorst, 1856).

—, *Evenings at the Microscope; Or, Researches Among the Minuter Organs and Forms of Animal Life* (London: Society for Promoting Christian Knowledge, 1859).

—, *Actinologia Britannica: A History of the British Sea-Anemones and Corals* (London, 1860).

—, *A Year at the Shore* (London: Alexander Strahan, 1865).

—, *The Romance of Natural History* (London: Nisbet, 1871).

Gould, S. J., *Time's Arrow, Time's Cycle: Myth and Metaphor in the Discovery of Geological Time* (Cambridge, MA: Harvard University Press, 1987).

—, *Leonardo's Mountain of Clams and the Diet of Worms: Essays on Natural History* (London: Jonathan Cape, 1998).

Grafe, F., 'Ein Wilderer – Jean Painlevé, 1902 bis 1989', *Cinema, CineZoo* (1997), pp. 9–19.

Graham, V. E., *The Imagery of Proust* (Oxford: Basil Blackwell, 1966).

Greenblatt, S., *Marvellous Possessions: The Wonder of the New World* (Chicago, IL: The University of Chicago Press, 1991).

Grimble, A., *A Pattern of Islands* (London: John Murray, 1952).

Groeben, C., 'The Stazione Zoologica Anton Dohrn as a Place for the Circulation of Scientific Ideas: Vision and Management', in K. L. Anderson and C. Tony (eds), Information for Responsible Fisheries: Libraries as Mediators; proceedings of the 31st Annual Conference, held in Rome, Italy (10–14 October 2005).

—, and I. Müller (eds), *The Naples Zoological Station at the Time of Anton Dohrn* (Naples: Stazione di Napoli, 1975)

Groom, G., 'The Late Work', in D. W. Druick (ed.), *Odilon Redon: Prince of Dreams 1840–1916* (Chicago, IL: The Art Institute of Chicago, 1994).

Guardini, R., *Zu Rainer Maria Rilkes Deutung des Daseins, Schriften für die geistige Überlieferung IV* (Berlin: Küpper, 1941).

Gudger, E. W., 'The Five Great Naturalists of the 16th Century: Belon, Rondelet, Salviani, Gesner and Aldrovandi: A Chapter in the History of Ichthyology', *Isis*, 22 (1934), pp. 21–40.

H. Heine's Pictures of Travel, trans. C. G. Leland, 9th rev. edn (Philadelphia, PA: Schaefer & Koradi, 1882)

Haberling, W., *Johannes Müller: Das Leben des Rheinischen Naturforschers* (Leipzig: Akademische Verlagsgesellschaft, 1924).

Hadot, P. *The Veil of Isis* (Cambridge, MA: Harvard University Press, 2006).

Hantschk, A. and P. Kruspel, 'Mit Skizzenblock und Taucherglocke: Ein Wiener Maler unter Wasser', *Divemaster*, 1 (2000), p. 57–60.

Haeckel, E., *Die Radiolarien (Rhizopoda radiaria). Eine Monographie* (Berlin: Reimer Verlag, 1862).

—, *Generelle Morphologie der Organismen* (Berlin, 1866).

—, *Das Leben in den grössten Meerestiefen. Vortrag, gehalten am 2. März 1870 im akademischen Rosensaale zu Jena* (Berlin: Lüderitz'sche Verlagsbuchhandlung, 1870).

—, *Natürliche Schöpfungsgeschichte* (Berlin: Reimer Verlag, 1870).

—, *Die Kalkschwämme (Calcispongiae). Eine Monographie* (Berlin: G. Reimer Verlag, 1872).

—, *Das System der Medusen I: Craspedoten* (Jena: G. Fischer, 1879).

—, *Das System der Medusen II: Acraspeden* (Berlin, 1880).

—, *The History of Creation*, trans. E. Ray Lankester (New York: Appleton and Company, 1880).

—, *Die Tiefsee-Medusen der Challenger-Reise und der Organismus der Medusen. Zweiter Theil einer Monographie der Medusen* (Jena, 1881).

—, *Report on the Deep-Sea Medusae dredged by HMS Challenger during the years 1873–76* (London, 1882).

—, *Report on the Radiolaria, collected by HMS Challenger during the years 1873–76, I. Part: Porulosa (Spumellaria and Acantharia). II. Part Osculosa (Nesselaria and Phaeodaria)* (London: Eyre & Spottiswoode, 1887).

—, *II. Part Osculosa (Nesselaria and Phaeodaria)* (London, 1887).

—, *Report on the Siphonophorae collected by HMS Challenger during the years 1873–76* (London: Eyre & Spottiswoode 1888).

—, *Kunstformen der Natur* (Leipzig: Verlag des Bibliographischen Instituts, 1904).

Hagner, M. and B. Wahrig-Schmidt, *Johannes Müller und die Philosophie* (Berlin: Akademie Verlag, 1992).

—, and H.-J. Rheinberger (eds), *Die Experimentalisierung des Lebens. Experimentalsysteme in den biologischen Wissenschaften 1850/1950* (Berlin: Akademie-Verlag, 1993).

Hahn, D., 'Tourbillons et turbulences. Zu einer Ästhetik des Experiments in Étienne-Jules Mareys Machines à fumée', in É.-J. Mareys (ed.), *Machines à fumée, Ilinx – Berliner Beiträge zur Kulturwissenschaft. Nr. 1, Wirbel, Ströme, Turbulenzen* (Hamburg: Philo Fine Arts, 2009), pp. 43–69.

Halley, E., 'The Art of Living under Water: Or, a Discourse Concerning the Means of Furnishing Air at the Bottom of the Sea, in any Ordinary Depths', *Philosophical Transactions of the Royal Society*, 29 (1716), pp. 492–9.

Harper, J., *The Sea-Side and Aquarium; Or, Anecdote and Gossip on Marine Zoology* (Edinburgh: William P. Nimmo, 1858).

Hart, C., *The Prehistory of Flight* (Berkeley, CA: University of California Press, 1985).

Harter, U., 'Le Paradis artificiel. Aquarien, Leuchtkästen und andere Welten hinter Glas', *Vorträge aus dem Warburg Haus*, 6 (2002), pp. 77–124.

—, 'Odilon Redons Tiefseebilder', in E. Schlebrügge (ed.), *Das Meer im Zimmer. Von Tintenschnecken und Muscheltieren* (Graz: Landesmuseum Joanneum, 2005), pp. 115–21.

Harting, P., 'Description de quelques fragments de deux céphalopodes gigantesques', *Verh. Akad. Wet., Amst*, 9:1 (1860), p. 2.

Hartwig, G., *Das Leben des Meeres. Eine Darstellung für gebildete Stände* (Glogau: Karl Flemming Verlag, 1862).

—, *The Sea and its Living Wonders: A Popular Account of the Marvels of the Deep and of the Progress of Maritime Discovery from the Earliest Ages to the Present Time* (London: Longman, 1873).

Harvey, E. N., *A History of Luminescence from the Earliest Times until 1900* (Philadelphia, PA: American Philosophical Society, 1957).

Harvey, M., 'How I Discovered the Great Devil-Fish', *Wide World Magazine* (1899), pp. 732–40.

Harvey, W. H., *The Sea-Side Book; Being an Introduction the the Natural History of the British Coasts* (London: John van Voorst, 1857).

Hauff, B., 'Der Ichthyosaurus des Berliner Zoo-Aquariums aus dem Museum Hauff in Holzmaden', *Bongo,* 11 (1986), pp. 115–20.

Hauptmann, J., M. van Zuylen and F. Starr (eds), *Beyond the Visible: The Art of Odilon Redon* (New York: The Museum of Modern Art, 2005).

Hauser, S. E., 'Der subaquatische Bilderkosmos. Eine kurze Geschichte des Aquarium- und Unterwasserfildms von 1890 bis heute ... ', in V. Weigel (ed.), *Unter wasser über wasser. Vom Aquarium. zum Videobild* (Wilhelmshaven: Kerber Art, 2009), pp. 18–35.

Hazéra, H. and D. Leglu, 'Jean Painlevé Reveals the Invisible. Interview in Libération 1986', in A. M. Bellows and M. McDougall (ed.), *Science is Fiction: The Films of Jean Painlevé* (San Francisco, CA: Brico Press, 2000), pp. 170–9.

Hedgpeth, J. W., 'De mirabili maris: Thoughts on the Flowering of Seashore Books', *The Royal Society of Edinburgh, Proceedings, Section B (Biology),* 72 (1971–2), pp. 107–14.

Heinroth, O., *Führer durch das Aquarium nebst Terrarium und Insektarium im Zoologischen Garten zu Berlin.* (Berlin: Zoologischer Garten zu Berlin, 1915).

Helmreich, S., 'An Anthropologist Underwater: Immersive Soundscapes, Submarine Cyborgs, and Transductive Ethnography', *American Ethnologist*, 34:4 (2007), pp. 621–41.

Heraeus, S., *Traumvorstellung und Bildidee. Surreale Strategien in der französischen Graphik des 19. Jahrhunderts* (Berlin: Reimer, 1998).

—, 'Artists and the Dream in Nineteenth-Century Paris: Towards a Prehistory of Surrealism', *History Workshop Journal*, 48 (Autumn 1999), pp. 151–68.

—, 'The Dream as an Artistic Strategy', in M. Hollein and M. Stuffmann (eds), *As in a Dream: Odilon Redon* (Ostfildern: HatjeCantz, 2007), pp. 71–8.

Herder, J. G., 'On the Cognition and Sensation of the Human Soul', in Philosophical Writings, trans. and ed. by Michael N. Foster (Cambridge, MA: Cambridge University Press, 2002), p. 197.

Herzmanovsky-Orlande, F. von, *Der Briefwechsel mit Alfred Kubin* (Salzburg: Residenz Verlag, 1983).

Heßler, M. and D. Mersch (eds), *Konstruierte Sichtbarkeiten. Wissenschafts- und Technikbilder seit der Frühen Neuzeit* (München: Fink Verlag, 2006).

— (ed.), *Logik des Bildlichen. Zur Kritik der ikonischen Vernunft* (Bielefeld: Transcript-Verlag, 2009).

Heuss, T., *Anton Dohrn* (Stuttgart und Tübingen: Wunderlich, 1948).

Heuvelmans, B., *In the Wake of Sea-Serpents* (New York: Hill and Wang, 1968).

Hevesi, L., *Acht Jahre Sezession: März 1897 – Juni 1905. Kritik, Polemik, Chronik* (Wien, 1906).

Hibberd, S., *The Fern Garden* (London: Groombridge & Sons, 1869).

—, *Rustic Adornments for Homes of Taste* (1856; London: Trafalger Square 1987).

Hoberg, A., Alfred Kubin 1877–1959 (München: Edition Spangenberg, 1990).

Hof, im U., 'Enlightenment – Lumieres – Illuminismo – Aufklaerung. Die 'Ausbreitung eines besseren Lichts im Zeitalter der Vernunft', in M. Svilar (ed.), '*Und es ward Licht*': zur Kulturgeschichte des Lichts (Bern, Frankfurt: Peter Lang Verlag, 1983).

Hogendorp Prosperetti, L. van , '"Conchas Legere": Shells as Trophies of Repose in Northern European Humanism', *Art History,* 29:3 (2006), pp. 387–413.

Holländer, H., 'Es rauscht in den Schachtelhalmen, verdächtig leuchtete das Mee ... ', in A. Preiß and B. Brock (eds), *Ikonographia: Anleitung zum Lesen von Bildern* (München: Klinkhardt & Biermann, 1990), pp. 179–202.

Hornell, J., 'Notes on Animal Colouration', *Journal of Marine Zoology,* 1 (1893), pp. 3–8.

Hugo, V., *Toilers of the Sea*, ed. E. Rhys, trans. W. Moy Thomas (London and Toronto: J. M. Dent, 1911).

Humboldt, A. von, *Views of Nature, or Contemplations on the Sublime Phenomena of Creation*, trans. E. C. Otté and H. G. Bohn (London: Henry G. Bohn, 1850), p. xi.

—, *Ansichten der Natur mit wissenchaftlichen Erläuterungen*, 3rd edn (1807; Stuttgart and Augsburg: Cotta'scher Verlag, 1859)

Humphreys, H. N., *River Gardens; Bring an Account of the Best Methods or Cultivation Fresh-Water Plants in Aquaria* (London: Sampson Low, 1857).

Huxley, T. H., 'On Some Organisms Living at Great Depths in the North Atlantic Ocean', *Quarterly Journal of Microscopical Science* (1868), pp. 203–12.

Huysman, J.-K., *A Rebours* (1884).

Ingenhousz, J., *Experiments Upon Vegetables: Discovering their Great Power of Purifying the Common Air in the Sunshine and of Injuring it in the Shade at Night* (London: P. Elmsly and H. Payne, 1779).

Jäger, G., *Das Leben im Wasser und das Aquarium* (Stuttgart: Kosmos, Gesellschaft der Naturfreunde, 1908).

James, T., *Dream, Creativity, and Madness in Nineteenth-Century France* (Oxford: Clarendon Press, 1995).

Jardine N., A. Secord and E. C. Spary (eds), *Cultures of Natural History* (Cambridge: Cambridge University Press, 2000 (1996).

Jarofke, D., 'Tiere der Vorzeit an der Fassade unseres Aquariums', *Bongo*, 8 (1984), pp. 19–40.

Johnston, G., *A History of British Sponges and Lithophytes* (Edinburgh: W. H. Lizars, 1842).

Jones, T. R., *The Aquarian Naturalist: A Manuar for the Seaside* (London: Van Voorst, 1858).

Jonston, J., *Historia naturalis de piscibus et cetis libri V* (Frankfurt am Main: Matth. Merian, 1650).

Journal of Researches into the National History and Geology of the Countries Visited During the Voyage of HMS Beagle Round the World, under the Command of Capt. Fitz R. N. (1860 1st edn; London: John Murray, 1913), p. 162.

Jung, M., *Das Handbuch zur Tauchgeschichte. Techniken, Geräte, Berufe, Erfindunge* (Stuttgart: Naglschmid, 1999).

Kegel, B., *Der Rote* (Hamburg: Marebuchverlag, 2009).

Kemp, M., *Visualizations: The Nature Book of Art and Science* (Oxford: Oxford University Press, 2000).

Kenny, N., *Curiosity in Early Modern Euorpe: Word Histories* (Wiesbaden: Harrassowitz-Verlag, 1998).

Kenseth, J., 'Age of the Marvelous: An Introduction', in J. Kenseth (ed.), *The Age of the Marvelous* (Hanover: Hood Museum of Art, Dartmouth College, 1991), pp. 25–60.

Kent, W. S., 'Note on a Gigantic Cephaolpod from Cenception Bay, Newfoundland', *Proceedings of the Zoological Society London* (1874), pp. 178–82.

—, 'A Further Communication Upon Certain Gigantic Cephalopods Recently Encountered off the Coast of Newfoundland', *Proceedings of the Zoological Society London* (1874), pp. 489–94.

Kepler, J., *Astronomia Nova seu de Motu Stellae Martis* (1609).

Kernbauer, E., 'Formationen des Meeres bei Victor Hugo', in T. Brandstetter, K. Harrasser and G. Friesinger (eds), *Grenzflächen des Meeres* (Wien: Turia and Kants, 2010), pp. 63–86.

Kingsley, C., *Glaucus; Or, the Wonders of the Shore* (Cambridge: Macmillan and Co., 1859).

Kinzer, J., 'In ewiger Nacht und Kälte. Sehen, Riechen, Hören, Sex: Die Lebens- und Überlebensstrategien der Tiefseebewohner wirken grotesk, sind aber effizient', *mare*, 13 (1999), pp. 36–44.

Kircher, A., *Mundus Subterraneus, in XII libros digestus* (Amsterdam, 1665).

Kirk, T. W., 'On the Occurrence of a Giant Cuttlefish on the New Zealand Coast', *Transactions of the New Zealand Institute* (1880), pp. 310–13.

Kisling, Jr V. N. (ed.), *Zoo and Aquarium History: Ancient Animal Collections to Zoological Gardens* (Boca Raton, London, New York and Washington: CRC Press, 2001).

Klee, P., *Tagebücher 1898–1916. Textkritische Neuedition*, 2:390 (Stuttgart: Hatje, Teufen, Niggli, 1988).

Kleeberg, B., *Theophysis. Ernst Haeckels Philosophie des Naturganzen* (Köln, Weimar: Böhlau, 2005).

Klein, J. T., *Historiae piscium naturalis promovendae missus I–V* (Lipsiae, 1740–9).

Klingel, G., *The Ocean Island (Inagua)* (New York: Dodd, Mead & Company, 1940).

Klös, H.-G., *Von der Menagerie zum Tierparadies. 125 Jahre Zoo Berlin* (Berlin: Reimer-Verlag, 1969).

— (ed.), *Festschrift 70 Jahre Aquarium* (Berlin: Zoologischer Garten, 1983).

—, and U. Klös, *Der Berliner Zoo im Spiegel seiner Bauten 1841–1989* (Berlin: Heeneman, 1990).

—, H. Frädrich and U. Klös, *Die Arche Noah an der Spree. 150 Jahre Zoologischer Garten Berlin. Eine tiergärtnerische Kulturgeschichte von 1844–1994* (Berlin: FAB Verlag, 1994).

—, and J. Lange, *Tierwelt hinter Glas: das Zoo-Aquarium Berlin* (Berlin: arani-Verlag, 1988).

Knox, J., 'Sounding the Depths: Jean Painlevé's Sunken Cinema', *Senses of Cinema*, 25 (March 2003) [accessed 28 March 2015].

Kockerbeck, C., *Ernst Haeckels 'Kunstformen der Natur' und ihr Einfluß auf die deutsche bildende Kunst der Jahrhundertwende* (Frankfurt am Main: Peter Lang Verlag, 1986).

Koella, R., 'Kubins symbolistische Zeichnungen', in R. Koella and D. Schwarz (eds), *Alfred Kubin, Ausstellung im Kunstmuseum Winterthur* (Winterthur: Kunstmuseum, 1986), pp. 31–43.

Kollmann, J., 'Die Cephalopoden der zoologischen Station des Dr. Dohrn', *Zeitschrift für Wissenschaftliche Zoologie* (1876), pp. 1–23.

—, 'Aus dem Leben der Cephalopoden', *Vierteljahresschrift für wissenschaftliche Philosophie* (1877), pp. 136–55.

Konvitz, J. W., 'Changing Concepts of the Sea, 1550–1950: An Urban Perspective', in M. Sears and D. Merriman (eds), *Oceanography: The Past* (New York, Heidelberg and Berlin: Springer-Verlag, 1980), pp. 32–41.

Kort, P., 'Die Dinge wirklich werden lassen: Odilon Redon und Jean Carriès', in P. Kort and M. Hollein (eds), *Darwin. Kunst und die Suche nach den Ursprüngen*, Ausstellungskatalog (Köln: Wienand Verlag, 2009), pp. 154–70.

—, 'Arnold Böcklin, Max Ernst und die Debatten um Ursprünge und Übereben in Deutschland und Frankreich', in P. Kort and M. Hollein (eds), *Darwin. Kunst und die Suche nach den Ursprüngen*, Ausstellungskatalog (Köln: Wienand Verlag, 2009), pp. 24–53.

—, and M. Hollein (eds), *Darwin. Kunst und die Suche nach den Ursprüngen* (Köln: Wienand Verlag, 2009).

Kranz, I., 'Zur Felsengrotte im Heimaquarium', in B. Butis (ed.), *Stehende Gewässer. Medien der Stagnation* (Berlin and Zürich: Diaphanes, 2007), pp. 247–58.

Krauße, E., *Ernst Haeckel* (Leipzig: Teubner Verlag, 1984).

Krohn, W., 'Die ästhetischen Dimensionen der Wissenschaft', in W. Krohn (ed.), *Ästhetik in der Wissenschaft. Interdisziplinärer Diskurs über das Gestalten und Darstellen von Wissen* (Hamburg: Meiner-Verlag, 2006), pp. 3–38.

Krolzik, U., 'Das Wasser als theologisches Thema der deutschen Frühaufklärung bei Johann Albert Fabricius ... ', in H. Böhme (ed.), *Kulturgeschichte des Wassers* (Frankfurt am Main: Suhrkamp, 1988), pp. 189–207.

Kruse, M., 'Zur Interpretation von Rousseaus Cinquieme Reverie und Laforgues Aquarium', in W. Pabst, R. Grossmann and E. Schramm (eds), *Der Vergleich. Literatur- und sprach-*

wissenschaftliche Interpretationen. Festgabe für Hellmuth Petriconi zum 1. April 1955 (Hamburg: Cram, de Gruyter & Co, 1955), pp. 91–103.

Kubin, A., 'Malerei des Übersinnlichen', in A. Kubin, *Aus meiner Werkstatt. Gesammelte Prosa mit 71 Abbildungen* (1933; München: Nymphenburger Verlag, 1973), pp. 43–5.

—, 'Fragment eines Weltbildes', in A. Kubin, *Aus meiner Werkstatt. Gesammelte Prosa mit 71 Abbildungen* (1933; München: Nymphenburger Verlag, 1973).

—, 'Aus meinem Leben', in W. K. Müller-Thalheim (ed.), *Erotik und Dämonie im Werk Alfred Kubins* (München: Nymphenburger Verlagshandlung, 1970), pp. 63–106.

—, 'Aus halbvergessenem Lande. Über künstlerische Befruchtung', in A. Kubin, *Aus meiner Werkstatt. Gesammelte Prosa mit 71 Abbildungen* (1926; München: Nymphenburger Verlag, 1973), pp. 19–24.

—, *Aus meinem Leben. Gesammelte Prosa mit 73 Abbildungen* (München: edition spangenberg im Ellermann Verlag, 1974).

—, 'Mein Tag in Zwickledt: Ein Brief an Wilhelm Hausenstein (1921)', in A. Kubin, *Aus meinem Leben. Gesammelte Prosa mit 73 Abbildungen* (1921; München: edition spangenberg im Ellermann Verlag, 1974) pp. 87–93, on p. 92.

Kubodera, T. and K. Mori, 'First-Ever Observations of a Live Giant Squid in the Wild', *Proceedings of the Royal Society: Biological Sciences*, 272 (2005), pp. 2583–6.

—, 'Observations of Wild Hunting Behaviour and Bioluminescence of a Large Deep-Sea, Eight-Armed Squid, Taningia danae', *Proceedings of the Royal Society: Biological Sciences*, 274 (2007), pp. 1029–34.

Kunzig, R., *Mapping the Deep: The Extraordinary Story of Ocean Science* (New York and London: W. W. Norton & Company, 2000).

Lacaz-Dutons, H. de, 'Sur la photographie sous-marine', *Comptes Rendus Hebdomadaires de l'Academie des Sciences*, 117, pp. 286-9.

Lacépède, B. G. de, *Histoire naturelle des poissons* (Paris, 1798–1803).

Laforgue, J., 'L'Aquarium', *La Vogue*, 6 (29 May – 3 June 1886).

—, *Revue Blanche*, 49 (15 June 1895).

Lang, W., 'Mary Anning of Lyme, Collector an Vendor of Fossils', *Natural History Magazine*, 5:34 (1936), pp. 64–81.

Lange, J., 'Schauaquarien im Wandel der Zeit', *Bongo*, 13 (1987), pp. 135–56.

—, 'Hundert Jahre Zoo-Aquarium Berlin – Hundert Jahre Schauaquaristik', in B. Blaszkiewitz (ed.), *Picassofisch und Kompasssqualle. 100 Jahre Zoo – Aquarium Berlin* (Berlin: Lehmanns, 2013)

Lankester, E., *The Aquavivarium, Fresh and Marine Being an Account of the Principles and Objects Involved in the Domestic Culture of Water Plants and Animals* (London: R. Hardwicke, 1856).

Larson, B., *The Dark Side of Nature: Science, Society, and the Fantastic in the Work of Odilon Redon* (Pennsylvania, PA: Pennsylvania State University Press, 2005).

Latour, B., 'Drawing Things Together', in M. Lynch and S. W. Woolgar (eds), *Representation of Scientific Practice* (London and Cambridge, MA: The MIT-Press, 1990), pp. 19–68.

—, 'Arbeit mit Bildern: oder Die Umverteilung der wissenschaftlichen Intelligenz', in B. Latour, *Der Berliner Schlüssel. Erkundungen eines Liebhabers der Wissenschaften* (Berlin: Akademie-Verlag, 1996).

Lee, H., *Aquarium Notes: The Octopus; or, The 'Devil-Fisch' of Fiction and of Fact* (London: Champman and Hall, 1875).

—, 'Sea Monsters Unmasked', in *The Great International Fisheries Exhibition Literature* (London: William Clowes and Sons, 1883), p. 439.

LeGoff, J., *The Medieval Imagination*, trans. A. Goldhammer (Chicago, IL and London: University of Chicago Press, 1988).

Leonhard, K., 'Shell Collecting: On 17th Century Conchyology, Curiosity Cabinets and Stillife Painting', in K. A. E. Enenkel (ed.), *Early Modern Zoology: The Construction of Animals in Science, Literature and the Visual Arts* (Leiden: Brill, 2007), pp. 177–214.

—, 'Die Muschel als symbolische Form, oder: Wie Rembrandt nach Oxford kam', in B. Gockel (ed.), *Vom Objekt zum Bild. Piktorale Prozesse in Kunst und Wissenschaft 1600–2000* (Berlin: Akademie Verlag, 2011), pp. 123–56.

Lepenies, W., *Das Ende der Naturgeschichte: Wandel kultureller Selbstverständlichkeiten in den Wissenschaften des 18. und 19. Jahrhunderts* (München: Hanser Verlag, 1976).

Lewes, G. H., *Sea-Side Studies at Ilfracombe, Tenby, the Scilly Isles & Jersey* (Edinburgh and London: W. Blackwood, 1858).

Lichtenberg, G. C., 'Vermischte Nachrichten und die aerostatischen Maschinen', in L. C. Lichtenberg and F. Kries (ed.), *Georg Christophs Lichtenbergs physikalische und mathematische Schriften* (Göttingen: Dieterich-Verlag, 1804), pp. 321–3.

Lindner, G., *Muscheln und Schnecken derWeltmeere: Aussehen, Vorkommen, Systematik* (München: BLV, 1999).

Link, *Über die Asterien* (Leipzig, 1733).

Linnebach, A., *Arnold Böcklin und die Antike. Mythos, Geschichte, Gegenwart* (München: Hirmer Verlag, 1991).

—, 'Antike und Gegenwart. Zu Böcklins mythoogischer Bilderwelt', in G. Magnaguagno and J. Steiner (eds), *Arnold Böcklin, Giorgio de Chirico, Max Ernst: Eine Reise ins Ungewisse* (Bern: Benteli, 1997).

Lohff, B., 'Die Entdeckung der Welt des Planktons', in *Historisch Meereskundliches Jahrbuch* (Berlin and Hamburg: Reimer, 1992), pp. 35–44.

Lonitzer, A., *Naturalis Historiae Opus Novum* (Frankfurt am Main: Egenolphus, 1551–5).

Lorenz, K., *Er redete mit den Fischen, den Vögeln und dem Vieh* (München: Deutscher Taschenbuch Verlag, 1964).

—, *King Solomon's Ring* (1949; London: Routledge, 2002), pp. 14–15.

Lovén, S., 'On the Bathymetrical Distribution of Submarine Life on the Northern Shores of Scandinavia', *Report of the 14th Meeting of the British Association for the Advancement of Science (1844)*, 2 (1845), pp. 50–1.

Luedecke, H., *Vom Zaubervogel zum Zeppelin. Eine Geschichte der Luftfahrt und des Fluggedankens* (Berlin: Kurt Wolff, Verlag, 1936).

Lyell, C., *Principles of Geology: Being an Attempt to Explain the Former Changes of the Earth's Surface by Reference to Causes Now in Operations* (London, 1830–3).

Lynch, M., 'The Production of Scientific Images: Vision and Re-Vision in the History, Philosophy, and Sociology of Science', in L. Pauwels (ed.), *Visual Cultures of Science: Rethinking Representational Practices in Knowledge Building and Science Communication* (Lebanon, NH: University Press of New England, 2006), pp. 26–40.

—, and S. Woolgar (eds), *Representation of Scientific Pratice* (London and Cambridge, MA: The MIT-Press, 1990).

Lyons, S., 'Sea Monsters: Myth or Genuine Relic of the Past', in K. R. Benson and P. F. Rehbock (eds), *Oceanographic History: The Pacific and Beyond* (Seattle, WA and London: University of Washington Press, 2002), p. 60–70.

Maak, N., *Der Architekt am Strand. Le Corbusier und das Geheimnis der Seeschnecke* (München: Carl Hanser Verlag, 2010).

Macartney, J., 'Observations on Luminous Animals', *Phil. Trans.*, 100 (1810), pp. 258–93.

Magnus, O., *Carta marina et desciptio septembtrionalium terrarum ac mirabilium rerum in eis contentarum diligentissime elaborata septentrio Anno dei 1539* (Venedig, 1539).

—, *Historia de gentibus septentrionalibus, earundumque diversis statibus, conditionibus, moribus, ritibus..., Librum XXI, De piscibus monstrosis* (Rome, 1555).

Mangin, A., *L'Air et le Monde Aérien* (Tours: Alfred Mame, 1865).

—, *The Mysteries of the Ocean* (1864; London: T. Nelson and Sons, 1870).

Mantell, G. A., *Illustrations of the Geology of Sussex* (1827).

—, *Wonders of Geology; Or, a Familiar Exposition of Geological Phenomena: Being the Substance of a Course of Lectures Delivered at Brighton, From Notes taken by G. F. Richardson, Curator of the Mantellian Museum* (London: Rolfe & Fletcher, 1838).

Marey, E.-J., 'La Locomotion dans L'Eau', *La Nature*, 911 (15 November 1890), pp. 375–8.

Marks, A., *Inventar der Bibliothek Alfred Kubin im Kubin-Haus des Landes Oberösterreich in Zwickledt/Wernstein* (n.p., n.d.).

Marr, A., 'Introduction', in A. Marr and R. J. W. Evans (eds), *Curiosity and Wonder from the Renaissance to the Enlightenment* (Aldershot: Ashgate, 2006), pp. 1–20.

Marshall, W., *Die Tiefsee und ihr Leben. Nach den neuesten Quellen* (Leipzig: Hirt, 1888).

Marsili, L. F., *Histoire physique de la mer* (Amsterdam: Aux dépens de la Compagnie, 1725).

Martens, F., *Spitzbergische oder Groenländische Reise. Beschreibung gethan im jahre 1671* (Hamburg, 1675).

Martin, W., M. J. Russell, 'On the Origins of Cells: A Hypothesis for the Evolutionary Transitions from Abiotic Geochemistry to Chemoautotrophic Prokaryotes, and from Prokaryotes to Nucleated Cells', *Philosphical Transactions of the Royal Society London B*; DOI 10.1098/rstb.2002.1183.

Martyn, T., *Man and the Natural World. Changing Attitudes in England 1500–1800* (New York: Pantheon Books, 1983).

Marx, R. F., *Into the Deep. The History of Man's Underwater Exploration* (New York: Van Nostrand Reinhold, 1978).

—, *The History of Underwater Exploration* (New York: Dover Publications, 1978).

Materlinck, M., *Aquarium* (1889).

Matsen, B., *Descent: The Heroic Discovery of the Abyss* (New York: Pantheon Books, 2005).

Mattenklott, G., 'Schleierhaft: The Veil of Oblivion', *Daidalos*, 33 (15 September 1989), pp. 54–9.

Matyssek, A., *Rudolf Virchow, das Pathologische Museum: Geschichte einer wissenschaftlichen Sammlung um 1900* (Darmstadt: Steinkopff-Verlag, 2002).

Mauriès, P., *Shell Shock: Conchological Curiosities* (London: Thames and Hudson, 1994).

Maury, A., *Le Sommeil et les rêves* (Paris: Didier, 1861).

Maury, M. F., *Explanations and Sailing Directions to Accompany the Wind and Current Charts* (Philadelphia, PA: Biddle, 1854).

—, *The Physical Geography of the Sea* (New York: Harper & Brothers, 1855).

McKnight, S. A., 'Naturwissenschaft und Mystik bei Francis Bacon', in K. Vondung and K. L. Pfeiffer (eds), *Jenseits der entzauberten Welt: Naturwissenschaft und Mystik in der Moderne* (München and Paderborn: Wilhelm Fink Verlag, 2006), pp. 57–82.

Meder, J., *Die Handzeichnung. Ihre Technik und Entwicklung* (Wien: Kunstverlag Anton Schroll & Co., 1919).

Mehrtens, H., 'Irresponsible Purity: The Political and Moral Structure of Mathematical Sciences in the National Socialist State', in M. Renneberg and M. Walker (eds), *Science, Technology and National Socialism* (Cambridge: Cambridge University Press, 1993), pp. 324–38.

Meidinger, K. F. von, *Icones piscium Ausriae indigenorum* (Wien, 1785–94).

Meier-Gräfe, J., *Hans von Marées, sein Leben und Werk*, 3 vols (München and Leipzig: Piper Verlag, 1909–10)

Mellerio, A., *Odilon Redon, peintre, dessinateur et graveur* (Paris, 1923).

Melville, H., *The Whales; Or, Moby Dick*, 3 vols (London: Richard Bentley, 1851), vol. 2.

Merleau-Ponty, M., *Phenomenology of Perception*, trans. C. Smith (New York: Routledge & K. Paul, 1962).

Merriman, D., 'Speculations on Life at the Depths: A XIXth-Century Prelude', *Bulletin Institute Océanographic Monaco*, 2:2 (1968), pp. 377–84.

Mersch, D., 'Naturwissenschaftliches Wissen und bildliche Logik', in M. Heßler (ed.), *Konstruierte Sichtbarkeiten. Wissenschafts- und Technikbilder seit der frühen Neuzeit* (München: Wilhelm Fink Verlag, 2006), pp. 405–20.

Mersenne, M., *Questions inouyes ou récréation des Scavans* (Paris, 1634).

Meyen, F. J. F., 'Über das Leuchten des Meeres und Beschreibung einiger Polypen und anderer niederer Tiere', *Nova Acta Leop. Carol. Supp.* 16 (1834), pp. 125–8.

Meyer-Sickendiek, B., *Tiefe. Über die Faszination des Grübelns* (München: Wilhelm Fink Verlag, 2010).

Michaelis, G. A., *Über das Leuchten der Ostsee nach eigenen Beobachtungen, nebst einigen Bemerkungen über diese Erscheinung in andern Meeren* (Hamburg: Perthes und Besser, 1830).

Michelet, J., *The Sea* (New York: Rudd & Carleton, 1861).

Mills, E. L., 'Edward Forbes, John Gwyn Jeffreys, and British Dredging before the *Challenger* Expedition', *Journal of the Society for the Bibliography of Natural History*, 8:4 (1978), pp. 507–36.

—, *Biological Oceanography. An Early History, 1870–1960* (Ithaca, NY and London: Cornell University Press, 1989).

Möbius, K. A., 'Ostseeaquarien', *Zoologischer Garten*, 3 (1862), pp. 165–8.

—, *Das Aquarium des zoologischen Gartens zu Hamburg, für die Besucher desselben beschrieben* (Hamburg, 1864).

—, 'Einige Fingerzeige für die Bevölkerung und Erhaltung der Aquarien', *Zoologischer Garten*, 6 (1865), pp. 211–14.

—, 'Mittheilungen über das Aquarium des zoologischen Gartens in Hamburg', *Zoologischer Garten*, 7 (1866), pp. 173–7.

—, 'Der Bau des Eozoon canadense nach eigenen Untersuchungen verglichen mit dem Bau der Foraminiferen', *Paleontographica*, 25 (1878), pp. 175–92.

—, *The Oyster and Oyster-Culture*, US Commission Fish and Fisheries Report (1880) pp. 683–751.

Montfort, P. D. de, *Histoire naturelle, generale et particuliere des mollusques, animaux sans vertebres et a sang blanc, Addendum zu Georges-Louis Leclerc de Buffon*, 4 vols (Paris: Dufart, 1801–4).

Montfort, P. D. de, *Allgemeine und besondere Naturgeschichte der Weichwürmer (Mollusques)* als Fortsetzung der Buffonschen Naturgeschichte, 2 vols (Hamburg, Mainz: G. Vollmer, 1803).

More, A. G., 'Gigantic Squid on the West Coast of Ireland', *Annual Magazine of Natural History* 14:16 (1875a), pp. 123–4.

—, 'Notice of a Gigantic Cephlopod (Dinoteuthis Proboscideus) which was Stranded at Dingle, in Kerry, Two Hundred Years Ago', *Zoologist* (1875b), pp. 4526–32.

—, 'Some Account of the Gigantic Squid (Architeuthis dux) Lately Captured off Boffin Island, Connemara', *Zoologist*, 2:10 (1875), pp. 4569–71.

Morin, E., *The Cinema, or the Imaginary Man*, trans. L. Mortimer (1956; Minneapolis, MN and London: University of Minnesota Press, 2005).

Morse, M., *Virtualitie: Television, Media Art and Cyberculture* (Bloomington, IN: Indiana University Press, 1998).

Morton, M., 'Natur und Seele: Österreichische Reaktionen auf Ernst Haeckels evolutionären Monismus', in P. Kort and and M. Hollein (eds), *Darwin. Kunst und die Suche nach den Ursprüngen* (Frankfurt am Main: Wienand Verlag, 2009), pp. 126–52.

Moseley, H. H., *Notes by a Naturalist: An Account of Obsrvations Made during the Voyage of HMS 'Challenger' Round the World in the Years 1872–76* (London: John Murray, 1892).

Moser, S., *Ancestral Images: The Iconography of Human Origins* (Ithaca, NY: Cornell University Press, 1998).

Moure, N. D. W., *The World of Zarh Pritchard* (Los Angeles, CA: William A. Karges Fine Art, 1999).

Müller, I., *Die Geschichte der Zoologischen Station in Neapel von der Gründung durch Anton Dohrn (1872) bis zum Ersten Weltkrieg und ihre Bedeutung für die Entwicklung der modernen Biologischen Wissenschaften* (Düsseldorf, 1976).

—, and C. Groeben (eds), *The Naples Zoological Station at the time of Anton Dohrn* (Naples: Stazione Zoologica di Napoli,1975).

Müller, J., 'Bericht über einige neue Thierformen der Nordsee', *Archiv für Anatomie, Physiologie und Wissenschaftliche Medicin* (1846), pp. 100–10.

—, *Über die Larven und die Metamorphose der Ophiuren und Seeigel* (Berlin: Dümmler, 1848).

—, *Über die Larven und die Metamorphose der Echinodermen. Zweite Abhandlung* (Berlin: Dümmler, 1849).

Murray, J., *Experimental Researches on the Light & the Luminosity of the Sea, The Phenomena of the Chameleon ...* (Glasgow: W. R. McPhan, 1826).

—, and J. Hjort, *Depths of the Ocean – A General Account of the Modern Science of Oceanography Based Largely on the Scientific Researches of the Norwegian Steamer MICHAEL SARS in the North Atlantic* (London, 1912).

Nancy, J.-L., 'Das liegende Auge oder: Oberfläche, Öffnung und Bewegung des Wassers', *Berliner Gazette,* 'Wasserwissen', 2 February 2009, p. 5.

Neubert, F., *Einleitung in eine kritische Ausgabe von B. de Maillets Telliamed ou Entretiens d'un philosophe indien avec un missionnaire francois. Ein Beitrag zur Geschichte der französischen Aufklärungsliteratur* (Berlin: Emil Ebering, 1920).

'Neue Fischarten in der Tiefsee entdeckt', in: Frankfurter Allgemeine Zeitung Nr. 297, 22 December 2014, p. 7.

Neuhardt, G., 'Das Fenster als Symbol. Versuch einer Systematik der Aspekte', in E. T. Reimbold (ed.), *Symbolon. Jahrbuch für Symbolforschung* (Köln: Wienand, 1972), pp. 77–91.

Newton, I., *Opticks or, a Treatise of the Reflections, Refractions, Inflections and Colours of Light* (1719; London: W. and J. Innys, 1730).

Norman, C., *Noah's Flood: The Genesis Story in Western Thought* (New Haven, CT and London: Yale University Press, 1996).

Norman, J. R. and P. H. Greenwood Norman, *A History of Fishes* (London: Ernest Benn, 1963).

Norton, T., *Stars Beneath the Sea: The Extraordinary Lives of the Pioneers of Diving* (London: Century Press, 1999).

Novalis, 'Miscellaneous Observations no. 17', in Philosophical Writing, trans. and ed. M. M. Stoljar (Albany, NY: State University of New York Press, 1997),

Nyhart, L. K., 'Civic and Economic Zoology in Nineteenth-Century Germany: The "Living Communities of Karl Möbius"', *Isis*, 89 (1998), pp. 605–30.

—, 'Science, Art, and Authenticity in Natural History Displays', in S. de Chadaveira and N. Hopwood (eds), *Models: The Third Dimension of Science* (Stanford, CA: Stanford University Press, 2004), pp. 307–38.

O'Connor, T., 'Capture of an Enormous Cuttle-Fish off Boffin Island, on the Coast of Connemara', *Zoologist* (1875), pp. 4502–3.

Oesau, W., *Hamburgs Grönlandfahrt auf Walfischfang und Robbenschlag vom 17. bis 19. Jahrhundert* (Hamburg: J. J. Augustin, 1955).

Olalquiaga, C., *The Artificial Kingdom: On the Kitsch Experience* (New York: Pantheon Books, 1998).

Onians, J., "'I wonder ... '": A Short History of Amazement', J. Onians (ed.), *Sight & Insight: Essays on Art and Culture in Honour of E. H. Gombrich at 85* (London: Phaidon Press, 1994), pp. 11–34.

Packard, A. S., 'Colossal Cuttlefishes', *American Naturalist*, 7 (1873), pp. 87–94.

Painlevé, J., 'Mysteries and Miracles of Nature' (French 'Mystères et Miracles), *Vu* (29 March 1931), in A. M. Bellows and M. McDougall (ed.), *Science is Fiction: The Films of Jean Painlevé* (San Francisco, CA: Brico Press, 2000), pp. 119–23.

Panceri, P., *Atti delle Reale Accademia delle Scienze Fisiche e Matematiche di Napoli* (1873), pp. 1–12.

Papp, C. S. *Scientific Illustration: Theory and Practice* (Dubuque: W. C. Brown Company, 1968).

Parker, G.H., 'Chromatophores', *Biological Review* (1930), pp. 59–90.

'Parlour Aquaria', *Family Friend*, 2 (1856), pp. 192–7.

Partsch, K. J., *Die Zoologische Station in Neapel* (Göttingen: Vandenhoeck & Ruprecht, 1980).

Pas, C. van de, *Piscium vivae Icones* (Amsterdam: o.J.).

Paul, J., *Des Luftschiffers Giannozo Seebuch*, in the 2nd volume of the 'Komische Anhang' zum Roman *Titan*, 4 vols (1801; Berlin: Buchhandlung Matzdorff, 1800–1803).

Périer, J. A. N. and A. Berbrugger, *Exploration Scientifique de l'Algérie pendant les années 1840, 1841, 1842: Recherches de Physique Générale sur la Méditerranée* (Paris: Impr. Royale, 1845).

Péron, F., *Voyage de découvertes aux terres australes, execute Sur Les Covettes Le Geographe, Le Naturaliste, et La Goelette Le Casuarina, Pendant Les annees 1800, 1801, 1803 et 1804*, 2 vols (Paris: de L'Impremerie Royale, 1907).

—, 'Mémoire sur le nouveau genre Pyrosoma', *Annales du Musée d'Histoire Naturelle*, Paris, 4:437–46 (1804).

Perucchi-Petri, U., 'Jeunes Peintres, mes amis: Odilon Redon and the Nabis', in M. Hollein and M. Stuffmann (eds), *As in a Dream: Odilon Redon* (Ostfildern: HatjeCantz, 2007), pp. 103–11.

Peters, G., 'Zwei Hamburger, ein Barbier (1675) und ein Bürgermeister (1746), erkunden und beschreiben die Wale des Nördlichen Eismeers', *Historisch Meereskundliches Jahrbuch* (Berlin and Hamburg: Reimer, 1992), pp. 9–34.

Playfair, J., 'Biographical Account of the late James Hutton', *F. R. S. Edinburgh: Royal Society Edinburgh* (1805), pp. 39–99.

Power de Villepreux, J., *Observations physiques sur le poulpe de l'Argonauta Argo: commencees en 1832 et terminees en 1843* (Paris: Charles de Mourgues Freres, 1856).

—, 'The Aquarium', *The Popular Science Monthly* (May 1874), pp. 687–935.

Price, W., *Adventures in Paradise: Tahiti, Samoa, Fiji* (London: William Heinemann, 1956).

Pritchard, A., *History of Infusoria, Living and Fossil: Arranged According to 'Die Infusionsth-ierchen' of C. G. Ehrenberg* (London: Whitaker and Co., 1845).

Proctor, R., *Value-Free Science? Purity and Power in Modern Knowledge* (Cambridge, MA: Harvard University Press, 1991).

— (ed.), *René Binet. Natur und Kunst. Mit Beiträgen von Robert Proctor und Olaf Breidbach* (München: Prestel Verlag, 2007).

Proust, M., *In Search of Lost Time*, ed. C. Prendergast, 6 vols (1918; London: Penguin, 1987–9).

Quatrefages, A. de, *The Rambles of a Naturalist on the Coasts of France, Spain, and Sicily* (London: Longman, Brown, Green, Longmans & Roberts, 1857).

Ransonnet-Villez, Baron E. von, *Reise von Kairo nach Tor zu den Korallenbänken des Rothen Meeres* (Wien: Ueberreuter, 1863).

—, *Cylon, Skizzen seiner Bewohner, seines Thier- und Pflanzenlebens und Untersuchungen des Meeresgrundes nahe der Küste* (Braunschweig: Georg Westermann, 1868).

Ray, J., *Observations Topographical, Moral and Physiological, Made in a Journey Through Parts of the Low Countries, Germany, Italy and France: With a Catalogue of Plants not Native of England* (London, 1673).

—, *The Wisdom of God Manifested in the Works of the Creation* (London: Printed for Samuel Smith, 1691).

Rehbock, P., 'Huxley, Haeckel, and the Oceanographers: The Case of Bathybius Haeckelii', *Isis*, 66 (1975), pp. 504–33.

—, 'The Victorian Aquarium in Ecological and Social Perspective', in M. Sears and D. Merriman (eds), *Oceanography: The Past* (New York and Heidelberg: Springer-Verlag, 1980), pp. 522–39.

—, 'The Early Dredgers: "Naturalizing" in British Seas, 1830–1850', *Journal for the History of Biology*, 12:2 (1979), pp. 293–368.

Reyna, F., 'Las Fotografías Submarinas de Painlevé', *Estampa Año*, 8:367 (1935).

Rhein, P., *The Verbal and Visual Art of Alfred Kubin* (Riverside: Ariadne Press, 1989).

Rheinberger, H.-J., *Experiment, Differenz, Schrift. Zur Geschichte epistemischer Dinge* (Marburg: Baslisken-Presse, 1992).

—, *Experimentalsysteme und epistemische Dinge. Eine Geschichte der Proteinsynthese im Reagenzglas* (2001; Frankfurt am Main: Suhrkamp, 2006).

Reiter, W. L., 'Zerstört und vergessen: Die Biologische Versuchsanstalt und ihre Wissenschaftler/innen', *Österreichische Zeitschrift für Geschichtswissenschaften*, 10:4 (1999), pp. 585–614.

Rice, A., *Voyages of Discovery, Three Centuries of Natural History Exploration* (Richmond Hill, ON: Firefly Books, 2008).

Rice, A. L., 'The Oceanography of John Ross' Arctic Expedition of 1818: A Reappraisal', *Journal of the Society for the Bibliography of Natural History*, 7:3 (1975), pp. 291–313.

Richards, R. J., *The Tragic Sense of Life, Ernst Haeckel and the Struggle over Evolutionary Thought* (Chicago, IL and London: University of Chicago Press, 2008).

Richter, D., *Das Meer. Geschichte der ältesten Landschaft* (Berlin: Wagenbach, 2014).

Roberts, G., 'The Fossil Finder of Lyme Regis', *Chambers Journal of Popular Literature*, 7 (1857), pp. 382–4.

Rodenbach, G., *Les Vies Encloses. Poème* (Paris: Bibliothèque-Charpentier, 1896).

Roemer, B. van de, 'Neat Nature: The Relation between Nature and Art in a Dutch Cabinet of Curiosities from the early Eighteenth Century', *History of Science* (2004), pp. 47–84.

Rondelet, G., *Historia piscium universa* (Lyon: Matthias Bonhomme, 1554).

—., *Libri de Piscibus marinis, in quibus verae piscium effigies expressae sunt* (Lugduni, 1554).

Ross, Sir J. C., *A Voyage of Discovery made under the Orders of the Admiralty in His Majesty's Ships 'Isabella' and 'Alexander', for the purpose of exploring Baffin's Bay, and inquiring into the Possibility of a North-west Passage* (London: J. Murray, 1819).

—, *A Voyage of Discovery and Research in the Southern and Antarctic Regions, During the Years 1839–1843* (London: J. Murray, 1847).

Rossi, P., *The Dark Abyss of Time: The History of the Earth and the History of Nations from Hooke to Vico* (Chicago, IL: University of Chicago Press, 1984).

Roßmäßler, E. A., 'Der See im Glase', *Die Gartenlaube. Illustriertes Familienblatt* (1856), pp. 252–6.

Rozwadowski, H., *Fathoming the Ocean: The Discovery and Exploration of the Deep Sea* (Cambridge, MA and London: Harvard University Press, 2005).

Rudloff, D., *Unvollendete Schöpfung. Künstler im zwanzigsten Jahrhundert* (Stuttgart: Urachhaus, 1982).

Rudwick, M. J. S., *The Meaning of Fossils. Episodes in the History of Palaeontology* (London: Macdonald, 1972).

—, 'The Emergence of a Visual Language for Geological Science 1760–1840', *History of Science* (1976), pp. 149–95.

—, *Scenes from Deep Time, Early Pictorial Representations of the Prehistoric World* (Chicago, IL and London: University of Chicago Press, 1992).

—, *Bursting the Limits of Time: The Reconstruction of Geohistory in the Age of Revolution* (Chicago, IL and London: University of Chicago Press, 2005).

—, *Worlds Before Adam: The Reconstruction of Geohistory in the Age of Reform* (Chicago, IL and London: University of Chicago Press, 2008).

Ruskin, J., *The Elements of Drawing: In Three Letters to Beginners* (New York: Wiley & Halsted, 1857).

Salvia, A. La, 'Der Krake mit dem seidenen Blick. Lautréamont, Dalí, Breton', in E. Puyplat, A. La Salvia and H. Heinzelmann (eds), *Salvador Dali: Facetten eines Jahrhundertkünstlers* (Würzburg: Königshausen & Neumann, 2005), pp. 107–18.

Scherren, H., *Ponds and Rock Pools* (London: The Religious Tract Society, 1906).

Scheuchzer, J. J., *Homo diluvii testis et theoscopos* (Zürich, 1726).

Schickore, J., *The Microscope and the Eye: A History of Reflections, 1740–1870* (Chicago, IL: University of Chicago Press, 2007).

Schlebrügge, E., 'Die Conchyliologie, diese angenehme Wissenschaft …', in E. Schlebrügge (ed.), *Das Meer im Zimmer. Von Tintenschnecken und Muscheltieren. Ausstellung im*

Landesmuseum Joanneum, Graz vom 13.:5. bis 1 November 2005 (Graz: Landesmuseum Johanneum 2005), pp. 123–41.

Schlee, S., *The Edge of an Unfamiliar World: A History of Oceanography* (New York: E. P. Dutton & Co., 1973).

—, 'The Controversial Dr. Beebe and his Brain Fish', *BBSR Newsletter*, 3:3 (1974), p. 2.

Schleiden, M. J., *Die Pflanze und ihre Leben* (Leipzig, 1858); *The Plant: A Biography in a Series of Popular Lectures*, trans. Arthur Henrey (London: Hippolyte Bailliere, 1848).

Schmalenbach, W., *Paul Klee: Fische, Einführung von Werner Schmalenbach. Werkmonographien zur Bildenden Kunst in Reclams Universal Bibliothek Band Nr. 31* (Stuttgart: Reclam, 1958).

Schmitt, E. (ed.), *Glas des Art Nouveau. Die Sammlung Gerda Koepff* (München and New York: Prestel, 1998).

Schmoll Gen. Eisenwerth, J. A., 'Der Blick durch das offene Fenster. Ein Motiv der deutschen Malerei des 19. Jahrhunderts ... ', in *Der frühe Realismus in Deutschland 1800–1850. Gemälde und Zeichnungen aus der Sammlung Georg Schäfer* (Schweinfurt: Ausstellung im Germanischen Nationalmuseum Nürnberg, 1967), pp. 132–44.

—, 'Fensterbilder. Motivketten in der europäischen Malerei', in L. Grote (ed.), *Beiträge zur Motivkunde des 19. Jahrhunderts* (München: Prestel Verlag, 1970), pp. 13–166.

Smollet, T., *The Expedition of Humphry Clinker*, ed. G. Saintsbury, 2 vols (1771; London: Gibbings 1895).

Schnier, J., 'Morphology of a Symbol: The Octopus', *American Image: A Psychoanalytic Journal for the Arts and Sciences*, 13:1 (1956), pp. 3–31.

Schöpf, J. D., *Travels in the Confederation* (1783–84), trans. and ed. A. J. Morrison (Philadelphia, PA: William J. Campell, 1911).

Schramm, H., L. Schwarte and J. Lazardzig (eds), *Spektakuläre Experimente. Praktiken der Evidenzproduktion im 17. Jahrhundert* (Berlin and New York: Springer, 2006).

Schubert, G., 'Aus dem Berliner Aquarium', *Der Zoologische Garten. Zeitschrift für Beobachtung, Pflege und Zucht der Thiere*, 21 (1880), pp. 92–6.

Sears, M. and D. Merriman (eds), *Oceanography: The Past* (New York, Heidelberg and Berlin: Springer Verlag, 1980).

Secord, J. A., 'Monsters at the Chrystal Palace', in S. de Chadarevian and N. Hopwood (eds), *Models: The Third Dimension of Science* (Stanford, CA: Stanford University Press, 2004), pp. 138–69.

Selbmann, R., *Eine Kulturgeschichte des Fensters von der Antike bis zur Moderne* (Berlin: Reimer Verlag, 2010).

Sereni, E., 'The Chromatophores of the Cephalopods', *Biological Bulletin, Woods Hole* (1930), pp. 247–68.

Sewell, R. B. S., 'Oceanographic Exploration, 1851–1951', *Science Progress*, 40 (1952), pp. 403–18.

Simmons, W. E., 'The Aquarium', *The Popular Science Monthly* (1874), pp. 687–95.

Simon, H.-R., 'Carl Chun und Ernst Haeckel als Herausgeber meeresbiologischer Tafelwerke', in C. Nissen (ed.), *Die zoologische Buchillustration: ihre Bibliographie und Geschichte*, 2 vols (Stuttgart: Hiersemann, 1978), pp. 327–37.

— (ed.), *Anton Dohrn und die Zoologische Station Neapel* (Frankfurt am Main: Edition Erbrich, 1980).

Simon, J., 'Jean Painlevé and Mark Dion', *Parkett*, 45 (1995), pp. 163–71.

Snyder, J., 'Visualization and Visibility', in C. A. Jones and P. Gallison (eds), *Picturing Science/Making Art* (New York: Routledge Chapman & Hall, 1998).

Sonrel, L., *Bottom of the Sea* (London: Sampson Low, Son and Marston, 1870).

Sowerby, G. B., *Popular British Conchology: A Familiar History of the Molluscs Inhabiting the British Isles* (London: Lovell Reeve, 1854).

—, *The Aquarium: A Popular Account of Marine and Fresh-Water Animals and Plants* (London: Routledge, 1865).

Spry, W. J. J., *The Cruise of her Majestys Ship 'Challenger': Voyages Over Many Seas, Scenes in Many Lands* (London: Sampson Low, Marston, Searle, & Rivington, 1876).

Stafford, B. M., *Voyage into Substance, Art, Science, Nature and the Illustrated Travel Account, 1740–1840* (Cambridge, MA: MIT Press, 1984).

—, 'Images of Ambiguity: Eighteenth-Century Microscopy and the Neither/Nor', in D. P. Miller and P. H. Reill (eds), *Visions of Empire. Voyages, Botany, and Representations of Nature* (Cambridge: Cambridge University Press, 1996), pp. 230–57.

Steche, O., 'Carl Chun', *Mitteilungen der Gesellschaft für Erdkunde zu Leipzig* (Leipzig 1914), pp. 43–90.

Steinach, E., 'Studien über die Hautfärbung und über den Farbenwechsel der Cephalopoden', *Arch. f. g. Physiol*, 87 (1901), pp. 1–37.

Steinberg, P. E., *The Social Construction of the Ocean* (Cambridge: Cambridge University Press, 2001).

Stott, R., 'Through a Glass Darkly: Aquarium Colonies and Nineteenth-Century Narratives of Marine Monstrosity', *Gothic Studies*, 2:3 (2000), pp. 305–27.

—, *Theatres of Glass: The Woman who Brought the Sea to the City* (London: Short Books, 2003).

—, *Darwin and the Barnacle: The Story of One Tiny Creature and History's Most Spectacular Scientific Breakthrough* (New York: W. W. Norton & Company, 2004).

Strehlow, H., 'Zur Geschichte des Berliner Aquariums Unter den Linden', *Der Zoologische Garten*, 57 (1987), pp. 26–40.

Taves, B., 'A Pioneer under the Sea: Library Restores Rare Film Footage', *LC Information Bulletin*, 55:15 (16 September 1996).

—, 'With Williamson Beneath the Sea', *Journal of Film Preservation*, 52 (1996), pp. 54–61.

Tennyson, A., *Poems, Chiefly Lyrical* (London: Effingham Wilson, 1830).

Thomson, C. W., *The Depths of the Sea: An Account of the General Results of the Dredging Cruises of HMSS 'Porcupine' and 'Lightning' during the summers of 1868, 1869, and 1870* (London: MacMillan, 1873).

—, *Voyage of the "Challenger": The Atlantic*, 2 vols (London: MacMillan, 1877).

—, and J. Murray (eds), *Report on the Scientific Results of the Voyage of HMp Challenger During the Years 1873–76* (London 1880–95).

Thwaite, A., *Glimpses of the Wonderful: The Life of Philip Henry Gosse 1810–1888* (London: Faber & Faber, 2002).

Thynne, A., 'On the Increase of Madrepores', *Annals of Natural History*, 18 (June 1859), pp. 449–61.

Tilesius von Tilenau, W. G., *Resultate seiner, während der drei Jahren der Krusensteruschen Entdeckungsreise gesammelten Erfahrungen* (St Peterburg, 1813).

Torrens, H., 'Mary Anning of Lyme: The Greatest Fossilist the World ever Knew', *British Journal of the History of Science*, 28 (1995), pp. 257–84.

Tradescant, J., *Musaeum Tradescantianum: Or, a Collection of Rarities Preserved at South-Lambeth, Near London* (London: John Grismond, 1656).

Treviranus, G. R., *Die Erscheinungen und Gesetze des organischen Lebens*, 2 vols (Bremen: Heyse-Verlag, 1833).

Tsutatani, N., 'A Floating Vision: Thoughts on Odilon Redon's *Guradian Spirit of the Waters* (1878)', in A. Yamamoto (ed.), *Odilon Redon – le Souci de l'Absolu* (Nagoya: Chunichi Shimbun, 2002), pp. 194–200.

Ullrich, W., *Die Geschichte der Unschärfe* (Berlin: Wagenbach-Verlag, 2009).

Ulmer, R. (ed.), *Art Nouveau: Symbolismus und Jugendstil in Frankreich* (Stuttgart and New York: Arnoldsche 1999).

Valéry, P., 'Blicke aufs Meer', in J. Schmidt-Radefeltd (ed.), *Paul Valéry. Werke*. Frankfurter Ausgabe in 7 Bänden (1930, French; Frankfurt: Insel, 1989), pp. 501–9.

—, *Eupalinos: Or The Architect*, trans. W. M. Stewart (1923; London and Oxford: Oxford University Press, H. Milford, 1932).

, 'Man and the Sea Shell', in P. Valéry, *Aesthetics, Collected Works*, trans. R. Manheim, 15 vols (New York: Pantheon Books, 1964), vol. 13, pp. 3–31.

Vanney, J.-R., *Le Mysètre des Abysses. Histoire et découvertes des profondeurs océaniques* (Paris: Fayard, 1993)

Verdi, R., *Klee and Nature* (London: Zwemmer Ltd, 1984).

Verhaeren, E., *L'Aquarium* (1889).

Verne, J., *Twenty Thousand Leagues under the Seas; Or, The Marvellous and Exciting Adventures of Pierre Aronnax, Conseil his Servant, and Ned Land, a Canadian Harpooner*, trans. from the French (Boston, MA: Geo. M. Smith & Co., 1873).

Verrill, A. E., 'Occurrence of Gigantic Cuttle-Fishes on the Coast of Newfoundland', *American Journal of Science and Arts* (1874), pp. 158–61.

—, 'The Giant Cuttle-Fishes of Newfoundland and the Common Squids of the New England Coast', *American Naturalist* (1874), pp. 167–74.

—, 'The Colossal Cephalopods of the North Atlantic', *American Naturalist* (1875), pp. 21–36.

—, *The Cephalopods of the North-Eastern Coast of America* (New Haven, CT: Connecticut Academy of Sciences, 1880).

—, 'Occurrence of Architeuthis', *American Journal of Science* (1881), pp. 171–2.

Viviani, D., *Phosphorescentia maris* (Genua: Joannis Giossi, 1805).

Vogt, C., 'Aus der zoologischen Station in Neapel', *Die Gartenlaube*, 21 (1880), pp. 340–4.

Vortriede, W., *Novalis und die französischen Symbolisten. Zur Entstehungsgeschichte des dichterischen Symbols* (Stuttgart: Kohlhammer, 1963).

Voss, J., 'Variieren und Selektieren: Die Evolutionstheorie in der englischen und deutschen illustrierten Presse im 19. Jahrhundert', in P. Kort and and M. Hollein (eds), *Darwin. Kunst und die Suche nach den Ursprüngen* (Frankfurt am Main: Wienand Verlag, 2009), pp. 246–57.

—, 'Die Entdeckung der Unordnung. Wie Charles Darwin den Zufall zeichnete', in N. Adamowsky, R. Felfe and H. Böhme (eds), *Ludi naturae. Spiele der Natur in Kunst und Wissenschaft* (München: Wilhelm Fink Verlag, 2011), pp. 257–84.

Voßkamp, W. (ed.), *Utopieforschung. Interdisziplinäre Studien zur neuzeitlichen Utopie* (Frankfurt am Main: Suhrkamp, 1985).

Waldenfels, B., 'Überraschte Wahrnehmung. Eine phänomenologische Betrachtung', in M. Frölich, R. Middel and K. Visarius (eds), *Zeichen und Wuner. Über das Staunen im Kino* (Marburg: Schüren, 2001), pp. 9–29.

Waldeyer, W., 'Rede des Geh. Rath Prof. W. Waldeyer zum 25-jährigen Bestehen der Zoologischen Station', in H.-R. Simon (ed.), *Anton Dohrn und die Zoologische Station Neapel* (Frankfurt am Main: Ed. Erbrich, 1980), pp. 77–81.

Wallich, G. C., *The North Atlantic Sea-Bed: Comprising a Diary of the Voyage on Board HMS 'Bulldog' in 1860* (London: J. van Voorst, 1862).

Ward, F., *Marvels of Fish Life: As Revealed by the Camera*, 2nd edn (London and New York: Cassell and Company, 1912).

—, *Animal Life Under Water* (London: Cassell, 1919).

Ward, J., *Weimar Surfaces: Urban Visual Culture in 1920s Germany* (Berkeley, CA: University of California Press, 2001).

Ward, N. B., 'Reports on the Subject of the Growth of Plants in Closed Glass Vessels', *Report of the British Association for the Advancement of Science*, 1 (1837), pp. 501–5.

—, *On the Growth of Plants in Closely Glazed Cases* (London: Van Voorst, 1842).

Warington, R., 'Notice of Observations on the Adjustment of the Relations Between the Animal and Vegetable Kingdoms, by which the Viral Functions of Both are Permanently Maintained', *Quarterly Journal of the Chemical Society*, 3 (1851), pp. 52–4.

—, 'On Preserving the Balance Between the Animal and Vegetable Organisms in Sea-Water', *Annals of Natural History*, 2:12 (1853), pp. 319–24.

—, 'Memoranda of Observations Made in Small Aquaria, in which the Balance Between the Animal and Vegetable Organisms was Permanently Maintained', *Annals of Natural History*, 2:14 (1854), pp. 366–73.

—, 'On the Injurious Effects of an Excess or Want of Heat and Light on the Aquarium', *Annals of Natural History*, 2:16 (1855), pp. 313–15.

—, 'On the Aquarium', *Proceedings of the Royal Institution of Great Britain*, 2 (1857), pp. 403–8.

Watermann, B. and O. J. Wrzesinski, *Bibliographie zur Geschichte der deutschen Meersforschung. Chronologische Titelaufzählung (1557–1989)* (Hamburg: Deutsche Gesellschaft für Meeresforschung, 1989).

Watson, L., *Sea Guide to Whales of the World* (London and Sydney: Hutchinson, 1981).

Weber, M., 'Wissenschaft als Beruf', in *Gesammelte Aufsätze zur Wissenschaftslehre* (Tübingen: Mohr Verlag, 1922), pp. 524–55.

—, *Gesammelte Aufsätze zur Religionssoziologie I,* photomechanischer Nachdruck der Erstauflage von 1920 (Tübingen: Mohr Verlag, 1988).

Weidner, H., 'Die Anfänge meeresbiologischer und ökologischer Forschung in Hamburg durch Karl August Möbius (1825–1908) und Heinrich Adolph Meyer (1822–1889)', in *Historisch Meereskundliches Jahrbuch* (Berlin and Hamburg: Dietrich Reimer, 1994), pp. 69–84.

Weigel, V., 'Vom Aquarium- zum Videobild. Einführende Bemerkungen zu den Werken ... ' in Weigel, V. (ed.), *unter wasser über wasser. Vom Aquarium. zum Videobild* (Wilhelmshaven: Kerber Art, 2009), pp. 36–49.

Weinberg, S., P. L. Dogué and J. Neuschwander, *Unterwasserfotografie: Hundert Jahre Geschichte, Technik, Faszination* (Gilching: Verlag Photographie, 1993).

Welker, R. H., *Natural Man: The Life of William Beebe* (Bloomington, IN: Indiana University Press, 1975).

Wells, H.G., *The Sea Raiders* (Weekly Sun Literary Supplement, 6 December 1896).

Whiston, W., *A New Theory of the Earth* (London, 1708).

Wichmann, S., *Jugendstil Floral Funktional in Deutschland und Österreich und den Einflußgebieten* (Herrsching: Schuler Verlagsgesellschaft, 1984).

'Wie er- und behält man den Ocean auf dem Tische, oder das Marine-Aquarium', *Die Gartenlaube*, 38 (1855), pp. 503–5.

Wild, J. J., *Thalassa: An Essay on the Depth, Temperature and Currents of the Ocean* (London: Marcus Ward & Co., 1877).

Willemoes-Suhm, R. von, *Die Challenger-Expedition. Zum tiefsten Punkt der Weltmeere, 1872–1876. Rudolf von Willemoes-Suhms Briefe von der Challenger-Expedition mit Auszügen aus dem Reisebericht des Schiffsingenieurs W. J. J. Spry* (Stuttgart: Thienemann, 1984).

Williams, R., *Notes on the Underground: An Essay on Technology, Society, and the Imagination* (Cambridge, MA and London: The MIT Press, 2008).

Williamson, J. E., *Twenty Years under The Sea* (Boston, MA: Ralph. T. Hale & Company, 1936).

Willughby, F., *De Historia piscium Libri IV* (1686).

Winter, F. W., 'Carl Chun', *Berichte der Senckenbergischen Naturforschenden Gesellschaft in Frankfurt am Main,* 45 (1914), pp. 176–83.

Wolfzettel, F., *Jules Verne: Eine Einführung* (Zürich: Artemis Verlag, 1988).

Wood, J. G., *The Fresh and Salt Water Aquarium* (London: Routledge, 1868).

Woolf, V., *A Room of One's Own* (London: Hogarth Press, 1929).

Wüst, G., 'The Major Deep-Sea Expeditions and Research Vessels, 1873–1960', in M. Sears (ed.), *Progress in Oceanography* (London and New York: Pergamon Press, 1964), pp. 1–52.

Yonge, C. M., 'Victorians by the Sea Shore', *History Today*, 25:9 (1975), pp. 602–9.

Zachariae, A. W., *Die Elemente der Luftschwimmkunst* (Wittenberg: Zimmermann, 1807).

—, *Fluglust und Fluges Beginnen: Hierbei mein schon fliegendes Blatt und auf diesem Kupferstich der Bauriss zu meinem Flugkahne nebst Abbildung von dessen Luftbahn* (Leipzig: C. Cnobloch, 1821).

Zajonc, A., *Catching the Light: The Entwined History of Light and Mind* (New York and London: Oxford University Press, 1995).

Zedler, J. H., *Großes Vollständiges Universal-Lexikon aller Wissenschaften und Künste, 1731–54*, 64 vols (Leipzig: Johann Heinrich Zedler)

Zegher, C. de and C. Armstrong (eds), *Ocean Flowers. Impressions from Nature* (New York, Princeton, NJ and Oxford: Princeton University Press, 2004).

Zinn, D. J., 'Alexander Agassiz (1835–1910) and the Financial Support of Oceanography in the United States', in M. Sears and D. Merriman (eds), *Oceanography: The Past* (New York, Heidelberg and Berlin: Springer Verlag, 1980), pp. 83–93.

NOTES

Introduction: Perspectives on the Epistemology and Aesthetics of Oceanic Wonders

1. A. Mangin, *The Mysteries of the Ocean* (1864; London: T. Nelson and Sons, 1870), p. 197.
2. Mangin, *Mysteries*, p. ix.
3. Mangin, *Mysteries*, p. 173.
4. J. Michelet, *The Sea* (New York: Rudd & Carleton, 1861), p. 131.
5. K. Benson, H. Rozwadowski and D. van Keuren (eds), *The Machine in Neptune's Garden: Historical Perspectives on Technology and the Marine Environment* (Sagamore Beach, MA: Science History Publications, 2004), pp. xiii, xiv.
6. ('put the mediated under conditions which they constitute and create'; all translations are mine unless otherwise stated), L. Engell and J. Vogl, 'Vorwort', in L. Engell, J. Vogl, O. Fahle and B. Neitzel (eds), *Kursbuch Medienkultur: Die maßgeblichen Theorien von Brecht bis Baudrillard* (Stuttgart: Deutsche Verlags-Anstalt, 1999), p. 10.
7. Compare to M. Heßler and D. Mersch, 'Einleitung: Bildlogik oder Was heißt visuelles Denken?', in M. Heßler and D. Mersch (eds), *Logik des Bildlichen. Zur Kritik der ikonischen Vernunft* (Bielefeld: Transcript-Verlag, 2009), pp. 8–62, especially pp. 14–18.
8. F. Belly, 'La Poésie de l'exposition', *Revue contemporaine*, 23 (1855), p. 162.
9. M. Weber, 'Wissenschaft als Beruf', in *Gesammelte Aufsätze zur Wissenschaftslehre* (Tübingen: Mohr-Verlag, 1922), pp. 524–55, on p. 554.
10. M. Weber, *Gesammelte Aufsätze zur Religionssoziologie I*, photomechanischer Nachdruck der Erstauflage von 1920 (Tübingen: Mohr-Verlag, 1988), p. 564.
11. See also S. A. McKnight, 'Naturwissenschaft und Mystik bei Francis Bacon', in K. Vondung and K. L. Pfeiffer (eds), *Jenseits der entzauberten Welt: Naturwissenschaft und Mystik in der Moderne* (München and Paderborn: Wilhelm Fink Verlag, 2006), pp. 57–82, on p. 57.
12. McKnight, 'Naturwissenschaft und Mystik', p. 57; and M. Berman, *The Reenchantment of the World* (Ithaca, NY: Cornell University Press, 1981).
13. R. M. Burns provides an excellent overview in *The Great Debate on Miracles: From Joseph Glanvill to David Hume* (Lewisburg, PA: Bucknell University Press, 1981).
14. H. Blumenberg, *Der Prozeß der theoretischen Neugierde* (Frankfurt am Main: Suhrkamp, 1966).
15. W. J. Bouwsma, 'Work on Blumenberg', *Journal for the History of Ideas*, 48:2 (April – June, 1987), pp. 347–54, on p. 347–8.
16. See also A. Marr, 'Introduction', in A. Marr and R. J. W. Evans (eds), *Curiosity and Wonder*

from the Renaissance to the Enlightenment (Aldershot: Ashgate 2006), pp. 1–20, on p. 8.

17. See also N. Kenny, *Curiosity in Early Modern Europe: Word Histories* (Wiesbaden: Harrassowitz-Verlag, 1998), p. 15.

18. L. Daston and K. Park, *Wonders and the Order of Nature* (New York: Zone Books, 1998), pp. 11, 17.

19. Daston and Park, *Wonders and the Order of Nature*, p. 19.

20. Daston and Park, *Wonders and the Order of Nature*, p. 367.

21. S. Greenblatt, *Marvellous Possessions: The Wonder of the New World* (Chicago, IL: The University of Chicago Press, 1991), p. 7.

22. Greenblatt, *Marvellous Possessions*, p. 17, 78.

23. K. Barck, 'Wunderbar', in K. Barck, M. Fontius, D. Schlenstedt, B. Steinwachs and F. Wolfzettel (eds), *Ästhetische Grundbegriffe. Historisches Wörterbuch in sieben Bänden* (Stuttgart and Weimar: J. B. Metzler, 2005), vol. 6, pp. 730–73, on p. 736.

24. C. Colón and D. de Colón, *Libro de la primera navegación y descubrimiento de las Indias: facsímil y transcripcion del manuscrito original de Fray Bartolomé de las Casas / introducción por Manuel Lucena Giraldo,* (Madrid: Guillermo Blázquez 2005), Lunes, 24 de diciembre 1492, p. 192.

25. Greenblatt, *Marvellous Possessions,* p. 73.

26. Greenblatt, *Marvellous Possessions*, pp. 74, 133–5.

27. J. LeGoff, *The Medieval Imagination*, trans. A. Goldhammer (Chicago, IL and London: University of Chicago Press, 1988).

28. Greenblatt, *Marvellous Possessions*, pp. 75, 24–25.

29. J. Onians, '"I wonder ... ": A Short History of Amazement', in J. Onians (ed.), *Sight & Insight: Essays on Art and Culture in Honour of E. H. Gombrich at 85* (London: Phaidon Press, 1994), pp. 11–34, on pp. 24, 26, 32.

30. Greenblatt, *Marvellous Possessions*, p. 13.

31. Compare to W. L. Reiter, 'Zerstört und vergessen: Die Biologische Versuchsanstalt und ihre Wissenschaftler/innen', *Österreichische Zeitschrift für Geschichtswissenschaften*, 10:4 (1999), 'Sprache Macht Geschichte', pp. 585–614; and A. Matyssek, *Rudolf Virchow, das Pathologische Museum: Geschichte einer wissenschaftlichen Sammlung um 1900* (Darmstadt: Steinkopff-Verlag, 2002).

32. A. Marr, 'Introduction', in A. Marr and R. J. W. Evans (eds), *Curiosity and Wonder from the Renaissance to the Enlightenment* (Aldershot: Ashgate 2006), pp. 1–20, on p. 17.

33. J. Kenseth, 'Age of the Marvelous: An Introduction', in J. Kenseth (ed.), *The Age of the Marvellous* (Hanover: Hood Museum of Art, Dartmouth College, 1991), pp. 25–60, on p. 55.

34. The Natural History Museum, map, London, 2002; the phrase refers to the exhibition in the central hall, which has been recently renamed Hintze Hall.

35. L. Federmair, 'Entzaubern – Verzaubern. Zu den außergewöhnlichen Reisen Jules Vernes', in B. Felderer (ed.), *Wunschmaschine Welterfindung: Eine Geschichte der Technikvisionen seit dem 18. Jahrhundert* (Wien, NY: Springer-Verlag, 1996), pp. 236–49, on p. 243.

36. M. Morse, *Virtualities: Television, Media Art and Cyberculture* (Bloomington, IN: Indiana University Press, 1998), p. 81.

37. R. Williams, *Notes on the Underground: An Essay on Technology, Society, and the Imagination* (Cambridge, MA and London: The MIT Press, 2008), p. 43.

38. Williams, *Notes on the Underground*, p. 45.

39. The idea of temporalization has been discussed in various directions; compare to W. Lepenies, *Das Ende der Naturgeschichte: Wandel kultureller Selbstverständlichkeiten in den Wissenschaften des 18. und 19. Jahrhunderts* (München: Hanser-Verlag, 1976); and

W. Voßkamp (ed.), *Utopieforschung. Interdisziplinäre Studien zur neuzeitlichen Utopie* (Frankfurt am Main: Suhrkamp, 1985).

40. J. Murray and J. Hjort, *Depths of the Ocean – A General Account of the Modern Science of Oceanography Based Largely on the Scientific Researches of the Norwegian Steamer MICHAEL SARS in the North Atlantic* (London: MacMillan, 1912), p. 22.

41. N. Adamowsky, *Das Wunder in der Moderne. Eine andere Kulturgeschichte des Fliegens* (Paderborn: Wilhelm Fink Verlag, 2010).

42. M. J. Schleiden, *Die Pflanze und ihre Leben* (Leipzig: Engelmann, 1858), p. 164 (7. Vorlesung; 7th Lecture); and compare to The Plant; A Biography in a Series of Popular Lectures, trans. A. Henrey (London: Hippolyte Bailliere, 1848).

43. The paradoxical reception and description of the sea, especially the Mediterranean Sea, is skillfully discussed in H. Baader and G. Wolf (ed.), *Das Meer, der Tausch und die Grenzen der Repräsentation* (Zürich: Diaphanes-Verlag, 2010).

44. J. Delumeau, *La Peur en Occident (XIVe-XVIIIe siècles). Une cité assiégé* (Paris: Fayard, Paris 1978).

1 Wondrous and Terrible Sea – Stations of an Unknown Modernity

1. H. Rozwadowski, *Fathcoming the Ocean: The Discovery and Exploration of the Deep Sea* (Cambridge, MA and London: Harvard University Press, 2005), p. 4.

2. J. W. Konvitz, 'Changing Concepts of the Sea', in M. Sears and D. Merriman (eds), *Oceanography: The Past* (New York, Heidelberg and Berlin: Springer-Verlag, 1980), pp. 32–41, on p. 32.

3. C. M. Yonge, 'Victorians by the Sea Shore', *History Today*, 25:9 (1975), pp. 602–9.

4. The notion of sea-bathing and the use of bathing-machines goes back to John Awsiter, Thoughts on Brightelmston, 1768 zurück. See D. E. Allen, *The Naturalist in Britain: A Social History* (Princeton, NJ: Princeton University Press, 1976), pp. 111–13; and D. Richter, *Das Meer. Geschichte der ältesten Landschaft* (Berlin: Wagenbach, 2014), p. 152.

5. W. Cowper, 'Retirement' (1787); Poems by William Cowper of the Inner Temple, Esq. (London: J. Johnson, 1782).

6. 'A Trip to Scarborough' was written by R. B. Sheridan (1751–1816) and it was staged for the first time in 1777.

7. T. Smollet, *The Expedition of Humphry Clinker*, ed. G. Saintsbury, 2 vols (1771; London: Gibbings 1895).

8. D. J. Zinn summarizes this development: '[B]y about 1730 the Europeans and, in particular, the English, had developed an interest in visiting the seashore for health and recreation. By the mid-nineteenth century attendance at seaside resorts became a popular summer diversion for English families ... [T]his led to the publication of several manuals and guides concerning the diversity, kinds and habitats of intertidal und subtidal marine organisms', in M. Sears and D. Merriman (eds), *Oceanography*, pp. 83–93, on p. 87.

9. See J. Schickore, *The Microscope and the Eye: A History of Reflections, 1740–1870* (Chicago, IL: University of Chicago Press, 2007).

10. See J. W. Hedgpeth, 'De mirabili maris: Thoughts on the Flowering of Seashore Books', *The Royal Society of Edinburgh, Proceedings, Section B (Biology)*, 72:8 (1971–2), pp. 107–14, on p. 108.

11. A. Corbin, *The Lure of the Sea: The Discovery of the Seaside in the Western World 1750–1840*, trans. J. Phelps (Berkeley, CA: University of California Press, 1994), pp. 1–2.

12. See W. Detel, 'Das Prinzip Wasser bei Thales', in H. Böhme (ed.), *Kulturgeschichte des Wassers* (Frankfurt am Main: Suhrkamp, 1998), pp. 43–64, on pp. 43–5.

13. See Detel, 'Das Prinzip Wasser', p. 45.

14. The book of Revelation: 'Then I saw a new heaven and a new earth, for the first heaven and the first earth had passed away, and there was no longer any sea'. (Rev. 21:1).

15. A. Corbin, *The Lure of the Sea: The Discovery of the Seaside in the Western World 1750–1840*, trans. J. Phelps (Berkeley, CA: University of California Press, 1994), pp. 2, 4.

16. H. Böhme and G. Böhme, *Feuer, Wasser, Erde, Luft: Eine Kulturgeschichte der Elemente* (München: C. H. Beck, 1996), p. 60.

17. Sophocles, Antigone, V. 332–6.

18. Compare to H. Bredekamp, 'Wasserangst und Wasserfreude in Renaissance und Manierismus', in H. Böhme (ed.), *Kulturgeschichte des Wassers*, pp. 145–88, on p. 148.

19. P. E. Steinberg, *The Social Construction of the Ocean* (Cambridge: Cambridge University Press, 2001), p. 105.

20. Böhme and Böhme, *Kulturgeschichte der Elemente*, p. 60.

21. E. Forbes, *The Natural History of the European Seas* (1859; New York: Arno Press, 1977), p. 12.

22. Corbin, *Lure of the Sea*, pp. 6–8.

23. Compare to H. Bredekamp, 'Coral versus Trees: Charles Darwin's Early Sketches of Evolution', in P. H. Smith, A. R. W. Meyers and H. J. Cook, *Ways of Making and Knowing: The Material Culture of Empirical Knowledge* (Ann Arbor, MI: University of Michigan Press, 2014), pp. 357–76, on p. 367.

24. C. Gesner, *Historiae Animalium: Liber 4 qui est de Piscium & Aquatilium Animantium natura* (Tiguri: Froscherus, 1558).

25. A. Lonitzer, *Naturalis Historiae Opus Novum* (Frankfurt am Main: Egenolphus, 1551–5); G. Rondelet, *Libri de Piscibus marinis, in quibus verae piscium effigies expressae sunt* (Lugdini, 1554); P. Belon, *La Nature et Diversité des Poissons* (Paris, 1555); and see E. W. Gudger, 'The Five Great Naturalists of the 16th Century: Belon, Rondelet, Salviani, Gesner and Aldrovandi: A Chapter in the History of Ichthyology', *Isis*, 22 (1934), pp. 21–40.

26. See R. Ellis, *Monsters of the Sea* (London: Robert Hale, 1994).

27. G. Bachelard, *The Poetics of Space: The Classic Look at how we Experience Intimate Places*, trans. M. Jolas (Boston, MA: Beacon Press, 1994), p. 107.

28. J. Baltrusaitis, *Das phantastische Mittelalter. Antike und exotische Elemente der Kunst der Gotik* (Frankfurt am Main: Propyläen, 1985), p. 75.

29. See Baltrusaitis, *Phantastisches Mittelalter*, p. 79.

30. On the history of shell collecting see P. S. Dance, *Shell Collecting: An Illustrated History* (London: Faber & Faber, 1966); B. Dietz, 'Mobile Objects: The Space of Shells in Eighteenth-Century France', *British Society of the History of Science*, 39:3 (2006), pp. 363–82; K. Leonhard, 'Shell Collecting: On 17th Century Conchology, Curiosity Cabinets and Stillife Painting', in K. A. E. Enenkel (ed.), *Early Modern Zoology: The Construction of Animals in Science, Literature and the Visual Arts* (Leiden: Brill, 2007), pp. 177–214; P. Mauriès, *Shell Shock: Conchological Curiosities* (London: Thames & Hudson, 1994); W. S. S. van Bentheim Jutting, 'A Brief History of the Conchological Collections at the Zoological Museum of Amsterdam with Some Reflections on 18th Century Shell Cabinets and their Proprietors', *Bijdragen tot de Dierkunde* (Leiden: Brill, 1938), pp. 167–245; B. van de Roemer, 'Neat Nature: The Relation between Nature and Art in a Dutch Cabinet of Curiosities from the early Eighteenth Century', *History of Science* (2004), pp. 47–84; and L. van Hogendorp Prosperetti, '"Conchas

Legere": Shells as Trophies of Repose in Northern European Humanism', *Art History*, 29:3 (2006), pp. 387–413.

31. T. Martyn, *Man and the Natural World: Changing Attitudes in England 1500–1800* (New York: Pantheon Books, 1983), p. 284.

32. Compare to K. Leonhard, 'Die Muschel als symbolische Form, oder: Wie Rembrandt nach Oxford kam', in B. Gockel (ed.), *Vom Objekt zum Bild. Piktorale Prozesse in Kunst und Wissenschaft 1600–2000* (Berlin: Akademie Verlag, 2011), pp. 123–56, on p. 124.

33. J. Brugman and H. J. D. Lulofs (eds), *Aristotle, Generation of Animals: The Arabic Translation commonly ascribed to Yaḥyā ibn al-Biṭrīq* (Leiden: Brill, 1971) book 5, 15–16; see Leonhard, 'Die Muschel als symbolische Form', p. 124.

34. See F. Neubert, *Einleitung in eine kritische Ausgabe von B. de Maillets Telliamed ou Entretiens d'un philosophe indien avec un missionnaire francois. Ein Beitrag zur Geschichte der französischen Aufklärungsliteratur* (Berlin: Emil Ebering, 1920).

35. Aristoteles, *De generatione animalium*, 761 a 14–42.

36. Corbin, *Lure of the Sea*, pp. 8–9.

37. See U. Krolzik, 'Das Wasser als theologisches Thema der Frühaufklärung bei Johann Albert Fabricius', in Böhme (ed.), *Kulturgeschichte des Wassers*, pp. 189–207, on p. 203.

38. J. Ray, *The Wisdom of God Manifested in the Works of the Creation* (London: Printed for Samuel Smith, 1691).

39. A. Coenen, *Visboock* (Fish Book) (1577–81); A. Coenen, *Walvisboock* (Whale Book) (1584–86); and U. Aldrovandi, *De reliquis Animalibus exanguibus libri quatuor post mortem ejus editi ... de Mollibus, Crustaceis, Testaceis, et Zoophytis* (Bononiae, 1623). See F. Egmond, 'Curious Fish', in K. A. E. Enenkel and P. Smith (eds), *Early Modern Zoology*, pp. 245–72.

40. N. de Bruyn, *Libellus varia genera piscium complectens* (Amsterdam: Claes J. Visscher, 1630); A. de Collaert, *Piscium vivae icones in aes incisae et editae* (Antwerpen, 1610); A. Flamen, *Icones diversorum piscium tum maris tum amnium* (Paris, 1664); J. Jonston, *Historia naturalis de piscibus et cetis libri V* (Frankfurt am Main: Matth. Merian, 1650); L. F. Marsili, *Histoire physique de la mer* (Amsterdam: Aux dépens de la Compagnie, 1725); J. T. Klein, *Historiae piscium naturalis promovendae missus I–V* (Lipsiae, 1740–9); E. L. Billardon de Sauvigny, *Histoire naturelle des dorades de la Chine* (Paris, 1780); K. Freiherr von Meidinger, *Icones piscium Austriae indigenorum* (Wien, 1785–94); M. E. Bloch, *Allgemeine Naturgeschichte der Fische* (Berlin, 1782–95); and B. Germaine de Lacépède, *Histoire naturelle des poissons* (Paris, 1798–1803).

41. F. Colonna, *Aquatilium, et Terrestrium aliquot Animalium, aliarumq. Naturalium Rerum observations* (Rome, 1616); B. Besler, *Fasciculus rariorum et aspect dignorum varii generis quae collegit* (Nürnberg, 1616); R. Besler, *Gazophylacium rerum naturalium e regno vegetabili, animali et minerali* (Leipzig, 1642); and J. Tradescant, *Musaeum Tradescatianum: Or, a Collection of Rarities Preserved at South Lambeth, Near London* (London: John Grismond, 1656).

42. Compare to E. Schlebrügge, 'Die Conchyliologie, diese angenehme Wissenschaft', in E. Schlebrügge (ed.), *Das Meer im Zimmer. Von Tintenschnecken und Muscheltieren*. Ausstellung im Landesmuseum Joanneum, Graz vom 13:5. bis 1 November 2005 (Graz: Landesmuseum Joanneum, 2005), pp. 123–41.

43. K. Leonhard, 'Shell Collecting', pp. 177–214, on p. 181.

44. See C. C. Albritton, *The Abyss of Time: Changing Conceptions of the Earth's Antiquity after the Sixteenth Century* (San Francisco, CA: Freeman, Cooper, 1980); C. Norman, *Noah's Flood: The Genesis Story in Western Thought* (New Haven, CT and London:

Yale University Press, 1996); and A. Cutler, *The Seashell on the Mountaintop: A Story of Science, Sainthood, and the Humble Genius Who Discovered a New History of the Earth* (London: Heinemann, 2003).

45. See N. Adamowsky, H. Böhme and R. Felfe (eds), *Ludi naturae. Spiele der Natur in den Künsten und Wissenschaften* (München: Wilhelm Fink Verlag, 2011); and P. Findlen, 'Jokes of Nature and Jokes of Knowledge: The Playfulness of Scientific Discourse in Early Modern Europe', *Renaissance Quarterly*, 43:2 (Summer 1990), pp. 292–331.

46. C. Encelius, *De Re Metallica, hoc. est. De Origine, Varietate & Natura Corporum Metallicorum, Lapidum, Gemmarum, atq. Aliarum, quae ex Fodinis cruuntur, Rerum ad Medicinae Usum deseruientium, Libri III* (Frankfurt, 1557). See M. J. S. Rudwick, *The Meaning of Fossils: Episodes in the History of Palaeontology* (London: MacDonald, 1972), Chapter One.

47. C. Gesner, *De Rerum fossilium, Lapidum et Gemmarum maxime, figuris et similitudinibus Liber* (Tiguri, 1565).

48. Compare to Rudwick, *The Meaning of Fossils*, p. 39.

49. J. Ray, *Observations Topographical, Moral and Physiological, Made in a Journey Through Parts of the Low Countries, Germany, Italy and France: With a Catalogue of Plants not Native of England* (London, 1673).

50. See M. Rudwick, *Bursting the Limits of Time: The Reconstruction of Geohistory in the Age of Revolution* (Chicago, IL and London: University of Chicago Press, 2005), pp. 249–53.

51. J. G. Bruguière, 'Histoire naturelle des vers, 1 [text]. Encyclopédie Méthodique: Histoire Naturelle', Paris, 1:28–43 (1792), quoted from Rudwick, *Bursting the Limits of Time*, p. 257.

52. M. Deacon, *Scientists and the Sea, 1650–1900: A Study of Marine Science* (London: Academic Press, 1971), p. 5.

53. H. Böhme, 'Umriß einer Kulturgeschichte des Wassers. Eine Eineitung', in H. Böhme (ed.), *Kulturgeschichte des Wassers* (Frankfurt am Main: Suhrkamp, 1988), p. 10.

54. Compare to Deacon, *Scientists and the Sea*, pp. 3–69.

55. Deacon, *Scientist and the Sea*, p. 58.

56. See, for example, A. Kircher, *Mundus Subterraneus, in XII libros digestus* (Amsterdam, 1665); Galileo Galilei, *Dialogo di Galileo Galilei sopra i due Massimi Sistemi del Mondo Tolemaico e Copernicano* (Fiorenza, 1632); J. Kepler, *Astronomia Nova seu de Motu Stellae Martis* (1609); and R. Descartes, *Les Principes de la Philosophie* (1644; french Rouen, 1647), part 4, sections 49–53.

57. Quoted from Deacon, *Scientists and the Sea*, p. 74; a detailed account is given by T. Birch, *History of the Royal Society of London* (1756), vol. 1, pp. 29–30.

58. E. Halley, 'The Art of Living under Water: Or, a Discourse Concerning the Means of Furnishing Air at the Bottom of the Sea, in any Ordinary Depths', *Philosophical Transactions of the Royal Society*, 29 (1716), pp. 492–9, on p. 492.

59. R. Boyle, 'Other Inquiries Concerning the Sea', *Philosophical Transactions of the Royal Society*, 18 (October 1666), pp. 315–16.

60. Deacon, *Scientists and the Sea*, p. 129.

61. Compare to G. Peters, 'Zwei Hamburger, ein Barbier (1675) und ein Bürgermeister (1746), erkunden und beschreiben die Wale des Nördlichen Eismeers', *Historisch Meereskundliches Jahrbuch* (Berlin and Hamburg: Reimer 1992), pp. 9–34; and W. Oesau, *Hamburgs Grönlandfahrt auf Walfischfang und Robbenschlag vom 17. bis 19. Jahrhundert* (Hamburg: J. J. Augustin, 1955).

62. O. Magnus, *Historia de gentibus septentrionalibus, earundumque diversis statibus, condi-*

tionibus, moribus, ritibus ... Librum XXI, De piscibus monstrosis (Rome, 1555); and O. Magnus, *Carta marina et desciptio septembtrionalium terrarum ac mirabilium rerum in eis contentarum diligentissime elaborata septentrio Anno dei 1539* (Venedig, 1539).

63. F. Martens, *Spitzbergische oder Groenländische Reise. Beschreibung gethan im jahre 1671* (Hamburg, 1675).

64. Compare to L. Watson, *Sea Guide to Whales of the World* (London and Sydney: Hutchinson, 1981).

65. I. Newton, *Opticks or, a Treatise of the Reflections, Refrections, Inflections and Colours of Light* (1719; London: W. and J. Innys, 1730), p. 160; and E. Halley, 'The Art of Living Underwater: Or, a Discourse concerning the Means of Furnishing Air at the Bottom of the Sea, in Any Ordinary Depths', *Philosophical Transactions of the Royal Society* (1683–1775), 29 (1753), pp. 492–9.

66. L. F. Marsili, *Histoire physique de la mer* (Amsterdam: Aux dépens de la Compagnie, 1725).

67. Link, *Über die Asterien* (Leipzig, 1733).

68. Compare to A. Corbin, *The Lure of the Sea: The Discovery of the Seaside in the Western World 1750–1840*, trans. J. Phelps (Berkeley, CA: University of California Press, 1994), pp. 121–37; and B. M. Stafford, *Voyage into Substance, Art, Science, Nature and the Illustrated Travel Account, 1740–1840* (Cambridge, MA: MIT Press, 1984).

69. R. Ellis, *Monsters of the Sea* (London: Robert Hale, 1995), p. 197.

70. See C. C. Albritton, *The Abyss of Time: Changing Conceptions of the Earth's Antiquity after the Sixteenth Century* (San Francisco, CA: Freeman, Cooper 1980); A. Geikie, *The Founders of Geology* (London: Macmillan and Co., 1897); S. J. Gould, *Time's Arrow, Time's Cycle: Myth and Metaphor in the Discovery of Geological Time* (Cambridge, MA: Harvard University Press, 1987); P. Rossi, *The Dark Abyss of Time: The History of the Earth and the History of Nations from Hooke to Vico* (Chicago, IL: University of Chicago Press, 1984); and M. J. S. Rudwick, *Bursting the Limits of Time: The Reconstruction of Geohistory in the Age of Revolution* (Chicago, IL and London: University of Chicago Press, 2005), Chapter 5.

71. H. Rozwadowski, *Fathoming the Ocean: The Discovery and Exploration of the Deep Sea* (Cambridge, MA and London: Harvard University Press, 2005), p. 7.

72. A. Corbin, *The Lure of the Sea: The Discovery of the Seaside in the Western World, 1750–1840*, trans. J. Phelps (Berkeley, CA: University of California Press, 1994), pp. 108–9.

73. Corbin, *Lure of the Sea*, p. 110.

74. Compare to, for example, T. Burnet, *The Sacred Theory of the Earth* (London, 1684); and W. Whiston, *A New Theory of the Earth* (London, 1708).

75. Corbin, *Lure of the Sea*, p. 109.

76. J. J. Scheuchzer, *Homo diluvii testis et theoscopos* (Zürich, 1726).

77. G. Cuvier, 'Sur quelques quadrupèdes ovipares fossiles conservés dans les schistes calcaires', *Annales du Muséum d'Histoire Naturelle*, 13 (1809), pp. 410–37.

78. Compare to Gould, *Time's Arrow, Time's Cycle*, pp. 3–4.

79. See J. Barr, 'Why the World was Created in 4004 BC: Archbishop Usher and Biblical Chronology', *Bulletin of the John Rylands University Library*, 67 (1985), pp. 575–608.

80. Rossi, *The Dark Abyss of Time*, p. ix.

81. 'An unconformity is a fossil surface of erosion, a gap in time separating two episodes in the formation of rocks. Unconformities are direct evidence that the history of our earth includes several cycles of deposition and uplift'. Gould, *Time's Arrow, Time's Cycle*, p. 62.

82. J. Playfair, 'Biographical Account of the late James Hutton', *F. R. S. Edinburgh: Royal Society Edinburgh* (Edinburgh, 1805), pp. 39–99, on pp. 71–2.

83. C. Lyell, *Principles of Geology: Being an Attempt to Explain the Former Changes of the Earth's Surface by Reference to Causes Now in Operations* (London, 1830–3), p. 63.

84. Corbin, *Lure of the Sea*, p. 102.

85. Compare to M. J. S. Rudwick, *Worlds Before Adam: The Reconstruction of Geohistory in the Age of Reform* (Chicago, IL: University of Chicago Press, 2008), Chapter 2; and D. Cadbury, *The Dinosaur Hunters: A True Story of Scientific Rivalry and the Discovery of the Prehistoric World* (London: Fourth Estate, 2000).

86. G. A. Mantell, *Wonders from Geology; Or, a Familiar Exposition of Geological Phenomena: Being the Substance of a Course of Lectures delivered at Brighton, from notes taken by G. F. Richardson, Curator of the Mantellian Museum* (London: Rolfe & Fletcher, 1838).

87. Compare to D. Cadbury, *The Dinosaur Hunters*, Part One.

88. Mary Anning was one of the most successful and remarkable fossil hunters of her time; see H. Torrens, 'Mary Anning of Lyme: The Greatest Fossilist the World ever Knew', *British Journal of the History of Science*, 28 (1995), pp. 257–84; W. Lang, 'Mary Anning of Lyme, Collector and Vendor of Fossils', *Natural History Magazine*, 5:34 (1936), pp. 64–81; and G. Roberts, 'The Fossil Finder of Lyme Regis', *Chambers Journal of Popular Literature*, 7 (1857), pp. 382–4.

89. W. Conybeare and H. de la Beche, 'Notice of the Discovery of a New Fossil Animal', *Transactions of the Geological Society*, 5 (1821), pp. 559–94.

90. See W. Conybeare, 'On the Discovery of an Almost Perfect Skeleton of the Plesiosaurus', *Transactions of the Geological Society of London*, 1 (1824), quoted in Cadbury, *The Dinosaur Hunters,* pp. 107–8.

91. G. A. Mantell, *Illustrations of the Geology of Sussex* (1827), quoted in from Cadbury, *The Dinosaur Hunters,* p. 158.

92. Compare to M. J. S. Rudwick, *Scenes from Deep Time, Early Pictorial Representations of the Prehistoric World* (Chicago, IL and London: University of Chicago Press, 1992); S. Moser, *Ancestral Images: The Iconography of Human Origins* (Ithaca, NY: Cornell University Press, 1998); and H. Holländer, 'Es rauscht in den Schachtelhalmen, verdächtig leuchtete das Meer ... ', in A. Preiß and B. Brock (eds), *Ikonographia: Anleitung zum Lesen von Bildern* (München: Klinkhardt & Biermann 1990), pp. 179–202.

93. Compare to J. Voss, 'Variieren und Selektieren: Die Evolutionstheorie in der englischen und deutschen illustrierten Presse im 19. Jahrhundert', in P. Kort and M. Hollein (eds), *Darwin. Kunst und die Suche nach den Ursprüngen* (Köln: Wienand, 2009), pp. 246–57; see also D. Donald, *Picturing Animals in Britain, 1750–1850* (New Haven CT and London: Yale University Press, 2007).

94. Compare to Rudwick, *Scenes from Deep Time*, p. 47.

95. Compare to Rudwick, *Scenes from Deep Time*, p. 62.

96. See, for example, J. A. Secord, 'Monsters at the Chrystal Palace', in S. de Chadarevian and N. Hopwood (eds), *Models: The Third Dimension of Science* (Stanford, CA: Stanford University Press, 2004), pp. 138–69.

2 Wonder and Mystery

1. M. F. Maury, *The Physical Geography of the Sea* (New York: Harper & Brothers, 1855), p. 308. Maury also wrote: 'The wonders of the sea are as marvellous as the glories of the heavens'. Maury, *Physical Geography*, p. 314.

2. M. J. Schleiden, *Die Pflanze und ihre Leben* (Leipzig, 1858), p. 163.

3. 'The bosom of the ocean is full of mysteries; it conceals a whole world of curiously-

shaped animals, which the naturalist only superficially knows, and may, perhaps, never be able to fathom'. G. Hartwig, *The Sea and its Living Wonders: A Popular Account of the Marvels of the Deep and of the Progress of Maritime Discovery from the Earliest Ages to the Present Time* (London: Longman, 1873), p. 186.

4. See, for example, S. Schlee, *The Edge of an Unfamiliar World: A History of Oceanography* (New York: E. P. Dutton & Co., 1973); J.-R. Vanney, *Le Mystère des Abysses. Histoires et découvertes des profondeurs océaniques* (Paris: Fayard, 1993); and R. Kunzig, *Mapping the Deep: The Extraordinary Story of Ocean Science* (New York and London: W. W. Norton & Company, 2000).

5. Compare to P. Hadot, *The Veil of Isis* (Cambridge, MA: Harvard University Press, 2006). The veil which covers the mysteries of nature differs from the ancient tradition of spiritual inquiry linked with gestures of 'demasking' and 'unveiling', a tradition which subsequently evolved into the systematic scientific examination of deception and self-deception with the aim of uncovering the underlying truth. Compare to G. Mattenklott, 'Schleierhaft: The Veil of Oblivion', *Daidalos*, 33 (15 September 1989), pp. 54–9, on p. 54.

6. On Müller's life and research see E. Du Bois-Reymond, 'Gedächtnisrede auf Johannes Müller', in *Reden von Emil Du Bois-Reymond* (Leipzig: Veit & Co., 1887); W. Haberling, *Johannes Müller: Das Leben des Rheinischen Naturforschers* (Leipzig: Akademische Verlagsgesellschaft, 1924); F. Bidder, 'Vor Hundert Jahren im Laboratorium Johannes Müllers', *Münchner Medizinische Wochenschrift* (1934), pp. 60–4; and M. Hagner and B. Wahrig-Schmidt, *Johannes Müller und die Philosophie* (Berlin: Akademie-Verlag, 1992).

7. Quoted from B. Lohff, 'Die Entdeckung der Welt des Planktons', in *Historisch Meereskundliches Jahrbuch* (Berlin and Hamburg: Reimer, 1992), pp. 35–44, on p. 38.

8. Quoted from B. Lohff, 'Die Entdeckung der Welt', p. 37.

9. In the German text Müller describes "'das eine oder andere 'rätselhafte Thier', an dem das 'Sonderbarste war, dass an dem Theil des Körpers, der in die Arme ausläuft, ein Seestern befestigt war', oder aber auch die 'sonderbare Form' der Roccocolarve". J. Müller, *Über die Larven und die Metamorphose der Echinodermen. Zweite Abhandlung* (Berlin: Dümmler, 1849), pp. 3, 5. (translation EB).

10. Müller speaks of "'sehr räthselhafte(n)' Tentakeln und 'eigenthümliche(n) Armen einiger 'gestachelter, seeigelförmiger Echinodermen'". J. Müller, *Über die Larven und die Metamorphose der Ophiuren und Seeigel* (Berlin: Dümmler, 1848), pp. 14, 17. (translation EB).

11. J. Müller, 'Bericht über einige neue Thierformen der Nordsee', *Archiv für Anatomie, Physiologie und Wissenschaftliche Medicin,* (1846), pp. 100–10, on p. 101.

12. Müller talks about: der "'räthselhaften *Tomopteris onisciformis Eschscholtz*', sowie von 'einigen neuen ... sehr räthselhaften Seethieren'. Diese teilweise nur wenige, 'langen Tierchen (*vexillaria flabellum*) waren 'von so eigenthümlicher Form und Structur, dass es dermalen unmöglich ist, zu errathen, zu welcher Klasse es gehören mag'. Ihre Eingeweide 'liegen so verschlungen, und sind so sonderbarer Gestalt, dass eine Erklärung ihres Zusammenhanges nach der Analogie anderer Tiere unmöglich war'". Müller, 'Bericht über einige neue Thierformen', p. 106. (translation EB).

13. Müller, 'Bericht über einige neue Thierformen', p. 105.

14. Müller, *Über die Larven und die Metamorphose der Ophiuren*, p. 2.

15. Quoted from H.-R. Simon (ed.), *Anton Dohrn und die Zoologische Station Neapel* (Frankfurt am Main: Edition Erbrich, 1980), p. 49.

16. On the life and work of Ernst Haeckel see E. Krauße, *Ernst Haeckel* (Leipzig: Teubner Verlag, 1984); B. Kleeberg, *Theophysis. Ernst Haeckels Philosophie des Naturganzen*

(Köln, Weimar: Böhlau, 2005); and R. J. Richards, *The Tragic Sense of Life, Ernst Hae-ckel and the Struggle over Evolutionary Thought* (Chicago, IL and London: University of Chicago Press, 2008).

17.   I. Eibl-Eibesfeld, 'Ernst Haeckel – Der Künstler im Wissenschaftler', in E. Haeckel (ed.), *Ernest Haeckel, Kunstformen der Natur* (München: Prestel Verlag, 1998), pp. 19–31, on p. 19. For the aesthetic feature in Haeckel's work see K. Bayertz, 'Biology and Beauty: Science and Aesthetics in Fin de Siècle Germany', in R. Porter and M. Teich (eds), *Fin de Siècle and its Legacy* (Cambridge: Cambridge University Press, 1990), pp. 278–95.

18.   E. Haeckel, *Das System der Medusen I: Craspedoten* (Jena: G. Fischer, 1879), p. 7.

19.   E. Haeckel, *Die Kalkschwämme (Calcispongiae). Eine Monographie*, 3 vols (Berlin: G. Reimer Verlag, 1872); Haeckel, *Das System der Medusen I*; Haeckel, *Das System der Medusen II: Acraspeden* (Jena: G. Fischer, 1880); and Haeckel, *Die Radiolarien (Rhizopoda radiaria). Eine Monographie*, 2 vols (Berlin: Reimer Verlag, 1862).

20.   E. Haeckel, *Report on the Deep-Sea Medusae dredged by HMS Challenger During the Years 1873–76* (London: Longmans & Co., 1882); Haeckel, *Report on the Radiolaria, collected by HMS Challenger During the Years 1873–76, I. Part: Porulosa (Spumellaria and Acantharia). II. Part: Osculosa (Nesselaria and Phaeodaria)* (London: Eyre & Spottiswoode, 1887); and Haeckel, *Report on the Siphonophorae Collected by HMS Challenger During the years 1873–76* (London: Eyre & Spottiswoode, 1888).

21.   'Entwicklung' heißt von jetzt an das Zauberwort, durch das wir alle uns umgebenden Räthsel lösen, oder wenigstens auf den Weg ihrer Lösung gelangen können'. E. Haeckel, *Natürliche Schöpfungsgeschichte* (Berlin: Reimer Verlag, 1870), p. xviii.

22.   E. Haeckel, *The History of Creation*, trans. E. Ray Lankester, 2 vols (New York: Appleton and Company, 1880), vol. 1, p. 11.

23.   Haeckel, *The History of Creation*, vol. 2, p. 49.

24.   Haeckel, *The History of Creation*, vol. 1, p. 332.

25.   Haeckel, *The History of Creation*, vol. 1, p. 332.

26.   Haeckel, *The History of Creation*, vol. 2, p. 344.

27.   Haeckel, *The History of Creation*, vol. 1, p. 333

28.   Haeckel, *The History of Creation*, vol. 1, p. 371.

29.   Haeckel, *The History of Creation*, vol. 2, p. 31.

30.   Haeckel, *The History of Creation*, vol. 2, p. 66.

31.   Haeckel, *The History of Creation*, vol. 2, p. 165.

32.   Haeckel, *The History of Creation*, vol. 2, p. 60.

33.   Haeckel, *The History of Creation*, vol. 1, p. 200.

34.   Haeckel, *The History of Creation*, vol. 2, p. 40.

35.   W. Waldeyer, 'Rede des Geh. Rath Prof. W. Waldeyer zum 25jährigen Bestehen der Zoologischen Station', in H.-R. Simon (ed.), *Anton Dohrn und die Zoologische Station Neapel*, pp. 77–81, on p. 77. Also see E. Haeckel, *Das Leben in den grössten Meerestiefen. Vortrag, gehalten am 2. März 1870 im akademischen Rosensaale zu Jena* (Berlin: Lüderitz'sche Verlagsbuchhandlung, 1870), p. 5.

36.   L. Sonrel, *Bottom of the Sea* (London: Sampson Low, Son and Marston, 1870), p. 109.

37.   E. Forbes, *The Natural History of the European Seas* (1859; New York: Arno Press, 1977), p. 13.

38.   K. A. Möbius, 'Ostseeaquarien', *Zoologischer Garten*, 3 (1862), pp. 165–8, on p. 165.

39.   G. Jäger, *Das Leben im Wasser und das Aquarium* (Stuttgart: Kosmos, Gesellschaft der Naturfreunde, 1908), p. iii.

40. K. A. Möbius, 'Einige Fingerzeige für die Bevölkerung und Erhaltung der Aquarien', *Zoologischer Garten*, 6 (1865), pp. 211–14, on p. 212.

41. Jäger, *Leben im Wasser*, p. 53.

42. P. H. Gosse, *The Aquarium: An Unveiling of the Wonders of the Deep Sea* (London: John van Voorst, 1854), p. 226.

43. Jäger, *Leben im Wasser*, p. 87.

44. G. Hartwig, *Das Leben des Meeres. Eine Darstellung für gebildete Stände* (Glogau: Karl Flemming Verlag, 1862), p. 175.

45. Compare to Sonrel: '(I)n no part does the creative power reveal itself with more of grandeur and magnivicence than in the abysses of Ocean ... No region of the earth could give so vivid an idea of the exuberance of life (as the sea). For the most unexpected, a fecundity the most marvellous, challenge our admiration at every step we take through these wonderful regions'. *Bottom of the Sea*, p. 109.

46. G. Hartwig, *The Sea and its Living Wonders: A Popular Account of the Marvels of the Deep and of the Progress of Maritime Discovery from the Earliest Ages to the Present Time* (London: Longman, 1873), p. 186.

47. Hartwig, *The Sea*, p. 186.

48. Hartwig, *The Sea*, pp. 263–7.

49. Hartwig, *Leben des Meeres*, p. 72.

50. Hartwig, *Leben des Meeres*, p. 80.

51. See R. Proctor, *Value-Free Science? Purity and Power in Modern Knowledge* (Cambridge, MA: Harvard University Press, 1991); and H. Mehrtens, 'Irresponsible Purity: The Political and Moral Structure of Mathematical Sciences in the National Socialist State', in M. Renneberg and M. Walker (eds), *Science, Technology and National Socialism* (Cambridge: Cambridge University Press, 1993), pp. 324–38.

52. H.-J. Rheinberger, *Experimentalsysteme und Epistemische Dinge. Eine Geschichte der Proteinsynthese im Reagenzglas* (2001; Frankfurt am Main: Suhrkamp, 2006), pp. 20–2.

53. See A. W. Daum, *Wissenschaftspopularisierung im 19. Jahrhundert. Bürgerliche Kultur, naturwissenschaftliche Bildung und die deutsche Öffentlichkeit, 1848–1914* (München: Oldenbourg Verlag, 1998).

54. Compare to L. Daston, 'The Moral Economy of Science', *Osiris*, 10 (1995), pp. 3–24.

55. Compare to Daum, *Wissenschaftspopularisierung*, p. 250–1.

56. A. von Humboldt, *Views of Nature, or Contemplations on the Sublime Phenomena of Creation*, trans. E. C. Otté and H. G. Bohn (London: Henry G. Bohn, 1850), p. xi; and A. von Humboldt, *Ansichten der Natur mit wissenschaftlichen Erläuterungen*, 3rd edn (1807; Stuttgart and Augsburg: Cotta'scher Verlag, 1859), p. vii.

57. J. Voss, 'Die Entdeckung der Unordnung. Wie Charles Darwin den Zufall zeichnete', in N. Adamowsky, H. Böhme and R. Felfe (eds), *Ludi naturae. Spiele der Natur in Kunst und Wissenschaft* (München: Wilhelm Fink Verlag, 2011), pp. 257–84, on p. 264.

58. W. Marshall, *Die Tiefsee und ihr Leben. Nach den neuesten Quellen* (Leipzig: Hirt, 1888), p. 14.

59. Forbes, *Natural History*, pp. 4–5.

60. C. Darwin, *On the Origin of Species by Means of Natural Selection, or the Preservation of Favoured Races in the Struggle for Life, A Facsimile of the First Edition (London 1859) with an Introduction by Ernst Mayer* (Cambridge, MA: Harvard University Press, 1964), p. 164.

61. Darwin, *On the Origin of Species*, p. 490.

62. Compare to Daum, *Wissenschaftspopularisierung*, pp. 243–64, pp. 464–5.

63. Quoted from H. Weidner, 'Die Anfänge meeresbiologischer und ökologischer
     Forschung in Hamburg durch Karl August Möbius (1825–1908) und Heinrich Adolph
     Meyer (1822–1889)', *Historisch Meereskundliches Jahrbuch* (Berlin and Hamburg:
     Dietrich Reimer, 1994), pp. 69–84, on p. 74.

64. M. J. Schleiden, *Die Pflanze und ihre Leben* (Leipzig, 1858), pp. 172–4; quoted from
     A. Mangin, *The Mysteries of the Ocean* (1864; London: T. Nelson and Sons, 1870),
     pp. 197–200.

65. Sonrel, *The Bottom of the Sea*, p. 113.

66. J. D. Schöpf, *Travels in the Confederation* (1783–84), trans. and ed. A. J. Morrison, 2
     vols (Philadelphia, PA: William J. Campell 1911), vol. 2, pp. 285–6.

67. C. Buci-Glucksmann, *Der kartographische Blick der Kunst* (Berlin: Merve-Verlag,
     1997), pp. 32–3, 56.

68. Buci-Glucksmann, *Der kartographische*, p. 34.

69. Compare to Adamowsky, *Das Wunder in der Moderne. Eine andere Kulturgeschichte des
     Fliegens* (Paderborn: Wilhelm Fink Verlag, 2010), Chapter 3.3.2.

70. Compare to Adamowsky, *Wunder in der Moderne*, Chapter 4.1.

71. G. C. Lichtenberg, 'Vermischte Nachrichten und die aerostatischen Maschinen', in
     L. C. Lichtenberg and F. Kries (ed.), *Georg Christophs Lichtenbergs physikalische und
     mathematische Schriften* (Göttingen: Dieterich-Verlag, 1804), pp. 321–3.

72. C. G. Ehrenberg, 'Über die Natur und Bildung der Corallenbänke des roten Meeres',
     Abhandlungen der Königlichen Akademie der Wissenschaften zu Berlin, Erster Teil (Ber-
     lin: Commission F. Dummler 1834).

73. J. Paul, *Des Luftschiffers Giannozo Seebuch*, in 'Komische Anhang' zum Roman *Titan*, 4
     vols (1801; Berlin: Buchhandlung Matzdorff, 1800–3), vol. 2.

74. Aristotle, *Physics*, Book 4, Chapter 5.

75. Ambrosius, *Hexameron*, Day 5, Chapter 14; Aurelius Augustinus, *De Genesi ad Lit-
     teram Libri Duodecim*. Book III (6).

76. J. H. Zedler, *Großes Vollständiges Universal-Lexikon aller Wissenschaften und Künste,
     1731–54*, 64 vols (Leipzig: Johann Heinrich Zedler) vol. 9, p. 1354. This opinion was
     shared by Giambattista della Porta, Gerolamo Cardano, Friedrich Hermann Flayder,
     Tito Livio Burattini and Marin Mersenne. In 1634 the latter wrote optimistically: 'I
     simply add that it is perhaps hardly more difficult to fly than to swim; and that as we
     find the art of swimming very easy when we have learned it, although we should have
     held it to be impossible, so we shall consider the art of flying to be very easy when
     we have achieved it'. M. Mersenne, *Questions inouyes, ou recreation des scavans* (Paris,
     1634), pp. 1–5; quoted in C. Hart, *The Prehistory of Flight* (Berkely, CA: University of
     California Press, 1985), p. 130.

77. Compare to A. W. Zachariae, *Fluglust und Fluges Beginnen, hierbei mein schon fliegend-
     es Blatt und auf diesem in Kupferstich der Bauriß zu meinem Flugkahne nebst Abbildung
     von dessen Luftkahn* (Leipzig: C. Cnobloch, 1821).

78. A. W. Zachariae, *Die Elemente der Luftschwimmkunst* (Wittenberg: Zimmermann,
     1807), pp. 30–1.

79. Zachariae, *Elemente der Luftschwimmkunst*, p. 85.

80. Zachariae, *Elemente der Luftschwimmkunst*, p. 85.

81. A. Mangin, *L'Air et le Monde Aérien* (Tours: Alfred Mame, 1865), p. 70.

82. A. de Quatrefages, *The Rambles of a Naturalist on the Coasts of France, Spain, and Sicily*
     (London: Longman, Brown, Green, Longmans & Roberts, 1857), p. 175.

83. H. Luedecke, *Vom Zaubervogel zum Zeppelin. Eine Geschichte der Luftfahrt und des*

Fluggedankens (Berlin: Kurt Wolff Verlag, 1936), p. 269.

84. J. R. Forster, *Observations Made During a Voyage Around the World, on Physical Geography, Natural History and Ethic Philosophy* (London: G. Robinson, 1788), pp. 61–2.

85. Compare to E. N. Harvey, *A History of Luminescence from the Earliest Times until 1900* (Philadelphia, PA: American Philosophical Society, 1957).

86. Compare to Harvey, *History of Luminescence*, p. 513.

87. Compare to H. Böhme, *Das Geheimnis*, p. 66.

88. On the notion of light in the 18th century see U. im Hof, 'Enlightenment – Lumieres – Illuminismo – Aufklaerung. Die 'Ausbreitung eines besseren Lichts im Zeitalter der Vernunft', in M. Svilar (ed.), '*Und es ward Licht': zur Kulturgeschichte des Lichts* (Bern, Frankfurt: Peter Lang Verlag, 1983), pp. 115–36; on the history of light and consciousness cf. A. Zajonc, *Catching the Light: The Entwined History of Light and Mind* (New York and London: Oxford University Press, 1993).

89. H. Böhme, *Licht*, p. 112.

90. Compare to Böhme, *Licht*, pp. 117, 132.

91. H. Blumenberg, *Paradigmen zu einer Metaphorologie* (1960; Frankfurt am Main: Suhrkamp, 1998).

92. On board the *Beagle* Charles Darwins wrote: 'While sailing a little south of the Plata on one very dark night, the sea presented a wonderful and most beautiful spectacle. There was a fresh breeze, and every part of the surface, which during the day is seen as foam, now glowed with a pale light. The vessel drove before her bows two billows of liquid phosphorus, and in her wake she was followed by a milky train'. *Journal of Researches into the Natural History and Geology of the Countries Visited During the Voyage of HMS Beagle Round the World, under the Command of Capt. Fitz Roy R. N.* (1860; London: John Murray, 1913), p. 162.

93. See, for example, F. Péron, 'Mémoire sur le nouveau genre Pyrosoma', *Annales du Musée d'Histoire Naturelle*, Paris 4:437–46 (1804); C. Bernoulli, *Ueber das Leuchten des Meeres, mit besonderer Hinsicht auf das Leuchten thierischer Körper* (Göttingen: Heinrich Dieterich, 1803); D. Viviani, *Phosphorescentia maris* (Genua: Joannis Giossi, 1805); J. Macartney, 'Observations on Luminous Animals', *Phil. Trans.*, 100 (1810) pp. 258–93; W. G. Tilesius von Tilenau, *Resultate seiner, während der drei Jahren der Krusensteruschen Entdeckungsreise gesammelten Erfahrungen* (St Peterburg, 1813); G. R. Treviranus, *Die Erscheinungen und Gesetze des organischen Lebens*, 2 vols (Bremen: Heyse-Verlag, 1833), vols 1 and 2, pp. 432–47; J. Murray, *Experimental Researches on the Light & the Luminosity of the Sea, The Phenomena of the Chameleon ...* (Glasgow: W. R. McPhan, 1826); and F. J. F. Meyen, 'Über das Leuchten des Meeres und Beschreibung einiger Polypen und anderer niederer Tiere', *Nova Acta Leop. Carol. Supp.* 16 (1834) pp. 125–216, on pp. 125–8.

94. G. A. Michaelis, *Über das Leuchten der Ostsee nach eigenen Beobachtungen, nebst einigen Bemerkungen über diese Erscheinung in andern Meeren* (Hamburg: Perthes und Besser, 1830); and C. G. Ehrenberg, *Das Leuchten des Meeres. Neue Beobachtungen nebst Übersicht der Hauptmomente der geschichtlichen Entwicklung dieses merkwürdigen Phänomens* (Berlin: Königliche Akademie der Wissenschaften, 1835).

95. Michaelis, *Leuchten der Ostsee*, p. 18.

96. Ehrenberg, *Leuchten des Meeres*, p. 5, 117.

97. C. G. Ehrenberg, *Die Infusionsthierchen als vollkommene Organismen. Ein Blick in das tiefere organische Leben der Natur. Nebst einem Atlas von vierundsechzig colorirten Kupfertafeln, gezeichnet vom Verfasser* (Leipzig: L. Voss, 1838), p. 256; see also A. Pritchard,

History of Infusoria, Living and Fossil: Arranged According to, Die Infusionsthierchen of C. G. Ehrenberg (London: Whitaker and Co., 1845).

98. Mangin, *Mysteries*, p. 179–80.
99. A. von Chamisso, 'Reise um die Welt', in H. Tardel (ed.), *Chamisso's Werke*, 3 vols (Leipzig: Bibliographisches Institut Leipzig und Wien, 1907), vol. 3, p. 56 (voyage from Plymouth to Teneriffa).
100. H. Rozwadowski, *Fathoming the Ocean: The Discovery and Exploration of the Deep Sea* (Cambridge, MA and London: Harvard University Press, 2005), p. 99.
101. Compare to D. E. Allen, 'Tastes and Crazes', in N. Jardine and E. C. Spary (eds), *Cultures of Natural History* (Cambridge: Cambridge University Press, 2000), pp. 394–407.
102. Compare to P. Rehbock, 'The Early Dredgers: "Naturalizing" in British Seas, 1830–1850', *Journal for the History of Biology*, 12 (1979), pp. 293–368.
103. Rozwadowski, *Fathoming the Ocean*, p. 110.
104. C. Kingsley, *Glaucus; Or, the Wonders of the* Shore (Cambridge: Macmillan & Co., 1859), pp. 3, 39.
105. L. Barber, *The Heyday of Natural History: 1820–1870* (London: Cape, 1989), Chapter 9.
106. H.-J. Rheinberger, *Experimentalsysteme und epistemische Dinge. Eine Geschichte der Proteinsynthese im Reagenzglas* (2001; Frankfurt am Main: Suhrkamp, 2006), p. 27.
107. Compare to J. W. Hedgpeth, 'De mirabili maris: Thoughts on the Flowering of Seashore Books', *The Royal Societey of Edinburgh, Proceedings, Section B (Biology)*, 72 (1971–2), pp. 107–14, on p. 111. On the history of the Bathybius compare to T. H. Huxley, 'On Some Organisms Living at Great Depths in the North Atlantic Ocean', *Quarterly Journal of Microscopical Science*, 8 (1868), pp. 203–12; E. Haeckel, *Das Leben in den grössten Meerestiefen. Vortrag, gehalten am 2. März 1870 im akademischen Rosensaale zu Jena* (Berlin: Lüderitz'sche Verlagsbuchhandlung, 1870), pp. 28–39; J. Y. Buchanan, *Accounts Rendered of Work Done and Things Seen* (Cambridge: Cambridge University Press, 1919); and P. Rehbock, 'Huxley, Haeckel, and the Oceanographers: The Case of *Bathybius haeckelii*', *Isis*, 66 (1975), pp. 504–33.
108. L. Sonrel, *Bottom of the Sea* (London: Sampson Low, Son and Marston, 1870), pp. 117, 215.
109. Sonrel, *The Bottom of the Sea*, p. 161.
110. Rozwadowski, *Fathoming the Ocean*, p. 138, 148.
111. Quoted from Rozwadowski, *Fathoming the Ocean*, p. 101.
112. G. Jäger, *Das Leben im Wasser und das Aquarium* (Stuttgart: Kosmos, Gesellschaft der Naturfreunde, 1908), p. 344.
113. Quoted from E. Batchelder, *A Romance of the Sea Serpent or the Ichthyosaurus and a Collection of Ancient and Modern Authorities* (Cambridge: J. Bartlett, 1849), pp. 135–6.
114. H. Lee, 'Sea Monsters Unmasked', in *The Great International Fisheries Exhibition Literature* (London: William Clowes and Sons, 1883), p. 439.
115. Rozwadowski, *Fathoming the Ocean*, p. 138. Compare to S. Lyons, 'Sea Monsters: Myth or Genuine Relic of the Past', in K. R. Benson and P. F. Rehbock (eds), *Oceanographic History: The Pacific and Beyond* (Seattle, WA and London: University of Washington Press, 2002), p. 60–70, on p. 60.
116. Compare to R. Ellis, *Seeungeheuer: Mythen, Fabeln, Fakten* (Basel, Boston, MA and Berlin: Birkhäuser Verlag, 1997), pp. 57–60.
117. J. R. Norman and P. H. Greenwood, *A History of Fishes* (London: Ernest Benn, 1963), p. 365.
118. Compare to B. Heuvelmans, *In the Wake of Sea-Serpents* (New York: Hill and Wang, 1968).

119. M. F. Maury, *Explanations and Sailing Directions to Accompany the Wind and Current Charts* (Philadelphia, PA: Biddle, 1854), p. 154.

120. C. W. Thomson reports on this in *The Depths of the Sea: An Account of the General Results of the Dredging Cruises of HMSS 'Porcupine' and 'Lightning' during the Summers of 1868, 1869, and 1870* (London: Macmillan, 1874), pp. 31–2.

121. E. Forbes writes: 'As we descend deeper and deeper in this region its inhabitants become more and more modified, and fewer and fewer, indicating our approach towards an abyss where life is either extinguished, or exhibits but a few sparks to mark its lingering presence ... It is in the exploration of this vast deep sea region that the finest field for submarine discovery yet remains', *The Natural History of the European Seas* (1859; New York: Arno Press, 1977), p. 27. Compare to also Forbes, 'Report on the Mollusca and Radiata of the Ægean Sea, and on their Distribution, Considered as Bearing on Geology', *Report of the 13th Meeting of the British Association for the Advancement of Science, 1843*, 1 (1844), pp. 130–93.

122. S. Lovén, 'On the Bathymetrical Distribution of Submarine Life on the Northern Shores of Scandinavia', *Report of the 14th Meeting of the British Association for the Advancement of Science (1844)*, 2 (1845), pp. 50–1.

123. Compare to M. Deacon, *Scientists and the Sea, 1650–1900: A Study of Marine Science* (London: Academic Press, 1971), Chapter 13; D. Merriman, 'Speculations on Life at the Depths: A XIXth-Century Prelude', *Bulletin Institute Océanographic Monaco*, 2:2 (1968), pp. 377–84; Rozwadowski, *Fathoming the Ocean*; and Thomson, *Depths of the Sea*.

124. Sir J. C. Ross, *A Voyage of Discovery made under the Orders of the Admiralty in His Majesty's Ships 'Isabella' and 'Alexander', for the purpose of exploring Baffin's Bay, and inquiring into the Possibility of a North-west Passage* (London: J. Murray, 1819); and Sir J. C. Ross, *A Voyage of Discovery and Research in the Southern and Antarctic Regions, During the Years 1839–1843* (London: J. Murray 1847). However, A. L. Rice doubts that Ross reached 1000 fathoms and presumes that it rather was a matter of 500–600 fathoms. Compare to 'The Oceanography of John Ross' Arctic Expedition of 1818: A Reappraisal', *Journal of the Society for the Bibliography of Natural History*, 7:3 (1975), pp. 291–313. Compare to also M. F. Maury, *The Physical Geography of the Sea* (New York: Harper & Brothers, 1855), p. 200.

125. Compare to J. A. N. Périer and A. Berbrugger, *Exploration Scientifique de l'Algérie pendant les années 1840, 1841, 1842: Recherches de Physique Générale sur la Méditerranée* (Paris: Impr. Royale, 1845).

126. G. C. Wallich, *The North Atlantic Sea-Bed: Comprising a Diary of the Voyage on Board HMS 'Bulldog' in 1860* (London: J. van Voorst, 1862), p. 68. See also H. L. Burstyn and A. G. E. Jones, 'G. C. Wallich, MD: Megalomaniac or Mis-Used Oceanographic Genius?', *Journal of the Society for the Bibliography of Natural History*, 7:4 (1976), pp. 432–50.

127. On this compare to Thomson, *Depths of the Sea*, pp. 26–30. See also B. Finn, *Submarine Telegraphy: The Grand Victorian Technology* (London: Science Museum, 1973).

128. Rozwadowski, *Fathoming the Ocean*, p. 62; see also E. L. Mills, 'Edward Forbes, John Gwyn Jeffreys, and British Dredging before the *Challenger* Expedition', *Journal of the Society for the Bibliography of Natural History*, 8:4 (1978), pp. 507–36.

129. Forbes, *Natural History*, p. 11.

130. Compare to W. B. Carpenter, 'Preliminary Report of Dredging Operations in the Seas to the North of the British Islands, Carried on in HMp *Lightning*, by Dr Carpenter und Dr Wyville Thomson, Professor of Natural History in Queen's College, Belfast', *Proceedings of the Royal Society*, 17 (1868–9), pp. 168–200.

131. W. B. Carpenter, J. G. Jeffreys and C. W. Thomson, 'Preliminary Report of the Scientific Exploration of the Deep Sea in HMS *Porcupine*, During the Summer of 1869', *Proceedings of the Royal Society*, 1 (1869–70), pp. 397–492.

132. C. W. Thomson and J. Murray (eds), *Report on the Scientific Results of the Voyage of HMp Challenger During the Years 1873–76* (London 1880–95); C. W. Thomson, *Voyage of the "Challenger": The Atlantic*, 2 vols (London: MacMillan, 1877); H. H. Moseley, *Notes by a Naturalist: An Account of Obsrvations Made during the Voyage of HMS 'Challenger' Round the World in the Years 1872–76* (London: John Murray, 1892); J. J. Wild, *Thalassa: An Essay on the Depth, Temperature and Currents of the Ocean* (London: Marcus Ward & Co., 1877); andR. von Willemoes-Suhm, *Die Challenger-Expedition. Zum tiefsten Punkt der Weltmeere, 1872–1876. Rudolf von Willemoes-Suhms Briefe von der Challenger-Expedition mit Auszügen aus dem Reisebericht des Schiffsingenieurs W. J. J. Spry* (Stuttgart: Thienemann, 1984).

133. W. J. J. Spry, *The Cruise of her Majestys Ship 'Challenger': Voyages Over Many Seas, Scenes in Many Lands* (London: Sampson Low, Marston, Searle, & Rivington, 1876), p. 384.

134. Quoted from Deacon, *Scientists and the Sea*, p. 371.

135. For example, some of the sponges were sent to Eduard Oscar Schmidt (1823–86) in Strasbourg, and Haeckel got the radiolaria and deep-sea medusae.

136. Rozwadowski, *Fathoming the Ocean*, p. 173.

137. This list is not complete. For a detailed register see Deacon, *Scientists and the Sea*, Chapter 16; Y. Fukano and H. Strickland, *Selected References to Literature on Marine Expeditions, 1700–1960* (Boston, MA: Hall, 1972); R. B. S. Sewell, 'Oceanographic Exploration, 1851–1951', *Science Progress*, 40 (1952), pp. 403–18; B. Watermann and O. J. Wrzesinski, *Bibliographie zur Geschichte der deutschen Meersforschung. Chronologische Titelaufzählung (1557–1989)* (Hamburg: Deutsche Gesellschaft für Meeresforschung, 1989); and G. Wüst, 'The Major Deep-Sea Expeditions and Research Vessels, 1873–1960', in M. Sears (ed.), *Progress in Oceanography* (London and New York: Pergamon Press, 1964), pp. 1–52.

138. Compare to Jäger, *Leben im Wasser*, p. 169.

139. F. W. Winter, 'Carl Chun', *Berichte der Senckenbergischen Naturforschenden Gesellschaft in Frankfurt am Main*, 45 (1914), pp. 176–83, on p. 179.

140. H.-R. Simon, 'Carl Chun und Ernst Haeckel als Herausgeber meeresbiologischer Tafelwerke', in C. Nissen (ed.), *Die zoologische Buchillustration: ihre Bibliographie und Geschichte*, 2 vols (Stuttgart: Hiersemann Verlag, 1978), vol. 2, pp. 327–37, on p. 329.

141. See Simon, 'Carl Chun', p. 329.

142. Compare to C. de Zegher and C. Armstrong (eds), *Ocean Flowers. Impressions from Nature* (New York, Princeton, NJ and Oxford: Princeton University Press, 2004), pp. 69–71.

143. J. Meder, *Die Handzeichnung. Ihre Technik und Entwicklung* (Wien: Kunstverlag Anton Schroll & Co., 1919), p. 276.

144. J. Ruskin, *Elements of Drawing: In Three Letters to Beginners* (New York: Wiley & Halsted, 1857), Letter 2, 'Sketching from Nature', pp. 91–136, on p. 119.

145. See L. Daston and P. Galison, *Objektivität* (Frankfurt am Main: Suhrkamp, 2007); and P. Geimer (ed.), *Ordnungen der Sichtbarkeit. Fotografie in Wissenschaft, Kunst und Technologie* (Frankfurt am Main: Suhrkamp, 2002).

146. Compare to A. Rice, *Voyages of Discovery: Three Centuries of Natural History Exploration* (Richmond Hill, ON: Firefly Books, 2008), p. 323.

147. L. Daston and P. Galison, 'The Image of Objectivity', *Representations*, 37 (1992), pp. 67–106; and S. de Chadarevian, 'Sehen und Aufzeichnen in der Botanik des 19. Jah-

rhunderts', in M. Wetzel and H. Wolf (eds), *Der Entzug der Bilder. Visuelle Realitäten* (München: Wilhelm Fink Verlag, 1994), pp. 121–44.

148. D. Mersch, 'Naturwissenschaftliches Wissen und bildliche Logik', in M. Heßler (ed.), *Konstruierte Sichtbarkeiten. Wissenschafts- und Technikbilder seit der Frühen Neuzeit* (München: Wilhelm Fink Verlag, 2006), pp. 405–20; J. Snyder, 'Visualization and Visibility', in C. A. Jones and P. Gallison (eds), *Picturing Science/Making Art* (New York: Routledge Chapman & Hall, 1998), pp. 379–99; and J. Schickore, *The Microscope and the Eye: A History of Reflections, 1740–1870* (Chicago, IL: University of Chicago Press, 2007).

149. L. Dippel, *Das Mikroskop und seine Anwendung* (Braunschweig: Vieweg-Verlag, 1867), p. 456.

150. Mersch, 'Naturwissenschaftliches Wissen', p. 415.

151. Rice, *Voyages of Discovery*, p. 323.

152. O. Steche, 'Carl Chun', *Mitteilungen der Gesellschaft für Erdkunde zu Leipzig* (Leipzig, 1914), pp. 43–90, on p. 53.

153. 'Perhaps there is nothing more interesting or challenging for an illustrator than drawing creatures of the sea'. C. P. Papp, *Scientific Illustration: Theory and Practice* (Dubuque: W. C. Brown Company, 1968), p. 84.

154. Steche, 'Carl Chun', p. 53.

155. M. Heßler, Einleitung. Annäherung an Wissenschaftsbilder, in M. Heßler (ed.), *Konstruierte Sichtbarkeiten. Wissenschafts- und Technikbilder seit der Frühen Neuzeit* (München: Wilhelm Fink Verlag, 2006), pp. 11–40, on p. 28.

156. Heßler, Einleitung, p. 28.

157. P. Geimer, 'Einleitung', in Geimer (ed.) *Ordnung der Sichtbarkeit*, pp. 7–27, on p. 7. Compare to M. Lynch, 'The Production of Scientific Images: Vision and Re-Vision in the History, Philosophy, and Sociology of Science', in L. Pauwels (ed.), *Visual Cultures of Science: Rethinking Representational Practices in Knowledge Building and Science Communication* (Lebanon, NH: University Press of New England, 2006), pp. 26–40.

158. The idea of drawing things together in the drawing process is based on Bruno Latour's text 'Drawing Things Together', in M. Lynch and S. Woolgar (eds), *Representation in Scientific Practice* (London and Cambridge, MA: The MIT-Press, 1990), pp. 19–68.

159. Compare to B. Latour, 'Arbeit mit Bildern: oder Die Umverteilung der wissenschaftlichen Intelligenz', in B. Latour, *Der Berliner Schlüssel. Erkundungen eines Liebhabers der Wissenschaften* (Berlin: Akademie-Verlag, 1996), pp. 159–90, on p. 183.

160. Winter, 'Carl Chun', p. 182.

3 *Twenty Thousand Leagues under the Sea*: The Overture to a Passion

1. J. Verne, *Twenty Thousand Leagues under the Seas; Or, The Marvellous and Exciting Adventures of Pierre Aronnax, Conseil his Servant, and Ned Land, a Canadian Harpooner*, trans. from the French (Boston, MA: Geo. M. Smith & Co., 1873), p. 3.

2. Verne, *Twenty Thousand Leagues under the Seas*, pp. 9–12.

3. F. Wolfzettel, *Jules Verne: Eine Einführung* (Zürich: Artemis Verlag, 1988), p. 131.

4. Wolfzettel, *Jules Verne*, p. 131.

5. Wolfzettel, *Jules Verne*, p. 80.

6. 'Verne, J., *Twenty Thousand Leagues under the Seas; Or, The Marvellous and Exciting Adventures of Pierre Aronnax, Conseil his Servant, and Ned Land, a Canadian Harpooner*, trans. from the French (Boston, MA: Geo. M. Smith & Co., 1873), p. 56.

7. Wolfzettel, *Jules Verne*, p. 133.

8.   Verne, *Twenty Thousand Leagues under the Seas*, p. 209.
9.   Wolfzettel, *Jules Verne*, p. 133.
10. Verne, *Twenty Thousand Leagues under the Seas*, pp. 210–11.
11. Wolfzettel, *Jules Verne*, p. 15.
12. Compare to H. Böhme, 'Umriß einer Kulturgeschichte des Wassers. Eine Einleitung', in H. Böhme (ed.), *Kulturgeschichte des Wassers* (Frankfurt am Main: Suhrkamp, 1988), pp. 7–42, on p. 25.
13. Verne, *Twenty Thousand Leagues under the Seas*, pp. 75–7.
14. Aronnax's role model was the French natural scientist Henri Milne-Edwards.
15. Compare to M. de Certeau, 'Writing the Sea: Jules Verne', in M. de Certeau, *Heterologies: Discourses on the Other*, trans. B. Massumi (Minneapolis, MN: University of Minnesota Press, 1986) pp. 137–49.
16. Certeau, 'Writing the Sea'.
17. Compare to Wolfzettel, *Jules Verne*, p. 69.
18. J. Verne, *Twenty Thousand Leagues under the Seas; Or, The Marvellous and Exciting Adventures of Pierre Aronnax, Conseil his Servant, and Ned Land, a Canadian Harpooner*, trans. from the French (Boston, MA: Geo. M. Smith & Co., 1873), p. 51–2.
19. Wolfzettel, F., *Jules Verne: Eine Einführung* (Zürich: Artemis Verlag, 1988), p. 127.
20. See V. Dehs, *Jules Verne. Eine kritische Biographie* (Düsseldorf and Zürich: Patmos, Artemis & Winkler, 2005).
21. Compare to U. Harter, 'Le Paradis artificiel. Aquarien, Leuchtkästen und andere Welten hinter Glas', *Vorträge aus dem Warburg Haus*, 6 (2002), pp. 77–124.
22. See R. Geißler, *Plaudereien über Paris und die Weltausstellung* (Berlin: Theobald Grieben, 1868).
23. Verne, *Twenty Thousand Leagues under the Seas*, p. 76.
24. See J. A. Schmoll Gen. Eisenwerth, 'Fensterbilder. Motivketten in der europäischen Malerei', in L. Grote (ed.), *Beiträge zur Motivkunde des 19. Jahrhunderts* (München: Prestel Verlag, 1970), pp. 13–166; G. Neuhardt, 'Das Fenster als Symbol. Versuch einer Systematik der Aspekte', in E. T. Reimbold (ed.), *Symbolon. Jahrbuch für Symbolforschung* (Köln: Wienand, 1972), pp. 77–91; and R. Selbmann, *Eine Kulturgeschichte des Fensters von der Antike bis zur Moderne* (Berlin: Reimer Verlag, 2010).
25. Compare to S. E. Hauser, 'Der subaquatische Bilderkosmos. Eine kurze Geschichte des Aquarium- und Unterwasserfilms von 1890 bis heute', in V. Weigel (ed.), *Unter wasser über wasser* (Wilhelmshaven: Kerber Art, 2009), pp. 18–35, on pp. 30–1.
26. Compare to Selbmann, *Kulturgeschichte des Fensters*, p. 23.
27. Compare to Selbmann, *Kulturgeschichte des Fensters*, p. 20.
28. Compare to Schmoll Gen. Eisenwerth, 'Der Blick durch das offene Fenster', p. 139.
29. Selbmann, *Kulturgeschichte des Fensters*, p. 20.
30. First published in A. Tennyson, *Poems, Chiefly Lyrical* (London: Effingham Wilson, 1830), p. 154.
31. H. Lee, *Aquarium Notes: The Octopus; or, The 'Devil-Fisch' of Fiction and of Fact* (London: Champman and Hall, 1875), p. 7.
32. Lee, *Aquarium Notes*, p. 10.
33. H. Melville, *The Whale; Or, Moby Dick*, 3 vols (London: Richard Bentley, 1851), vol. 2, p. 138.
34. F. Péron, *Voyage de découvertes aux terres australes, execute Sur Les Covettes Le Geographe, Le Naturaliste, et La Goelette Le Casuarina, Pendant Les annees 1800, 1801, 1803 et 1804*, 2 vols (Paris: de L' Impremerie Royale, 1807), vol. 2, p. 18.

35. Compare to R. Cardinal, 'Victor Hugo: Somnambulist of the Sea', in P. Collier (ed.), *Artistic Relations: Literature and the Visual Arts in Nineteenth-Century France* (New Haven, CT and London: Yale University Press, 1994), pp. 210–20, on p. 213; and E. Kernbauer, 'Formationen des Meeres bei Victor Hugo', in T. Brandstetter, K. Harrasser and G. Friesinge (eds), *Grenzflächen des Meeres* (Wien: Turia and Kant, 2010), pp. 63–86, on p. 81–2.

36. V. Hugo, *Toilers of the Sea*, ed. Ernest Rhys, trans. W. Moy Thomas (London and Toronto: J. M. Dent, 1911), p. 292.

37. J. Michelet, *The Sea* (New York: Rudd & Carleton, 1861), pp. 194–201.

38. Michelet, *The Sea*, p. 194.

39. Michelet, *The Sea*, p. 196.

40. P. D. de Montfort, *Histoire naturelle, generale et particuliere des mollusques, animaux sans vertebres et a sang blanc, addendum to Georges-Louis Leclerc de Buffon, Histoire Naturelle, Generale et Particuliere*, 4 vols (Paris: Dufart, 1801–1804), vol. 1. Additionally see R. Caillois, *Der Krake. Versuch über die Logik des Imaginativen* (München: dtv, 1989), pp. 31–4.

41. The most incisive positive account on octopuses and squids probably is from U. Aldrovandi's *De reliquis Animalibus exanguibus libri quatuor post mortem ejus editi ... de Mollibus, Crustaceis, Testaceis, et Zoophytis* (Bologna, 1606). The first part adresses the squid as an animal that even surpasses eagle and lion; see Chapter 2, 'De Polypo Ordinis, Ratio'.

42. See also Pierre Denys de Montfort, *Allgemeine und besondere Naturgeschichte der Weichwürmer (Mollusques) als Fortsetzung der Buffonschen Naturgeschichte*, 2 vols (Hamburg, Mainz: G. Vollmer, 1803) vol. 2, on the kraken's lust to kill see pp. 68–70, on the giant squid see pp. 170–2.

43. Quoted from Caillois, *Der Krake*, p. 37.

44. Hugo does not refer to Michelet in *Toilers of the Sea*, but he mentions Denys de Montfort as key witness for the kraken as a monster full of hatred. However, the idea of the kraken as a deadly suction bell originates in Michelet's work.

45. Hugo, *Toilers of the Sea*, p. 296.

46. Hugo, *Toilers of the Sea*, p. 290.

47. Compare to J. Schnier, 'Morphology of a Symbol: The Octopus', *American Image: A Psychoanalytic Journal for the Arts and Sciences*, 13:1 (1956), pp. 3–31.

48. Hugo, *Toilers of the Sea*, p. 295.

49. Caillois, *Der Krake*, p. 63.

50. J. Verne, *Twenty Thousand Leagues under the Seas; Or, The Marvellous and Exciting Adventures of Pierre Aronnax, Conseil his Servant, and Ned Land, a Canadian Harpooner*, trans. from the French, 2 vols (Boston, MA: Geo. M. Smith & Co., 1873), vol. 2, pp. 272–3.

51. Verne, *Twenty Thousand Leagues under the Seas*, vol. 2, p. 276.

52. Caillois, *Der Krake*, p. 61.

53. Caillois, *Der Krake*, p. 103.

54. See, for example: H. G. Wells, *The Sea Raiders* (Weekly Sun Literary Supplement, 6 December 1896); F. T. Bullen, 'The Last Stand of the Decapods', in F. T. Bullen, *Deep-Sea Plunderings* (New York: Appleton, 1902), pp. 211–34; *Reap the Wild Wind* (1942) director Cecil B. DeMille; I. Fleming, *Dr. No* (London, Jonathan Cape, 1958); M. Crichton, *Sphere* (New York: Knopf, 1987); A. C. Clarke, 'The Shining Ones', *Argosy* (December 1964), pp. 77–93; P. Benchley, *Beast* (New York: Random House, 1991); *Pirates of the Caribbean, Dead Man's Chest* (2002), director Gore Verbinski; and B.

Kegel, *Der Rote* (Hamburg: Marebuchverlag, 2009).

55. See consul M. Sabin Berthelot's report to the *Académie des Sciences* 1861 concerning an encounter of the sailing ship *Alecton* with a giant squid. Throughout the whole report the denomination of squid and octopus is used interchangeably, even though it was definitely a squid which had been sighted. Verne probably knew this report as he lets Aronnax refer to it extensively. *Comptes rendus hebdomadaires des séances de l'Academie des Sciences*, 53 (Paris Juillet – Décembre 1861), pp. 1256–67, on pp. 1265–6.

56. However, there are three illustrations which clearly depict an octopus, namely the frontispiece *Vingt Mille Lieves sous Les Mers*, an illustration in Chapter 4 which shows a sponge diver, and in Chapter 9, which shows Nemo and Aronnax on their way to Atlantis.

57. Verne, *Twenty Thousand Leagues under the Seas*, p. 273.

58. The giant squid often is confounded with a giant octopus (kraken), because it first appeared in Norwegian literature under the denomination of *kraken*.

59. Compare to F. A. Aldrich, 'Some Aspects of the Systematics and Biology of Squid of the Genus Architeuthis Based on a Study of Specimens from Newfoundland Waters', *Bulletin of Marine Science*, 49:1–2 (1992), pp. 457–81; and R. Ellis, *The Search for the Giant Squid: The Biology and Mythology of the World's Most Elusive Sea Creature* (London: Lyons Press, 1998).

60. T. Kubodera and K. Mori, 'First-Ever Observations of a Live Giant Squid in the Wild', *Proceedings of the Royal Society: Biological Sciences*, 272 (2005), pp. 2583–6; and T. Kubodera and K. Mori, 'Observations of Wild Hunting Behaviour and Bioluminescence of a Large Deep-Sea, Eight-Armed Squid, Taningia danae', *Proceedings of the Royal Society: Biological Sciences*, 274 (2007), pp. 1029–34.

61. In addition see S. Lyons, 'Sea Monsters: Myth or Genuine Relic of the Past', in K. R. Benson and P. F. Rehbock (eds), *Oceanographic History: The Pacific and Beyond* (Seattle, WA and London: University of Washington Press, 2002), p. 60–70.

62. P. Belon, *L'Histoire naturelle des estranges poissons marins avec la vraie peincture et description du Dauphin et de plusieurs autres de son espèce observée par Pierre Belon du Mans* (Paris: Regnaud Chaudiere, 1551); C. Gesner, *Historia Animalium. Liber 4 qui est de Piscium & Aquatilium Animantium natura* (Zürich: Froschauer, 1558); and G. Rondelet, *Historia piscium universa* (Lyon: Matthias Bonhomme, 1554).

63. Compare to A. G. More, 'Some Account of the Gigantic Squid (Architeuthis dux) Lately Captured off Boffin Island, Connemara', *Zoologist*, 2:10 (1875), pp. 4569–71; and L. de Freycinet (ed.), *Voyage autour du Monde, entrepris par Ordre du Roi ... exécuté sur les corvettes de ... l'Uranie et la Physicienne, pendant les années 1817, 1818, 1819 et 1820*, 2 vols (Paris: Chez Pillet Aîné, 1824).

64. Compare to Berthelot, *Comptes rendus*, p. 1267.

65. A. Mangin, *The Mysteries of the Ocean* (1864; London: T. Nelson and Sons, 1870), p. 292.

66. P. Harting, 'Description de quelques fragments de deux céphalopodes gigantesques', *Verh. Akad. Wet., Amst.*, 9:1 (1860), p. 2.

67. Also compare to W. S. Kent, 'Note on a Gigantic Cephaolpod from Cenception Bay, Newfoundland', *Proceedings of the Zoological Society London* (1874), pp. 178–82; and W. S. Kent, 'A Further Communication Upon Certain Gigantic Cephalopods Recently Encountered off the Coast of Newfoundland', *Proceedings of the Zoological Society London* (1874), pp. 489–94.

68. Compare to A. E. Verrill, 'Occurrence of Gigantic Cuttle-Fishes on the Coast of Newfoundland', *American Journal of Science and Arts* (1874), pp. 158–61; A. E. Verrill, 'The Giant Cuttle-Fishes of Newfoundland and the Common Squids of the New England

Coast', *American Naturalist* (1874), pp. 167–74; A. E. Verrill, 'The Colossal Cephalopods of the North Atlantic', *American Naturalist* (1875), pp. 21–36; A. E. Verrill, *The Cephalopods of the North-Eastern Coast of America* (New Haven, CT: Connecticut Academy of Sciences, 1880); and A. E. Verrill, 'Occurrence of Architeuthis', *American Journal of Science* (1881), pp. 171–2.

69. For further reports on giant cuttlefish see: A. S. Packard, 'Colossal Cuttlefishes', *American Naturalist*, 7 (1873), pp. 87–94; A. G. More, 'Gigantic Squid on the West Coast of Ireland', *Annual Magazine of Natural History* 14:16 (1875a), pp. 123–4; A. G. More, 'Notice of a Gigantic Cephlopod (Dinoteuthis Proboscideus) which was Stranded at Dingle, in Kerry, Two Hundred Years Ago', *Zoologist*, 2:10 (1875b), pp. 4526–32; More, 'Some Account of the Gigantic Squid', pp. 4569–71; T. O'Connor, 'Capture of an Enormous Cuttle-Fish off Boffin Island, on the Coast of Connemara', *Zoologist*, 2:10 (1875), pp. 4502–3; T. W. Kirk, 'On the Occurrence of a Giant Cuttlefish on the New Zealand Coast', *Transactions of the New Zealand Institute* (1880), pp. 310–13; and M. Harvey, 'How I Discovered the Great Devil-Fish', *Wide World Magazine* (1899), pp. 732–40.

70. Mangin, *The Mysteries of the Ocean*, pp. 290–1.

71. R. Ellis, *Monsters of the Sea* (New York: Robert Hale, 1994), p. 121.

72. H. Holländer, 'Es rauscht in den Schachtelhalmen, verdächtig leuchtete das Meer …', in A. Preiß and B. Brock (eds), *Ikonographia: Anleitung zum Lesen von Bildern* (München: Klinkhardt & Biermann, 1990), pp. 179–202, on p. 200.

73. Verne, *Twenty Thousand Leagues under the Seas*, p. 273.

74. Verne, *Twenty Thousand Leagues under the Seas*, p. 273.

75. E. Haeckel, *Kunstformen der Natur* (Leipzig: Verlag des Bibliographischen Instituts, 1904), Foreward.

76. Compare to A. Brinckmann-Voss, P. Fioroni and S. von Boletzky, *Adolf Portmanns frühe Studien mariner Lebewesen* (Basel: Schwabe-Verlag, 1997), p. 6.

77. J. Cousteau and P. Diolè, *Octopus and Squid: The Soft Intelligence* (New York: Doubleday, 1973).

78. Caillois, *Der Krake*, p. 63.

79. A. D. d'Orbigny and A. É. d'Audebert de Ferussac, *Histoire naturelle, général et particulière, des céphalopodes acétabulifères*, 2 vols (Paris 1835–48), see, for example, the plates showing *Octopus aranea*, *Octopus lunelatus* or *Eladone moschatus*.

80. G. Hartwig, *Das Leben des Meeres. Eine Darstellung für gebildete Stände* (Glogau: Karl Flemming Verlag, 1862), p. 246.

81. W. Bölsche, *Liebesleben der Natur: Eine Entwicklungsgeschichte der Liebe* (Jena: Diederichs, 1898), p. 224.

82. G. Schubert, 'Aus dem Berliner Aquarium', *Der Zoologische Garten. Zeitschrift für Beobachtung, Pflege und Zucht der Thiere*, 21 (1880), pp. 92–6, on p. 93.

83. See, for example, J. Kollmann, 'Die Cephalopoden der zoologischen Station des Dr. Dohrn', *Zeitschrift für Wissenschaftliche Zoologie* (1876), pp. 1–23; J. Kollmann, 'Aus dem Leben der Cephalopoden', *Vierteljahresschrift für wissenschaftliche Philosophie* (1877), pp. 136–55; P. Baglioni, 'Zur Kenntnis der Leistungen einiger Sinnesorgane (Gesichtssinn, Tastsinn und Geruchssinn) und des Zentralnervensystems der Cephalopoden und Fische', *Zeitschrift für Biologie* (1909), pp. 255–86; V. Bauer, 'Einführung in die Physiologie der Cephalopoden. Mit besonderer Berücksichtigung der im Mittelmeer häufigen Formen', *Mittheilungen aus der Zoologischen Station zu Neapel*, 19:2 (1909), pp. 149–268; and B. B. Boycott, 'Learning in *Octopus vulgaris* and other cephalopods', *Publications of the Stazzione zoologica Napoli*, 25 (1954), pp. 67–93.

84. N. J. Berrill, *atlantische wunderwelt* (Frankfurt amMain: Büchergilde Gutenberg 1957), p. 167.
85. A. Grimble, *A Pattern of Islands* (London: John Murray, 1952), p. 71.
86. J. Hornell, 'Notes on Animal Colouration', *Journal of Marine Zoology*, 1 (1893), pp. 3–8; E. Steinach, 'Studien über die Hautfärbung und über den Farbenwechsel der Cephalopoden', *Arch. f. g. Physiol,* 87 (1901), pp. 1–37; C. Chun, 'Ueber die Natur und die Entwicklung der Chromatophoren bei den Cephalopoden', *Verhandlungen der deutschen zoologischen Gesellschaft* (1902), pp. 162–82; E. V. Cowdry, *The Colour Changes of Octopus Vulgaris* (Toronto: University Library, 1911); G. H. Parker, 'Chromatophores', *Biological Review* (1930), pp. 59–90, on pp. 62–6; E. Sereni, 'The Chromatophores of the Cephalopods', *Biological Bulletin, Woods Hole* (1930), pp. 247–68; and H. B. Cott, *Adaptive Coloration in Animals* (London: Methuen, 1940).
87. Sereni, 'The Chromatophores', p. 2.
88. G. Klingel, *The Ocean Island (Inagua)* (New York: Dodd, Mead & Company, 1940), p. 296.
89. N. J. Berrill, *The Living Tide* (New York: Dodd, Mead & Company, 1951), p. 52.
90. J.-W. Atz, 'The Timid Octopus: For Centuries Man has Credited This Relative of the Oyster With Fearsome Powers which It Does Not Possess', *Bulletin of the New York Zoological Society* (1940), pp. 48–54, on p. 48.
91. See, for example, W. Price, *Adventures in Paradise: Tahiti, Samoa, Fiji* (London: William Heinemann, 1956), p. 201.
92. Atz, 'The Timid Octopus', p. 54.
93. On Painlevé's life and work see A. M. Bellows and M. McDougall (eds), *Science is Fiction: The Films of Jean Painlevé* (San Francisco, CA: Brico Press, 2000); B. Berg, 'The Art of Science: 1902–1989', in Museum of Contemporary Art Sydney (ed.), *Liquid Sea* (Sydney: Museum of Contemporary Art, 2003), pp. 14–19; J. Knox, 'Sounding the Depths: Jean Painlevé's Sunken Cinema', *Senses of Cinema*, 25 (March 2003), www.sensesofcinema.com [accessed March 2015]; *les documents cinématographiques*, www.lesdocs.com [accessed March 2015]; F. Reyna, 'Las Fotografías Submarinas de Painlevé', *Estampa Año*, 8:367 (1935); and J. Simon, 'Jean Painlevé and Mark Dion', *Parkett*, 45 (1995), pp. 163–71.
94. Compare to E. Barnow, *Documentary: A History of the Non-Fiction Film* (New York and Oxford: Oxford University Press, 1993), pp. 72–4.
95. H. Hazéra and D. Leglu, 'Jean Painlevé Reveals the Invisible. Interview in Libération 1986', in A. M. Bellows and M. McDougall (ed.), *Science is Fiction: The Films of Jean Painlevé* (San Francisco, CA: Brico Press, 2000), pp. 170–9, on p. 174.
96. Compare to A. La Salvia, 'Der Krake mit dem seidenen Blick. Lautréamont, Dalí, Breton' in E. Puyplat, A. La Salvia and H. Heinzelmann (eds), *Salvador Dali: Facetten eines Jahrhundertkünstlers* (Würzburg: Königshausen & Neumann, 2005), pp. 107–18.
97. J. Painlevé, 'Mysteries and Miracles of Nature' (French 'Mystères et Miracles'), *Vu* (29 March 1931), in Bellows and McDougall (eds), *Science is Fiction*, pp. 119–23, on p. 119.
98. Painlevé explained later that this effct happened by chance: 'He was an old man who, out of vanity, refused to wear glasses. He was therefore obliged to stick his face right up against the script, close to the microphone, where one could hear his emphysema'. Quoted from B. Berg, 'Contradictory Forces: Jean Painlevé, 1902–89', in Bellows and McDougall (eds), *Science is Fiction*, pp. 2–47, on p. 41.

4 *Mise-en-Scène:* Invented Realities, or the Mediality of Wonders – The Sea in the Aquarium

1. Compare to T. James, *Dream, Creativity, and Madness in Nineteenth-Century France* (Oxford: Clarendon Press, 1995); C. Kockerbeck, *Ernst Haeckels 'Kunstformen der Natur' und ihr Einfluß auf die deutsche bildende Kunst der Jahrhundertwende* (Frankfurt am Main: Peter Lang Verlag, 1986); P. Kort and M. Hollein (eds), *Darwin. Kunst und die Suche nach den Ursprüngen* (Frankfurt am Main, Wienand Verlag, 2009); B. Larson, *The Dark Side of Nature: Science, Society, and the Fantastic in the Work of Odilon Redon* (Pennsylvania, PA: Pennsylvania State University Press, 2005); M. Stuffmann and M. Hollein (eds), *As in a Dream: Odilon Redon* (Ostfildern: Hatje & Cantz, 2007); R. Verdi, *Klee and Nature* (London: Zwemmer Ltd, 1984); and P. Wichmann, *Jugendstil Floral Funktional in Deutschland und Österreich und den Einflußgebieten* (Herrsching: Schuler Verlagsgesellschaft, 1984).

2. Compare to E. Schmitt (ed.), *Glas des Art Nouveau. Die Sammlung Gerda Koepff* (München and New York: Prestel Verlag, 1998); and R. Ulmer (ed.), *Art Nouveau: Symbolismus und Jugendstil in Frankreich* (Stuttgart and New York: Arnoldsche, 1999).

3. E. Haeckel, *Report on the Radiolaria, collected by HMS Challenger during the years 1873–76, I. Part: Porulosa (Spumellaria and Acantharia). II. Part Osculosa (Nesselaria and Phaeodaria)* (London: Eyre & Spottiswoode, 1887). On Binet und Haeckel see R. Proctor, *René Binet. Natur und Kunst. Mit Beiträgen von Robert Proctor und Olaf Breidbach* (München: Prestel Verlag, 2007); and E. Krauße, *Ernst Haeckel* (Leipzig: Teubner Verlag, 1984), pp. 363–7.

4. M. Foucault, *History of Sexuality: An Introduction*, 3 vols (New York: Random House, 1990), vol. 1, pp. 141–2.

5. Compare to H.-J. Rheinberger, *Experiment, Differenz, Schrift. Zur Geschichte epistemischer Dinge* (Marburg: Baslisken-Presse, 1992); and H.-J. Rheinberger and M. Hagner (eds), *Die Experimentalisierung des Lebens. Experimentalsysteme in den biologischen Wissenschaften 1850/1950* (Berlin: Akademie-Verlag, 1993).

6. P. Geimer, *Ordnungen der Sichtbarkeit. Fotografie in Wissenschaft, Kunst und Technologie* (Frankfurt am Main: Suhrkamp, 2002).

7. Compare to H. Schramm, L. Schwarte and J. Lazardzig (eds), *Spektakuläre Experimente. Praktiken der Evidenzproduktion im 17. Jahrhundert* (Berlin and New York: Springer Verlag, 2006).

8. W. Krohn, 'Die ästhetischen Dimensionen der Wissenschaft', in W. Krohn (ed.), *Ästhetik in der Wissenschaft. Interdisziplinärer Diskurs über das Gestalten und Darstellen von Wissen* (Hamburg: Meiner-Verlag, 2006), pp. 3–38, on p. 4.

9. Compare to B. M. Stafford, 'Images of Ambiguity: Eighteenth-Century Microscopy and the Neither/Nor, in D. P. Miller and P. H. Reill (eds), *Visions of Empire. Voyages, Botany, and Representations of Nature* (Cambridge: Cambridge University Press, 1996), pp. 230–57.

10. K. A. Möbius, 'Der Bau des Eozoon canadense nach eigenen Untersuchungen verglichen mit dem Bau der Foraminiferen', *Paleontographica*, 25 (1878), pp. 175–92, on pp. 175, 189.

11. Compare to L. K. Nyhart, 'Science, Art, and Authenticity in Natural History Displays', in S. de Chadaveira and N. Hopwood (eds), *Models: The Third Dimension of Science* (Stanford, CA: Stanford University Press, 2004), pp. 307–38, on p. 313.

12. Compare to H. Weidner, 'Die Anfänge meeresbiologischer und ökologischer Forschung in Hamburg durch Karl August Möbius (1825–1908) und Heinrich Adolph

Meyer (1822–1889)', in *Historisch Meereskundliches Jahrbuch* (Berlin and Hamburg: Dietrich Reimer, 1994), pp. 69–84, on pp. 70–2; L. Nyhart, 'Civic and Economic Zoology in Nineteenth-Century Germany: The "Living Communities" of Karl Möbius', *Isis*, 89 (1998), pp. 605–30, on pp. 624–6; and M. Glaubrecht, 'Karl August Möbius: Von Lebensgemeinschaften zur Artenvielfalt', *Naturwissenschaftliche Rundschau*, 61:5 (2008), pp. 230–6.

13. K. A. Möbius, 'Einige Fingerzeige für die Bevölkerung und Erhaltung der Aquarien', *Zoologischer Garten*, 6 (1865), pp. 211–14, on pp. 211–12.

14. K. A. Möbius, *The Oyster and Oyster-Culture*, US Commission Fish and Fisheries Report (1880), pp. 683–751, on p. 732.

15. [Karl August Möbius], 'Blicke in das Leben der Thiere im Aquarium', *Hamburger Nachrichten*, 11 May 1864.

16. A. Mangin, *The Mysteries of the Ocean* (1864; London: T. Nelson and Sons, 1870), pp. 201–2.

17. On the history of the aquarium see B. Brunner, *Wie das Meer nach Hause kam* (Berlin: Transit–Verlag, 2003); A. Frédol, *Le Monde de la Mer* (Paris: Librairie de L-Hachette, 1865); P. H. Gosse, *The Aquarium: An Unveiling of the Wonders of the Deep Sea* (London: John van Voorst, 1854); G. Jäger, *Das Leben im Wasser und das Aquarium* (Stuttgart: Kosmos, Gesellschaft der Naturfreunde, 1908); Jr. Vernon N. Kisling (ed.), *Zoo and Aquarium History: Ancient Animal Collections to Zoological Gardens* (Boca Raton, London, New York and Washington: CRC Press, 2001); Mangin, *The Ocean*, pp. 195–205; J. Power de Villepreux, 'The Aquarium', *The Popular Science Monthly* (May 1874), pp. 687–935; P. Rehbock, 'The Victorian Aquarium in Ecological and Social Perspective', in M. Sears and D. Merriman (eds), *Oceanography: The Past* (New York and Heidelberg: Springer-Verlag, 1980), pp. 522–39; G. B. Sowerby, *The Aquarium: A Popular Account of Marine and Fresh-Water Animals and Plants* (London: Routledge, 1865); and R. Stott, *Theatres of Glass: The Woman who Brought the Sea to the City* (London: Short Books, 2003).

18. Compare to Gosse, *The Aquarium*, p. 3.

19. 'The aquativ vivarium at the Zoological Gardens, Regent's Park', *Illustrated London News*, 28 May 1853, p. 420.

20. T. Gautier, *Paris et Les Parisiens*, ed. C. Lacoste-Veysseyre (Paris: La Boîte à Documents, 1996), p. 297.

21. Mangin, *The Ocean*, p. 203.

22. I owe this reference to Hartmut Böhme.

23. Compare to G. Groom, 'The Late Work', in D. W. Druik (ed.), *Odilon Redon: Prince of Dreams 1840–1916* (Chicago, IL: The Art Institute of Chicago, 1994), Ausstellungskatalog, pp. 305–52, on p. 321.

24. *Gartenlaube* 1865; quoted from Brunner, *Wie das Meer*, p. 105.

25. On the Hamburg Aquarium see K. A. Möbius, 'Mittheilungen über das Aquarium des zoologischen Gartens in Hamburg', *Zoologischer Garten*, 7 (1866), pp. 173–7. On the history of the Berlin Aquarium see E. Biella, 'Das Berliner Aquarium Unter den Linden/Schadowstraße: Zur Konzeption des Rundganges und Bedeutung des Grottenstils als Ausstellungsarchitektur', *Der Bär von Berlin: Jahrbuch des Vereins für die Geschichte Berlins*, 49 (2000), pp. 63–80; A. Brehm, 'Von der Baustätte des Berliner Aquariums', *Die Gartenlaube. Illustrirtes Familienblatt* (1868), pp. 620–3; A. Brehm, 'Das Berliner Aquarium', *Westermann's Jahrbuch der Illustrirten Deutschen Monatshefte. Ein Familienbuch für das gesammte geistige Leben der Gegenwart* (1870), pp. 138–69; O. Heinroth,

Führer durch das Aquarium nebst Terrarium und Insektarium im Zoologischen Garten zu Berlin (Berlin: Zoologischer Garten zu Berlin, 1915); Kisling, *Zoo and Aquarium History*; H.-G. Klös, H. Fädrich and U. Klös, *Die Arche Noah an der Spree. 150 Jahre Zoologischer Garten Berlin. eine tiergärtnerische Kulturgeschichte von 1844–1994* (Berlin: FAB-Verlag, 1994); H.-G. Klös, *Von der Menagerie zum Tierparadies. 125 Jahre Zoo Berlin* (Berlin: Reimer-Verlag, 1969); H.-G. Klös and U. Klös, *Der Berliner Zoo im Spiegel seiner Bauten 1841–1989* (Berlin: Heenemann-Verlag, 1990); H.-G. Klös and J. Lange, *Tierwelt hinter Glas: das Zoo-Aquarium Berlin* (Berlin: arani-Verlag, 1988); H.-G. Klös, (ed.), *Festschrift 70 Jahre Aquarium* (Berlin: Zoologischer Garten, 1983); J. Lange, 'Schauaquarien im Wandel der Zeit', *Bongo*, 13 (1987), pp. 135–56; G. Schubert, 'Aus dem Berliner Aquarium', *Der Zoologische Garten. Zeitschrift für Beobachtung, Pflege und Zucht der Thiere*, 21 (1880), pp 92–6; and H. Strehlow, 'Zur Geschichte des Berliner Aquariums Unter den Linden', *Der Zoologische Garten*, 57 (1987), pp. 26–40.

26. Compare to I. Kranz, 'Zur Felsengrotte im Heimaquarium', in B. Butis (ed.), *Stehende Gewässer. Medien der Stagnation* (Berlin and Zürich: Diaphanes, 2007), pp. 247–58, on p. 249.

27. H. Beta, 'Der Sohn des, alten Brehm', *Die Gartenlaube. Illustrirtes Familienblatt*, 17 (1869), pp. 20–2.

28. Klös and Lange, *Tierwelt hinter Glas*, pp. 20–2.

29. Compare to D. Jarofke, 'Tiere der Vorzeit an der Fassade unseres Aquariums', *Bongo*, 8 (1984), pp. 19–40.

30. Compare to B. Hauff, 'Der *Ichthyosaurus* des Berliner Zoo-Aquariums aus dem Museum Hauff in Holzmaden', *Bongo*, 11 (1986), pp. 115–20.

31. Compare to J. Lange, 'Hundert Jahre Zoo-Aquarium Berlin – Hundert Jahre Schauaquaristik', in B. Blaszkiewitz (ed.), *Picassofisch und Kompasssqualle. 100 Jahre Zoo – Aquarium Berlin* (Berlin: Lehmanns, 2013), pp. 44–83, on p. 52.

32. Klös and Lange, *Tierwelt hinter Glas*, p. 22.

33. On the history of the home aquarium: J. W. Atz, 'The Balanced Aquarium Myth', *Natural History*, 58 (1949), pp. 72–7, 96; L. Barber, *The Heyday of Natural History: 1820–1870* (London: Cape, 1989), Chapter 8; P. H. Gosse, *Handbook to the Marine Aquarium* (London: Van Voorst, 1855); J. Harpe, *The Sea-Side and Aquarium; Or, Anecdote and Gossip on Marine Zoology* (Edinburgh: William P. Nimmo, 1858); S. Hibberd, *Rustic Adornments for Homes of Taste* (1856; London: Trafalgar Square 1987); E. Lankester, *The Aquavivarium, Fresh and Marine Being an Account of the Principles and Objects Involved in the Domestic Culture of Water Plants and Animals* (London: R. Hardwicke, 1856); and G. H. Lewes, *Sea-Side Studies at Ilfracombe, Tenby, the Scilly Isles, & Jersey* (Edinburgh and London: W. Blackwood, 1858).

34. J. Ingenhousz, *Experiments Upon Vegetables: Discovering Their Great Power of Purifying the Common Air in the Sunshine and of Injuring it in the Shade at Night* (London: P. Elmsly and H. Payne, 1779).

35. Rehbock, 'The Victorian Aquarium', p. 523.

36. C. DesMoulins, 'Note sur les moyens d'empecher la corruption dans les bocaux où l'on conservedes animaux aquatiques vivants', *Actes de la Société Linnéenne de Bordeaux*, 4 (1830), pp. 257–72.

37. J. de Villepreux Power, *Observations physiques sur le poulpe de l'Argonauta Argo: commencees en 1832 et terminees en 1843* (Paris: Charles de Mourgues Frères, 1856).

38. J. G. Dalyell, *Rare and Remarkable Animals of Scotland, Represented from Living Objects: With Practical Observations on their Nature* (London: Van Voorst, 1847–8).

39. G. Johnston, A *History of British Sponges and Lithophytes* (Edinburgh: W. H. Lizars, 1842), p. 215.

40. Compare to Stott, *Theatres of Glass*.

41. Compare to Stott, *Theatres of Glass*, pp. 135–43.

42. Compare to Stott, *Theatres of Glass*, p. 109.

43. Compare to Stott, *Theatres of Glass*, p. 115.

44. A. Thynne, 'On the Increase of Madrepores', *Annals and Magazine of Natural History*, 18 (June 1859), pp. 449–61.

45. A. Thwaite, *Glimpses of the Wonderful: The Life of Philip Henry Gosse 1810–1888* (London: Faber & Faber, 2002).

46. Compare to P. F. Rehbock, 'The Early Dredgers: "Naturalizing" in British Seas, 1830–1850', *Journal for the History of Biology*, 12:2 (1979), pp. 293–368.

47. N. Bagshaw Ward, 'Reports on the Subject of the Growth of Plants in Closed Glass Vessels', *Report of the British Association for the Advancement of Science*, 1 (1837), pp. 501–5, on p. 505.

48. N. Bagshaw Ward, *On the Growth of Plants in Closely Glazed Cases* (London: Van Voorst, 1842), p. 58.

49. S. Hibberd, *The Fern Garden* (London: Groombridge and Sons, 1869); quoted from D. E. Allen, *The Victorian Fern Craze* (London: Hutchinson, 1969), p. 62.

50. R. Warington, 'Notice of Observations on the Adjustment of the Relations Between the Animal and Vegetable Kingdoms, by which the Viral Functions of Both are Permanently Maintained', *Quarterly Journal of the Chemical Society*, 3 (1851), pp. 52–4, on pp. 52–3.

51. Warington, 'Notice of Observations', p. 53.

52. P. H. Gosse, On Keeping Marine Animals and Plants Alive in Unchanged Seawater, *Annuals of Natural History*, 19:2 (1852), pp. 263–8; and R. Warington, 'On Preserving the Balance Between the Animal and Vegetable Organisms in Sea-Water', *Annals of Natural History*, 2:12 (1853), pp. 319–24.

53. R. Warington, 'Memoranda of Observations Made in Small Aquaria, in which the Balance Between the Animal and Vegetable Organisms was Permanently Maintained', *Annals of Natural History*, 2:14 (1854), pp. 366–73; R. Warington, 'On the Injurious Effects of an Excess or Want of Heat and Light on the Aquarium', *Annals of Natural History*, 2:16 (1855), pp. 313–15; and R. Warington, 'On the Aquarium', *Proceedings of the Royal Institution of Great Britain*, 2 (1857), pp. 403–8.

54. Quoted from Stott, *Theatres of Glass*, p. 132; Compare to Thwaite, *Glimpses of the Wonderful*, pp. 185–7.

55. Lewes, *Sea-Side Studies*, p. 115.

56. E. A. Roßmäßler, 'Der See im Glase', *Die Gartenlaube. Illustriertes Familienblatt* (1856), pp. 252–6, on p. 252.

57. 'Wie er und behält man den Ocean auf dem Tische, oder das Marine-Aquarium', *Die Gartenlaube*, 38 (1855), pp. 503–5, on p. 505.

58. Gosse, *The Aquarium*, p. 226.

59. Compare to M. Lynch and S. W. Woolgar (eds), *Representation in Scientific Practice* (London and Cambridge, MA: The MIT-Press, 1990).

60. C. Olalquiaga, *The Artificial Kingdom: On the Kitsch Experience* (New York: Pantheon Books, 1998), p. 17.

61. P. Valéry, *Eupalinos: Or The Architect*, trans. W. M. Stewart (1923; London and Oxford: Oxford University Press, H. Milford, 1932); and P. Valéry, 'Man and the Sea Shell', in P. Valéry, *Aesthetics, Collected Works*, trans. R. Manheim, 15 vols (New York:

Pantheon Books, 1964), vol. 13, pp. 3–31.

62. Compare to N. Maak, *Der Architekt am Strand. Le Corbusier und das Geheimnis der Seeschnecke* (München: Carl Hanser Verlag, 2010), p. 27.

63. See, for example, G. C. Bateman, *Fresh-Water Aquaria: Their Construction, Arrangement, and Management* (London: Upcott Gill, 1870); W. E. Damon, *Ocean Wonders: A Companion for the Sea Side* (New York: D. Appleton & Company, 1879); W. Furneaux, *Life in Ponds and Streams* (London: Longmans, Green & Co., 1896); H. N. Humphreys, *River Gardens; Bring an Account of the Best Methods or Cultivation Fresh-Water Plants in Aquaria* (London: Sampson Low, 1857); H. D. Butler, *The Family Aquarium; Or, Aqua Vivarium. A New Pleasure for the Domestic Circle* (New York: Dick & Fitzgerald, 1858); J. Harper, *The Sea-Side and Aquarium; Or, Anecdote and Gossip on Marine Zoology* (Edinburgh: William P. Nimmo, 1858); 'Parlour Aquaria', *Family Friend*, 2 (1856), pp. 192–7; W. E. Simmons, 'The Aquarium', *The Popular Science Monthly* (1874), pp. 687–95; G. B. Sowerby, *The Aquarium: A Popular Account of Marine and Fresh-Water Animals and Plants* (London: Routledge, 1865); 'The Aquarium Simplified', *Home Friend*, 1 (1856), pp. 130–2; and J. G. Wood, *The Fresh and Salt Water Aquarium* (London: Routledge, 1868).

64. Lewes, *Sea-Side Studies,* pp. 9–10.

65. Harper, *The Sea-Side*, p. 164.

66. G. B. Sowerby, *Popular British Conchology: A Familiar History of the Molluscs Inhabiting the British Isles* (London: Lovell Reeve, 1854), pp. 1–2.

67. T. R. Jones, *The Aquarian Naturalist: A Manual for the Seaside* (London: Van Vorst, 1858), p. 419.

68. Gosse, *The Aquarium*, p. 230.

69. Gosse, *The Aquarium*, p. 29.

70. Gosse, *The Aquarium*, p. 55.

71. P. H. Gosse, *A Year at the Shore* (London: Alexander Strahan, 1865), p. 222.

72. Gosse, *The Aquarium*, pp. 104–6.

73. Gosse's instructions are very precise: 1 Æquoreal Pipefish (*Syngnathus æquoreus*); 1 Rough Doris (*Doris pilosa*), 2 Magus Top (*Trochus magus*), 1 Nerit (*Natica Alderi*), 1 Squin (*Pecten opercularis*), 1 Pholas (*Pholas parva*), 1 Pisa (*Pisa tetraodon*), 1 Cleanser Crab (*Portunus depurator*), 1 Ebalia (*Ebalia Pennantii*), 1 Hermit (small) (*Pagurus* ?), 3 Lobster-prawn (*Athanas nitescens*), 1 Brittle-star (*Ophiocoma rosula*), 1 Eyed Cribella (*Cribella oculata*), 2 Scarlet Sunstar (*Solaster papposa*), 1 Birdsfoot Star (*Palmipes membranaceus*), 3 Gibbous Starlet (*Asterina gibbosa*), 1 Purple tipped Urchin (*Echinus miliaris*), 7 Scarlet Madrepore (*Ralanophyllia regia*), 3 Cloak Anemone (*Adamsia palliata*).

74. Gosse, *Year at the Shore*, p. 236.

75. Stott, *Theatres of Glass*, pp. 126–8.

76. E. Gosse, *Father and Son* (London: Penguin Books, 1907), pp. 125–6.

77. P. H. Gosse, *Evenings at the Microscope; Or, Researches Among the Minuter Organs and Forms of Animal Life* (London: Society for Promoting Christian Knowledge, 1859), p. 3 (Preface).

78. H.-R., Simon, *Anton Dohrn und die Zoologische Station Neapel* (Frankfurt am Main: Edition Erbrich, 1980), pp. 50, 81.

79. D. E. Allen, *The Naturalist in Britain: A Social History* (Princeton, NJ: Princeton University Press, 1976), p. 115.

80. Gosse, *Evenings at the Microscope*, p. 244. See also R. Stott, *Darwin and the Barnacle: The Story of One Tiny Creature and History's Most Spectacular Scientific Breakthrough*

(New York: W. W. Norton & Company, 2004).

81. C. Darwin, *A Monograph of the Sub-Class Cirripedia with Figures of all the Species* (London: The Ray Society, 1851).

82. Compare to R. Stott, 'Through a Glass Darkly: Aquarium Colonies and Nineteenth-Century Narratives of Marine Monstrosity', *Gothic Studies*, 2:3 (2000), pp. 305–27, on pp. 313–14.

83. B. M. Stafford, 'Images of Ambiguity: Eighteenth-Century Microscopy and the Neither/Nor', in D. P. Miller and P. H. Reill (eds), *Visions of Empire. Voyages, Botany, and Representations of Nature* (Cambridge: Cambridge University Press, 1996), p. 230.

84. F. Burckhardt (ed.), *Charles Darwin's Letters: A Selection* (Cambridge: Cambridge University Press, 1996), p. 109; quoted from Stott, 'Through a Glass Darkly', p. 310.

85. Gosse, *The Aquarium*, p. 30.

86. C. Kingsley, *Glaucus; Or, the Wonders of the Shore* (Cambridge: Macmillan and Co., 1859), pp. 27, 83, 74.

87. Compare to Stott, 'Aquarium Colonies', p. 312.

88. Kingsley, *Glaucus*, pp. 89–90.

89. P. H. Gosse, *Actinologia Britannica: A History of the British Sea-Anemones and Corals* (London, 1860), p. 15.

90. Lewes, *Sea-Side Studies*, p. 242.

91. The expression comes from Le Corbusier, who derived it from his readings of Valéry. It does not denote lyrical objects, but, in the classical sense of poeisis, describes things which create something. Compare to Maak, *Der Architekt am Strand*, p. 60.

92. W. Krohn (ed.), *Ästhetik in der Wissenschaft. Interdisziplinärer Diskurs über das Gestalten und Darstellen von Wissen* (Hamburg: Meiner-Verlag, 2006), p. 3.

93. J.-L. Nancy, 'Das liegende Auge oder: Oberfläche, Öffnung und Bewegung des Wassers', *Berliner Gazette*, 'Wasserwissen', 4 February 2009, on http://berlinergazette.de/das-liegende-auge/ [accessed 12 December 2014].

94. Consequently and in accordance with Anne Friedberg aquaria can be addressed as 'machines of virtual transport'. In the same way panoramas brought the landscape to the city, aquaria introduced the sea to urban space and added elucidated bassins and dark auditoria to metropolitan illumination. Compare to A. Friedberg, *Window Shopping: Cinema and the Postmodern* (Berkeley, CA: University of California Press, 1993), pp. 4, 22.

95. Compare to S. J. Gould, *Leonardo's Mountain of Clams and the Diet of Worms: Essays on Natural History* (London, Jonathan Cape Ltd, 1998), pp. 57–73.

96. Compare to S. E. Hauser, 'Der subaquatische Bilderkosmos. Eine kurze Geschichte des Aquarium- und Unterwasserfildms von 1890 bis heute ... ', in V. Weigel (ed.), *Unter wasser über wasser. Vom Aquarium. zum Videobild* (Wilhelmshaven: Kerber Art, 2009), pp. 18–35.

97. Dohrn was highly interested in Marey's experiments and his idea to introduce new representational techniques to scientific research. Dohrn also supplied Marey with marine animals for his observational work. Compare to T. Heuss, *Anton Dohrn* (Stuttgart und Tübingen: Wunderlich, 1948), p. 284; and M. Braun, *Picturing Time: The Work of Etienne-Jules Marey (1830–1904)* (Chicago, IL and London: University of Chicago Press, 1992), pp. 158–60.

98. E.-J. Marey, 'La Locomotion dans L'Eau', *La Nature*, 911 (15 November 1890), pp. 375–8.

99. D. Hahn, 'Tourbillons et turbulences. Zu einer Ästhetik des Experiments in Étienne-Jules Mareys Machines à fumée', in É.-J. Mareys (ed.), *Machines à fumée, Ilinx – Berliner Beiträge zur Kulturwissenschaft. Nr. 1, Wirbel, Ströme, Turbulenzen* (Hamburg: Philo

Fine Arts, 2009), pp. 43–69, on p. 55.

100. The expression comes from Claude Bernard; G. Didi-Huberman uses it in his article: 'Expérimenter pour voir. Experimentieren, um zu sehen', *Kongress-Akten der Deutschen Gesellschaft für Ästhetik, Band 2 Experimentelle Ästhetik*: http://www.dgae.de/kongress-akten-band-2.html, p. 1 [accessed 18 December 2014].

101. Hahn, 'Tourbillons et turbulences', p. 63.

102. E. Morin, *The Cinema, or the Imaginary Man*, trans. L. Mortimer (1956; Minneapolis, MN and London: University of Minnesota Press, 2005).

103. Braun, *Picturing Time*, pp. 276–8.

104. Compare to J. Ward, *Weimar Surfaces: Urban Visual Culture in 1920s Germany* (Berkeley, CA: University of California Press, 2001), pp. 193–223.

105. I owe this hint to Moritz Jacobi.

106. K. Lorenz, *King Solomon's Ring* (1949; London: Routledge, 2002), p.14–15.

107. Lorenz, *King Solomon's Ring*, p. 10.

108. H. Bredekamp, 'Coral versus Trees: Charles Darwin's Early Sketches of Evolution', in P. H. Smith, A. R. W. Meyers and H. J. Cook, *Ways of Making and Knowing: The Material Culture of Empirical Knowledge* (Ann Arbor, MI: University of Michigan Press, 2014), pp. 357–76.

109. M. Foucault, *The Order of Things* (New York: Random House, 1994), p. 251.

110. Compare to B. Meyer-Sickendiek, *Tiefe. Über die Faszination des Grübelns* (München: Wilhelm Fink Verlag, 2010), pp. 53, 84–6, 92, 135–7.

111. J. G. Herder, 'On the Cognition and Sensation of the Human Soul', in *Philosophical Writings*, trans. and ed. by M. N. Foster (Cambridge, MA: Cambridge University Press, 2002), p. 197.

112. Herder, 'On the Cognition and Sensation', p. 168.

113. Compare to H. Böhme, 'Geheime Macht im Schoß der Erde. Das Symbolfeld des Bergbaus zwischen Sozialgeschichte und Psychohistorie', in H. Böhme (ed.), *Natur und Subjekt* (Frankfurt am Main: Suhrkamp, 1988), pp. 67–144.

114. Novalis, 'Miscellaneous Observations no. 17', in *Philosophical Writing*, trans. and ed. M. M. Stoljar (Albany, NY: State University of New York Press, 1997), p. 25.

115. *H. Heine's Pictures of Travel*, trans. C. G. Leland, 9th rev. edn (Philadelphia, PA: Schaefer & Koradi, 1882), pp. 149–50.

116. C. Baudelaire, 'Man and the Sea', in C. Baudelaire, *Flowers of Evil*, with an introduction by J. Culler and trans. by J. McGowern (Oxford: Oxford University Press, 1993), pp. 33–4.

117. Compare to W. Vortriede, *Novalis und die französischen Symbolisten. Zur Entstehungsgeschichte des dichterischen Symbols* (Stuttgart: Kohlhammer, 1963), p. 77.

118. Compare to R. Cardinal, 'Victor Hugo: Somnambulist of the Sea', in P. Collier (ed.), Artistic Relations: Literature and the Visual Arts in Nineteenth-Century France (New Haven, CT and London: Yale University Press, 1994), pp. 210–20, on p. 212.

119. V. Hugo, *Toilers of the Sea*, ed. E. Rhys, trans. W. Moy Thomas (London and Toronto: J. M. Dent, 1911), p. 131.

120. Compare to S. Friede, 'Die Welt als Aquarium – Spuren eines Schlüsselmotivs in Gides Paludes, Prousts Recherche und Robbe-Grillets Les Gommes', *Romanistische Zeitschrift für Literaturgeschichte*, 27:1–2 (2003), pp. 161–88; J.-K. Huysman, *A Rebours* (1884); J. Laforgue, 'L'Aquarium', *La Vogue*, 6 (29 May – 3 June 1886); M. Materlinck, *Aquarium* (1889); E. Verhaeren, *L'Aquarium* (1889); and A. Gide, *Paludes* (1895).

121. Hugo, *Toilers of the Sea*, p. 27.

122. Compare to S. Heraeus, 'Artists and the Dream in Nineteenth-Century Paris: Towards

a Prehistory of Surrealism', *History Workshop Journal*, 48 (Autumn 1999), pp. 151–68; and T. James, *Dream, Creativity, and Madness in Nineteenth-Century France* (Oxford: Clarendon Press, 1995).

123. S. Heraeus, 'The Dream as an Artistic Strategy', in M. Hollein and M. Stuffmann (eds), *As in a Dream: Odilon Redon* (Ostfildern: HatjeCantz, 2007), pp. 71–8, on p. 72.

124. A. Maury, *Le Sommeil et les rêves* (Paris: Didier, 1861), p. 113.

125. J. Michelet, *The Sea* (New York: Rudd & Carleton, 1861), p. 113.

126. J. Laforgue, *Revue Blanche*, 49 (15 June 1895); and M. Kruse, 'Zur Interpretation von Rousseaus *Cinquieme Reverie* und Laforgues Aquarium', in R. Grossmann, W. Pabst and E. Schramm (eds), *Der Vergleich. Literatur- und sprachwissenschaftliche Interpretationen. Festgabe für Hellmuth Petriconi zum 1. April 1955* (Hamburg: Cram, de Gruyter & Co, 1955), pp. 91–103.

127. Laforgue, 'L'Aquarium'.

128. The phrase of the 'All-Einigen' signifies an omniscient and wise unconscious. The term was coined by Eduard von Hartmann in his well known *Philosophie des Unbewußten*, published 1968, which Laforgues had studied intensively.

129. G. Rodenbach, *Les Vies Encloses. Poème* (Paris: Bibliothèque-Charpentier, 1896).

130. Still, the idea of an aquarium of the soul remained virulent. In *A Room of One's Own* for example Virginia Woolf (1882–1941) describes her thoughts as little fish and herself as the resident of a miraculous glass cabinet. V. Woolf, *A Room of One's Own* (London: Hogarth Press, 1929), Chapter 1.

131. H. C. Andersen, *The Dryad* (1868) trans. J. Hersholt; published online by the Hans Christian Andersen Centre: andersen.sdu.dk/vaerk/hersholt/TheDryad_e.html [accessed 12 February 2015].

132. M. Proust, *In Search of Lost Time*, ed. C. Prendergast, 6 vols (1918; London: Penguin, 1987–9), vol. 2: *In the Shadow of Young Girls in Flower*, p. 270.

133. Proust, *In Search of Lost Time*, vol. 3: *The Guermantes Way* (French 1920–1), p. 31.

134. Compare to D. W. Druick (ed.), *Odilon Redon: Prince of Dreams 1840–1916, Ausstellungskatalog* (Chicago, IL: The Art Institute of Chicago, 1994); U. Harter, 'Odilon Redons Tiefseebilder', in E. Schlebrügge (ed.), *Das Meer im Zimmer. Von Tintenschnecken und Muscheltieren* (Graz: Landesmuseum Joanneum, 2005), pp. 115–21; B. Larson, *The Dark Side of Nature: Science, Society, and the Fantastic in the Work of Odilon Redon* (Pennsylvania, PA: Pennsylvania State University Press, 2005); M. van Zuylen, S. Figura and J. Hauptmann (eds), *Beyond the Visible: The Art of Odilon Redon* (New York: The Museum of Modern Art, 2005); and Stuffmann and Hollein (eds), *As in a Dream*.

135. Compare to P. Kort, 'Die Dinge wirklich werden lassen: Odilon Redon und Jean Carriès', in P. Kort and M. Hollein (eds), *Darwin. Kunst und die Suche nach den Ursprüngen*, Ausstellungskatalog (Köln: Wienand, 2009), pp. 154–70, on p. 154; and N. Tsutatani, 'A Floating Vision: Thoughts on Odilon Redon's *Guradian Spirit of the Waters* (1878)', in A. Yamatoto (ed.), *Odilon Redon – le Souci de L'Absolu* (Nagoya: Chunichi Shimbun, 2002), pp. 194–200, on p. 195.

136. A. Clavaud, *Sur la prétendue Parthénogènese du Chara crinita* (Sonderdruck, 1878); and A. Clavaud, 'Sur la Nitella stelligera', *Actes Soc. linn. de Bordeaux,* 25 (1865), pp. 348–52.

137. Compare to Tsutatani, 'A Floating Vision'.

138. D. W. Druick and P. K. Zegers, 'In the Public Eye', in Druick (ed.), *Odilon Redon*, on p. 137.

139. Quoted from U. Perucchi-Petri, 'Jeunes Peintres, mes amis: Odilon Redon and the Nabis', in Stuffmann and Hollein (eds), *As in a Dream*, pp. 103–11, on p. 111.

140. Compare to H. Esswein, *Alfred Kubin: Der Künstler und sein Werk* (München: Georg Müller, 1911); H. Bisanz, *Alfred Kubin. Zeichner, Schriftsteller und Philos-*

oph (München: Edition Spangenberg, 1977); P. Rhein, *The Verbal and Visual Art of Alfred Kubin* (Riverside: Ariadne Press, 1989); A. Hoberg, *Alfred Kubin 1877–1959* (München: Edition Spangenberg, 1990); P. Assmann (ed.), *Alfred Kubin (1877–1959)* (Linz: Oö. Landesgalerie, 1995); and M. Morton, 'Natur und Seele: Österreichische Reaktionen auf Ernst Haeckels evolutionären Monismus', in P. Kort and and M. Hollein (eds), *Darwin. Kunst und die Suche nach den Ursprüngen* (Frankfurt am Main: Wienand Verlag, 2009), pp. 126–52.

141. Compare to A. Marks, *Inventar der Bibliothek Alfred Kubin im Kubin-Haus des Landes Oberösterreich in Zwickledt/Wernstein* (no place, no publisher, no date of publication); and W. Bölsche, *Das Liebesleben der Natur. Eine Entwicklungsgeschichte der Liebe*, 2 vols (Jena: Diederichs, 1898).

142. A. Kubin, 'Aus halbvergessenem Lande. Über künstlerische Befruchtung', in A. Kubin, *Aus meiner Werkstatt. Gesammelte Prosa mit 71 Abbildungen*, ed. U. Riemerschmidt (1926; München: Nymphenburger Verlagsbuchhandlung, 1973), pp. 19–24, on p. 22.

143. At that time Redon was very well known for his evocations of underwater landscapes. Compare to G. Groom, 'The Late Work', in Druick (ed.), *Odilon Redon*, pp. 314–15.

144. A. Kubin, 'Mein Tag in Zwickledt: Ein Brief an Wilhelm Hausenstein (1921)', in A. Kubin, *Aus meinem Leben. Gesammelte Prosa mit 73 Abbildungen* (1921; München: edition spangenberg im Ellermann Verlag, 1974), pp. 87–93, on p. 92.

145. Morton, 'Natur und Seele', p. 138.

146. Esswein, *Alfred Kubin*, p. 45.

147. A. Kubin, 'Malerei des Übersinnlichen', in A. Kubin, *Aus meiner Werkstatt. Gesammelte Prosa mit 71 Abbildungen* (1933; München: Nymphenburger Verlag, 1973) pp. 43–5, on p. 44.

148. A. Kubin, 'Malerei des Übersinnlichen', p. 42.

149. Compare to P. Assmann, 'Annäherungen an die Farbe im Werk Alfred Kubins', in P. Assmann, *Alfred Kubin*, pp. 164–206, on p. 188.

150. C. Brockhaus, 'Das zeichnerische Frühwerk Alfred Kubins bis 1904', in H. A. Peters (ed.), *Alfred Kubin. Das zeichnerische Frühwerk bis 1904* (Baden-Baden: Kunsthalle Baden-Baden, 1977), pp. 6–28, on p. 28.

151. F. von Herzmanovsky-Orlando, *Der Briefwechsel mit Alfred Kubin* (Salzburg: Residenz Verlag, 1983), p. 10.

152. Compare to R. Koella, 'Kubins symbolistische Zeichnungen', in R. Koella and D. Schwarz (eds), *Alfred Kubin, Ausstellung im Kunstmuseum Winterthur* (Winterthur: Kunstmuseum, 1986), pp. 31–43, on p. 40.

153. A. Kubin, 'Fragment eines Weltbildes', in U. Riemerschmidt (ed.), *Aus meiner Werkstatt*, pp. 35–8, on p. 38.

154. R. Verdi, *Klee and Nature* (London: Zwemmer Ltd, 1984), pp. 62–3.

155. P. Klee, *Tagebücher 1898–1916. Textkritische Neuedition*, 2:390 (Stuttgart: Hatje, Teufen, Niggli, 1988), p. 123.

156. C. Vogt, 'Aus der zoologischen Station in Neapel', *Die Gartenlaube*, 21 (1880), pp. 340–4, on p. 341.

157. J. Meier-Gräfe, *Hans von Marées, sein Leben und Werk* (München and Leipzig: Piper Verlag, 1909–10), p. 66.

158. 'For Anton Dohrn arts and science were still the two sides of – one – human culture'. C. Groeben, 'The Stazione Zoologica Anton Dohrn as a Place for the Circulation of Scientific Ideas: Vision and Management', in K. L. Anderson and C. Tony (eds), Information for Responsible Fisheries: Libraries as Mediators; proceedings of the 31st Annual Conference, held in Rome, Italy (10–14 October 2005); Compare to C.

Groeben and I. Müller (eds), *The Naples Zoological Station at the time of Anton Dohrn* (Naples: Stazione Zoologica di Napoli, 1975); K. J. Partsch, *Die Zoologische Station in Neapel* (Göttingen: Vandenhoeck & Ruprecht, 1980); and H.-R. Simon, *Anton Dohrn und die Zoologische Station Neapel* (Frankfurt am Main: Edition Erbrich, 1980).

159. Quoted from T. Heuss, *Anton Dohrn* (Stuttgart und Tübingen: Wunderlich, 1948), p. 215.

160. Compare to P. Kort, 'Arnold Böcklin, Max Ernst und die Debatten um Ursprünge und Übereben in Deutschland und Frankreich', in Kort and Hollein (eds), *Darwin*, pp. 24–53; A. Linnebach, *Arnold Böcklin und die Antike. Mythos, Geschichte, Gegenwart* (München: Hirmer Verlag, 1991); and A. Linnebach, 'Antike und Gegenwart. Zu Böcklins mythoogischer Bilderwelt', in G. Magnaguagno and J. Steiner (eds), *Arnold Böcklin, Giorgio de Chirico, Max Ernst: Eine Reise ins Ungewisse* (Bern: Benteli, 1997), pp. 195–203.

161. L. Hevesi, 'Böcklins, Meeresidylle', in L. Hevesi, *Acht Jahre Sezession. März 1897 – Juni 1905. Kritik, Polemik, Chronik* (Wien, 1906), pp. 364–8, on p. 366.

162. A. Böcklin, *Neben meiner Kunst: Flugstudien, Briefe und Persönliches von und über Arnold Böcklin*, ed. F. Runke and C. Böcklin (Berlin: Vita-Verlag, 1909), pp. 55–7; and A. Frey, *Anrold Böcklin, Nach den Erinnerungen seiner Zürcher Freunde*, 2nd edn (Stuttgart and Berlin: Cotta'sche Buchhandlung, 1912), pp. 133–4.

163. Klee in a letter to Lily Stumpf from 28th March 1902, in: F. Klee (ed.), *Paul Klee. Briefe an die Familie 1893–1940*, vol. 1: 1893–1906 (Köln: Du Mont, 1979), p. 220.

164. C. Giedion-Welcker, *Paul Klee mit Selbstzeugnissen und Bilddokumenten* (Reinbek bei Hamburg: Rowohlt, 1961), p. 20; and D. Rudloff, *Unvollendete Schöpfung. Künstler im zwanzigsten Jahrhundert* (Stuttgart: Urachhaus, 1982), p. 65.

165. Compare to W. Schmalenbach, *Paul Klee: Fische, Einführung von Werner Schmalenbach. Werkmonographien zur Bildenden Kunst in Reclams Universal Bibliothek Band Nr. 31* (Stuttgart: Reclam, 1958). pp. 4–6.

166. Verdi, *Klee and Nature*, p. 64.

167. Compare to Rudloff, *Unvollendete Schöpfung*, p. 66.

5 *Mise-en-Action*: *Asleep in the Deep*

1. K. Benson, H. Rozwadowski and D. van Keuren (eds), *The Machine in Neptune's Garden: Historical Perspectives on Technology and the Marine Environment* (Sagamore Beach, MA: Science History Publications, 2004), p. xiii.

2. C. van Dover, *Deep-Ocean Journeys: Discovering New Life at the Bottom of the Sea* (Redwood City, CA: Addison-Wesley, 1996), p. 16.

3. S. Helmreich, 'An Anthropologist Underwater: Immersive Soundscapes, Submarine Cyborgs, and Transductive Ethnography', *American Ethnologist*, 34:4 (2007), pp. 621–41.

4. M. Merleau-Ponty, *Phenomenology of Perception*, trans. C. Smith (New York: Routledge & K. Paul, 1962), p. 58.

5. M. Merleau-Ponty, *Phenomenology of Perception*, trans. C. Smith (New York: Routledge & K. Paul, 1962), p. 147.

6. A. Hantschk and P. Kruspel, 'Mit Skizzenblock und Taucherglocke: Ein Wiener Maler unter Wasser', *Divemaster*, 1 (2000), pp. 57–60.

7. Baron E. von Ransonnet-Villez, *Reise von Kairo nach Tor zu den Korallenbänken des Rothen Meeres* (Wien: Ueberreuter, 1863), p. 17.

8. Ransonnet-Villez, *Reise von Kairo*, p. 18.

9. In 'The Art of Living under Water: Or, a Discourse Concerning the Means of Furnishing Air at the Bottom of the Sea, in any Ordinary Depths', Halley writes: 'And by

the Glass Window, so much Light was transmitted that, when the Sea was clear, and especially when the Sun shone, I could see perfectly well to Write or Read, much more to fasten or lay hold on any thing under us, that was to be taken up', in *Philosophical Transactions of the Royal Society*, 29 (1716), pp. 492–9.

10. R. F. Marx, *The History of Underwater Exploration* (New York: Dover Publications, 1978), p. 61.

11. Compare to M. Jung, *Das Handbuch zur Tauchgeschichte. Techniken, Geräte, Berufe, Erfindunge* (Stuttgart: Nagelschmid, 1999); J. Bevan, *The Infernal Diver: The Lives of John and Charles Deane, Their Invention of the Diving Helmet, and its First Application to Salvage, Treasure Hunting, Civil Engineering and Military Uses* (London: Submex, 1996); and Z. Cowan, *Early Divers: Underwater Adventures in the 17th and 18th Centuries* (Norfolk: Treasure World, 1992).

12. J. Forest, 'Henri Milne Edwards', *Journal of Crustacean Biology*, 16:1 (1996), p. 207–13; and T. Norton, *Stars Beneath the Sea: The Extraordinary Lives of the Pioneers of Diving* (London: Century Press, 1999), pp. 27–42.

13. The equipment had been designed by Milne-Edwards's friend Gustave Paulin, commandant of the Parisian fire-brigade and was originally meant for firefighters. Compare to Norton, *Stars Beneath the Sea*, pp. 28–9.

14. A. de Quatrefages *The Rambles of a Naturalist on the Coasts of France, Spain, and Sicily*, 2 vols (London: Longman, Brown, Green, Longmans & Roberts, 1857), vol. 2, p. 15.

15. 'large Panopeas of the Mediterranean' or Panopées. Quatrefages, *Rambles*, vol. 2, p. 15.

16. Henry Milne-Edwards, quoted from Norton, *Stars Beneath the Sea*, p. 40.

17. Jung, *Das Handbuch zur Tauchgeschichte*, p. 137; and I. Müller, *Die Geschichte der Zoologischen Station in Neapel von der Gründung durch Anton Dohrn (1872) bis zum Ersten Weltkrieg und ihre Bedeutung für die Entwicklung der modernen Biologischen Wissenschaften* (Düsseldorf; no publisher, 1976), pp. 142–6.

18. ASZN, BA 438, quoted from Müller, *Die Geschichte der Zoologischen Station in Neapel*, p. 142.

19. Compare to ASZN, BA 438, quoted from Müller, *Die Geschichte der Zoologischen Station in Neapel*, pp. 143–6.

20. Compare to T. Ecott, *Neutral Buoyancy: Adventures in a Liquid World* (London: Michael Joseph, 2001), pp. 48, 178.

21. Compare to Ecott, *Neutral Buoyancy*, pp. 14–17.

22. Ecott, *Neutral Buoyancy*, p. 11.

23. Compare to Hantsch and Kruspel, 'Mit Skizzenblock und Taucherglocke', p. 59.

24. W. Ullrich, *Die Geschichte der Unschärfe* (Berlin: Klaus Wagenbach, 2009), pp. 86–7.

25. See, for example, *The Valley in the Sea* (1862), Indianapolis Museum of Art.

26. S. Weinberg, P. L. Dogué and J. Neuschwander, *Unterwasserfotografie. Hundert Jahre Geschichte, Technik, Faszination* (Gilching: Verlag Photographie, 1993), pp. 20–2.

27. The first underwater photography is credited to William Thompson (1822–79). Compare to V. Adam, 'William Thompson – 100 Years of Underwater Photography?', *Focus*, 49 (September 1993), pp. 4–8; and N. Baker, 'William Thompson – The World's First Underwater Photographer', *Historical Diving Times*, 19 (Summer 1997), pp. 8–16.

28. On Boutan see: Weinberg, Dogué and Neuschwander, *Unterwasserfotografie*; and Norton, *Stars Beneath the Sea*, pp. 161–76.

29. L. Boutan, 'Submarine Photography', *The Century Magazine* (May 1898), pp. 42–9, on p. 43.

30. H. de Lacaz-Dutons, 'Sur la photographie sous-marine', *Comptes Rendus Hebdomadaires de l'Academie des Sciences*, 117, pp. 286–9.

31. L. Boutan, *La Photographie sous-marine et les progrès de la photographie* (1900; Paris: Jean-Michel Place, 1987), p. 189.
32. Boutan, 'Submarine Photography', p. 48.
33. Boutan, 'Submarine Photography', pp. 44, 47.
34. F. Ward, *Marvels of Fish Life: As Revealed by the Camera*, 2nd edn (London and New York: Cassell and Company, 1912), p. xi.
35. Ward, *Marvels of Fish Life*, p. xi.
36. E. Bippus, 'Skizzen und Gekritzel. Relationen zwischen Denken und Handeln in Kunst und Wissenschaft', in M. Heßler (ed.), *Logik des Bildlichen* (Bielefeld: Transcript-Verlag, 2009), p. 76–93, on p. 76
37. Ward, *Marvels of Fish Life*, p. 162.
38. F. Ward, *Animal Life Under Water* (London: Cassell and Company, 1919), p. 1.
39. Compare to V. Allemandy, *Wonders of the Deep: The Story of the Williamson Submarine Expeditio* (London, Jarrold & Sons, 1916); T. Burgess, *Take Me under the Sea: The Dream Merchants of the Deep* (Salem, MA: The Ocean Archives, 1994); T. Norton, *Stars Beneath the Sea: The Extraordinary Lives of the Pioneers of Diving* (London: Century Press, 1999), pp. 177–98; B. Taves, 'A Pioneer under the Sea: Library Restores Rare Film Footage', *LC Information Bulletin*, 55:15 (16 September 1996); B. Taves, 'With Williamson Beneath the Sea', *Journal of Film Preservation*, 52 (1996), pp. 54–61; S. Weinberg, P. L. Dogué and J. Neuschwander, *Unterwasserfotografie: Hundert Jahre Geschichte, Technik, Faszination* (Gilching: Verlag Photographie, 1993), pp. 72–6; and J. E. Williamson, *Twenty Years under the Sea* (Boston, MA: Ralph T. Hale & Company, 1936).
40. N. D. W. Moure, *The World of Zarh Pritchard* (Los Angeles, CA: William A. Karges Fine Art, 1999); and Burgess, *Take Me under the Sea*.
41. Burgess, *Take Me under The Sea*, p. 170.
42. Williamson, *Twenty Years under The Sea*, pp. 29–30.
43. R. William, *Notes on the Underground: An Essay on Technology, Society, and the Imagination* (Cambridge, MA and London: The MIT Press, 2008), p. 7.
44. Allemandy, *Wonders of the Deep*, p. 47.
45. Allemandy, *Wonders of the Deep*, p. 45.
46. Allemandy, *Wonders of the Deep*, p. 87.
47. Allemandy, *Wonders of the Deep*, p. 92.
48. The film is also known as *Thirty Leagues under the Sea* or *In the Tropical Seas*.
49. Quoted from Burgess, *Take Me under The Sea*, p. 186.
50. *20,000 Leagues under The Sea* (1916), director Stuart Paton, photography Eugene Gaudio, music Alexander Rannie and Brian Benison, Special Effects, Williamson Inventions.
51. Quoted from T. Ecott, *Unter Wasser. Abenteuer in einer anderen Welt* (Berlin: Argon, 2002), p. 181.
52. Quoted from Norton, *Stars Beneath the Sea*, p. 191.
53. Submarine Eye (1917), Girl of the Sea (1920), Wet Gold (1921), Wonders of the Sea (1922), The Uninvited Guest (1924).
54. Quoted from Burgess, *Take Me under the Sea*, p. 180.
55. Quoted from Burgess, *Take Me under the Sea*, p. 212.
56. Williamson, *Twenty Years under the Sea*, p. 320.
57. On Beebe's life and work see R. H. Welker, *Natural Man: The Life of William Beebe* (Bloomington, IN: Indiana University Press, 1975); B. Matsen, *Descent: The Heroic Discovery of the Abyss* (New York: Pantheon Books, 2005); and T. M. Berra, *William Beebe: An Annotated Bibliography* (Hamden: Archon Books, 1977).

58. W. Beebe, 'Kingdom of the Helmet', in W. Beebe, J. Tee-Van, G. Hollister, J. Crane and O. Barton, *Half Mile Down* (New York: Hartcourt Brace and Company 1934), pp. 66–86, on p. 86.

59. Quoted from T. Norton, *Stars Beneath the Sea: The Extraordinary Lives of the Pioneers of Diving* (London: Century Press, 1999), pp. 64–5; and Matsen, *Descent*, pp. 16, 53, 64–5.

60. Matsen, *Descent*, p. 16.

61. Beebe, 'Kingdom of the Helmet', p. 85.

62. Compare to Beebe, 'Kingdom of the Helmet', p. 64.

63. Beebe mentions his favourite book frequently throughout his work.

64. W. Beebe, *Nonsuch: Land of Water* (London and New York: Puntams, 1932), p. 194

65. Beebe, *Nonsuch*, p. 42.

66. Beebe, *Nonsuch*, p. 176.

67. W. Beebe, 'A Wonderer under Sea', *National Geographic*, 62 (December 1932), pp. 741–58, on 741.

68. Released in *Modern Mechanix* (September 1935): blog.modernmechanix.com [accessed 12 December 2014].

69. Beebe, 'A Wonderer under Sea', p. 741.

70. Beebe, 'A Wonderer under Sea', p. 746.

71. Matsen, *Descent*, p. 52.

72. 'Undersea Classroom Reveals Ocean Secrets' (July 1934), *Modern Mechanix* [accessed 12 December 2014].

73. 'Build Your Own One-Man Submarine!' (September 1933); 'Inventor Gets Thrill in Homemade Submarine' (June 1933); 'Under-Sea Tractor-Sphere Roams Ocean Floor' (January 1935); 'Submarine Auto' (September 1936); and 'Fish is One-Man Submarine' (December 1938), *Modern Mechanix* [accessed 12 December 2014].

74. The analogy of the deep sea and outer space was widespread at the time. Beebe often compared the depths of the sea with the unknown regions of the universe; journalists quite frequently referred to him as 'the man from Mars' when reporting on his submarine adventures.

75. W. Beebe, 'A Half Mile Down: Strange Creature, Beautiful and Grotesque as Figments of Fancy, Reveal Themselves at Windows of the Bathysphere', *The National Geographic Magazine* (1934), pp. 661–704; and O. Barton, *The World Beneath the Sea: The Story of the Deepest Dive Ever Made by Man* (New York: Thomas Y. Crowell, 1953).

76. W. Beebe, 'Down to Davy Jones' Locker', *New York Times Magazine*, 13 July 1930, p. 1; W. Beebe, 'Diving to a Depth of a Quarter of a Mile', *Illustrated London News*, 11 April 1931, pp. 594–5; W. Beebe, 'A Round Trip to Davy Jones' Locker', *National Geographic Magazine*, 59:6 (1931), pp. 644–78; W. Beebe, 'The Depths of the Sea. Strange Forms a Mile Below the Surface', *National Geographic Magazine*, 61:1 (1932), pp. 65–88; W. Beebe, 'Descent into Perpetual Night', *New York Times Magazine*, 9 October 1932, pp. 1; W. Beebe, 'Exploration of the Deep Sea', *Science*, 76 (14 October 1932), p. 344; W. Beebe, 'Thoughts on Diving', *Harpers,* 166 (April 1933), pp. 582–6; W. Beebe, 'Going Down', *McCall's Magazine*, 14, March 1933, pp. 51–2; and W. Beebe, 'A New Deep Sea Fish Story', *Royal Gazette and Colonist Daily*, 11 October 1933, pp. 1–2.

77. Compare to Matsen, *Descent*, pp. 140–51.

78. M. Kemp, *Visualizations: The Nature Book of Art and Science* (Oxford: Oxford University Press, 2000), p. 2.

79. Beebe, *Half Mile Down*, p. 676.

80. W. Beebe, 'A First Round Trip to Davy Jones' Locker', in W. Beebe, *Adventuring with Beebe: Selections from the Writings of William Beebe* (London: The Bodley Head, 1956),

pp. 54–84, on p. 62.

81. B. Waldenfels, 'Überraschte Wahrnehmung. Eine phänomenologische Betrachtung', in M. Frölich, R. Middel and K. Visarius (eds), *Zeichen und Wuner. Über das Staunen im Kino* (Marburg: Schüren, 2001), pp. 9–29, on p. 25.
82. Beebe, 'Davy Jones's Locker', pp. 62–3.
83. Beebe, 'Davy Jones's Locker', pp. 67–9.
84. Compare to J. Kinzer, 'In ewiger Nacht und Kälte. Sehen, Riechen, Hören, Sex: Die Lebens und Überlebensstrategien der Tiefseebewohner wirken grotesk, sind aber effizient', *mare*, 13 (1999), pp. 36–44, on pp. 36–8.
85. Beebe, *Half Mile Down*, p. 685.
86. Beebe, *Half Mile Down*, p. 691.
87. Compare to Beebe, *Half Mile Down*, p. 691.
88. The Bermuda Biology Station for Research (ed.), *Bermuda 100 Years The First Century: Celebrating 100 years of Marine Science. BBSR 1903–2003* (Bermuda, 2003), p. 28.
89. S. Schlee, 'The Controversial Dr. Beebe and his Brain Fish', *BBSR Newsletter*, 3:3 (1974), p. 2.
90. Bermuda Biology Station, *The First Century*, p. 29.
91. www.noaav.gov [accessed 9 December 2014].
92. W. Martin and M. J. Russell, 'On the Origins of Cells: A Hypothesis for the Evolutionary Transitions from Abiotic Geochemistry to Chemoautotrophic Prokaryotes, and from Prokaryotes to Nucleated Cells', *Philosphical Transactions of the Royal Society London B*; DOI 10.1098/rstb.2002.1183.
93. Beebe, *Half Mile Down*, p. 685.
94. Beebe, *Half Mile Down*, p. 685.
95. Beebe, *A Half Mile Down*, p. 691.
96. W. Beebe, *A Half Mile Down*, p. 704. The sea-heaven's speciality is, of course, that the stellar fish move and twinkle colourfully.
97. Beebe, 'Davy Jones's Locker', pp. 662–3.
98. Beebe, 'Davy Jones's Locker', p. 83.
99. Beebe, 'Davy Jones's Locker', p. 82.
100. Beebe, 'Davy Jones's Locker', p. 678.
101. Beebe, 'Davy Jones's Locker', p. 675.
102. Beebe, *Half Mile Down*, plate x.
103. Beebe, *Half Mile Down*, plate xi.
104. Beebe, *Half Mile Down*, plate xiii.
105. Beebe, *Half Mile Down*, plate I.
106. Beebe, *Half Mile Down*, plate xv.
107. Beebe, 'A Wonderer under Sea', p. 10.
108. Beebe, 'A Wonderer under Sea', p. 8.
109. 'Neue Fischarten in der Tiefsee entdeckt', *Frankfurter Allgemeine Zeitung Nr. 297*, 22 December 2014, p. 7.

Epilogue

1. Valéry, 'Blicke aufs Meer', in J. Schmidt-Radefeldt (ed.), *Paul Valéry. Werke. Frankfurter Ausgabe in 7 Bänden* (1930, French; Frankfurt: Insel, 1989), pp. 501–9, on p. 504.
2. R. Guardini, *Zu Rainer Maria Rilkes Deutung des Daseins, Schriften für die geistige Überlieferung IV* (Berlin: Küpper, 1941), p. 26.

INDEX

Agassiz, Louis, 44, 56
Aldrovandi, Ulisse, 21, 93
Andersen, Hans Christian, 109, 136
Anning, Joseph, 31
Anning, Mary, 31–2
Aristoteles, 18, 20, 23, 49, 52
  De generatione animalium, 18, 20

Barton, Otis, 165–6
Beebe, William, 162–73
Belon, Pierre, 18, 21, 89, 142
Berliner Aquarium, 109, 111, 131
  Unter den Linden, 109
Blumenberg, Hans, 5, 52
Böcklin, Arnold, 142–3
Böhme, Gernot, 17
Böhme, Hartmut, 17, 25, 51
Bölsche, Wilhelm, 93, 139
Bostelmann, Else, 163–4, 166–7, 171, 173
Boutan, Louis Marie Auguste, 151–4
Boyle, Robert, 9, 25–7
  Other Enquiries Concerning the Sea, 26
Brehm, Alfred, 109, 140

Caillois, Roger, 87, 93
Chamisso, Adalbert von, 49, 53
Chun, Carl, 61–2, 65, 70–2, 93
Colonna, Fabio, 21, 23
  Aquatilium, et Terrestrium aliquot Animalium (1616), 23
Conybeare, William, 31–2
Cousteau, Jaques-Yves, 93

Dalyell, Sir Graham John, 112, 115
Darwin, Charles, 36, 40, 44–5, 53–4, 56, 93, 101, 104, 123–4, 132, 138, 142–3, 173
  Origin of Species, 45, 53, 56, 104, 138

Daston, Lorraine, 5–6, 44, 65
David, Joseph, 151–2
De La Beche, Henry, 31–4
DesMoulins, Charles, 112, 115
Dessalines d'Orbigny, Alcide, 86, 93
Dohrn, Anton, 129, 142–3, 147, 149
Dohrn, Reinhard, 148

Ehrenberg, Christian Gottfried, 44, 49, 52, 56, 78
Eibl-Eibesfeld, Irenäus, 40, 131

Fabricius, Johann Albert, 21
Flaubert, Gustave, 76, 138
Forbes, Edward, 17, 57, 113
Freud, Sigmund, 133

Gesner, Conrad, 18–19, 23, 89, 142
  Fisch-Buch (Book of the Fishes) (1558), 18, 21
  Historia animalium (1558), 23, 26
Gosse, Philip Henry, 107, 114–15, 119–20, 122

Haeckel, Ernst, 40, 44, 92, 102, 106, 111, 139
  The History of Creation, 40
Haeckel, Ernst , 38
Halley, Edmond, 25, 27, 146
  The Art of Living Underwater (1716), 25
Heinroth, Dr Oskar, 110, 131
Hooke, Robert, 27, 30, 67
Hugo, Victor, 76, 84–7, 93, 135

Jäger, Gustav, 140
Johnston, George, 113–15

Keuren, David van, 2, 145
Kingsley, Charles, 54, 125

Klee, Paul, 101, 137, 141–4
Kubin, Alfred, 101, 137, 139–41

Laforgue, Jules, 135–7
Lee, Henry, 56, 84
Lichtenberg, Georg Christoph, 48, 133
Lister, Martin, 21, 24
London Aquarium, 107–8
  Fish House, 107–8
Lonicer, Adam, 18, 21
Lyell, Charles, 30, 124

Mangin, Arthur, 1, 50, 79, 82, 90–2, 108
Mantell, Gideon, 31–2, 34
Marey, Etienne-Jules, 129–30
Marsili, Louis Ferdinand Comte de, 21, 27
Martens, Friedrich, 26–7
  Spitzbergische oder Groenlandischen Reise Beschreibungen, 26
Maury, Alfred, 135
Meier-Graefe, Julius, 142–3
Melville, Herman, 84
  Moby Dick, 84
Meyer, Heinrich Adolph, 105–6
Michelet, Jules, 1, 76, 85–6, 136, 138–9
Milne-Edwards, Henry, 89, 147–9
Möbius, Karl August, 46, 104–6
Montagu, Edward, Earl of Sandwich, 25
Moray, Robert Sir, 25, 27
Müller, Johannes Peter, 38–9, 44

Neuville, Alphonse de, 80–2, 87
Newton, Isaac, 27, 30
  Newtonian, 54

Oldenburg, Henry, 25, 27

Painlevé, Jean, 94–8
  La Pieuvre, film, 94–6, 98
  Les Amours de la Pieuvre, film, 94–5, 98
Paris, 3, 73, 79, 95, 102, 107–9, 137–8, 140, 151–3, 160
Paton, Stuart, 80, 82
Pritchard, Zarh, 157, 160, 163

Ransonnet-Villez, Eugen Freiherr von, 145–7, 149–51
Ray, John, 21, 24
Redon, Odilon, 101, 109, 137–41
Riou, Édouard, 35, 79, 87, 92
Rondelet, Guillaume, 18, 21, 89, 142
Rooke, Lawrence, 25, 27
Roßmäßler, Emil Adolf, 44, 116
Rozwadowski, Helen, 2, 56, 145
Russell, Richard, 13

Schleiden, Matthias Jakob, 12, 37, 44, 47, 49

Thynne, Anna, 113–15

Valéry, Paul, 117, 121, 175
Verne, Jules, 35–6, 75–6, 78–85, 87–9, 91–2, 160
  Twenty Thousand Leagues under the Seas, 75, 78, 80, 84
Virchow, Rudolf, 8, 39
Vogt, Carl, 142

Ward, Francis, 151, 154–7
Warington, Robert, 114–15
Williamson, George, 158–9, 161
Williamson, John Ernest, 157–62

Zachariae, August Wilhelm, 49–50